FUNDAMENTAL PROBLEMS IN THE THEORY OF STELLAR EVOLUTION

INTERNATIONAL ASTRONOMICAL UNION
UNION ASTRONOMIQUE INTERNATIONALE

SYMPOSIUM No. 93

HELD AT KYOTO UNIVERSITY, KYOTO, JAPAN, JULY 22–25, 1980

FUNDAMENTAL PROBLEMS IN THE THEORY OF STELLAR EVOLUTION

EDITED BY

DAIICHIRO SUGIMOTO
Department of Earth Science and Astronomy,
College of General Education, University of Tokyo, Tokyo, Japan

DONALD Q. LAMB
Harvard Smithsonian Center for Astrophysics,
Cambridge, Massachusetts, U.S.A.

and

DAVID N. SCHRAMM
Astronomy and Astrophysics Center,
The University of Chicago, Chicago, Illinois, U.S.A.

D. REIDEL PUBLISHING COMPANY

DORDRECHT : HOLLAND / BOSTON : U.S.A. / LONDON : ENGLAND

Library of Congress Cataloging in Publication Data

Main entry under title:

Fundamental problems in the theory of stellar evolution.

At head of title: International astronomical union. Sponsored by IAU commission 35 on stellar constitution and by commission 42 on close binary stars.

Includes index.

1. Stars–Evolution–Congresses. I. Sugimoto, Daiichiro, 1937- . II. Lamb, Donald Q. III. Schramm, David N. IV. International astronomical union. V. International astronomical union. Commission 35. VI. International astronomical union. Commission 42.

QB806.F86 521'.58 81-4057

ISBN-13: 978-90-277-1274-5 e-ISBN-13: 978-94-009-8477-6

DOI: 10.1007/978-94-009-8477-6

Published on behalf of
the International Astronomical Union
by
D. Reidel Publishing Company, P.O. Box 17, 3300 AA Dordrecht, Holland

Softcover reprint of the hardcover 1st Edition 1981

Sold and distributed in the U.S.A. and Canada
by Kluwer Boston Inc.,
190 Old Derby Street, Hingham, MA 02043, U.S.A.

In all other countries, sold and distributed
by Kluwer Academic Publishers Group,
P.O. Box 322, 3300 AH Dordrecht, Holland

D. Reidel Publishing Company is a member of the Kluwer Group.

TABLE OF CONTENTS

Contributed papers

Session 4: MASS EXCHANGE ON CLOSE BINARY STARS AND THE EFFECT ON STELLAR EVOLUTION
Chairman; C. de LOORE

Review papers

Contributed papers

PREFACE

This volume presents the Proceedings of International Astronomical Union Symposium No.93 on Fundamental Problems in the Theory of Stellar Evolution. It contains the texts of all the invited papers, the abstracts of the contributed papers that were read by one of the attending author(s), and edited discussions. Only one abstract is included in this volume from each author who attended, and the abstracts of papers which were read on behalf of absent author(s) are not included. Those papers, which were read but are not included in the volume, are indicated by asterisks in the table of contents.

The meeting took place at the University Hall, Kyoto University, Kyoto, Japan from July 22 to 25, 1980, and was sponsored by IAU Commission 35 on Stellar Constitution and co-sponsored by the IAU Commission 42 on Close Binary Stars. Locally, the Symposium was hosted by the Research Institute for Fundamental Physics, Kyoto University with encouragement from the Astronomical Society of Japan. Financial support for the meeting was provided by the IAU, the Japan Society for the Promotion of Sciences, the Japan World Exposition Commemorative Fund, and the Yamada Science Foundation. Preparation for the Symposium and editing of the Proceedings were supported in part by Scientific Research Fund of Japanese Ministry of Education, Science and Culture (530603).

The Scientific and Local Organizing Committees carried most of the work toward the success of the meeting. Thanks are also due to our colleagues of Professor C. Hayashi's school who helped with the Symposium in various ways.

Daiichiro Sugimoto
Donald Q. Lamb
David N. Schramm

Tokyo
July 1980

D. Sugimoto, D. Q. Lamb, and D. N. Schramm (eds.), Fundamental Problems in the Theory of Stellar Evolution, xi.

SCIENTIFIC ORGANIZING COMMITTEE

G.S. Bisnovatyi-Kogan
P. Bodenheimer
S. Hayakawa
R. Kippenhahn
D.Q. Lamb
B. Paczyński
M. Plavec
D.N. Schramm
D. Sugimoto (Joint Chairman)
R.J. Tayler (Joint Chairman)

LOCAL ORGANIZING COMMITTEE

S. Hayakawa
R. Hoshi
N. Itoh
T. Matsuda
T. Nakano
K. Nakazawa
H. Sato (Chairman)
D. Sugimoto

LIST OF PARTICIPANTS

K. Aizu,	Rikkyo Univ., Tokyo, JAPAN
N. Arimoto,	Tohoku Univ., Sendai, JAPAN
G. Beaudet,	Univ. Montreal, Montreal, Quebec, CANADA
P. Bodenheimer,	Univ. California, Santa Cruz, CA, USA
W.S. Cameron,	NASA/GSFC, Greenbelt, MD, USA
V. Castellani,	Lab. Astrofis., Frascati, ITALY
V.M. Chechetkin,	Inst. Applied Mathematics, Moscow, USSR
K-Y. Chen,	Univ. of Florida, Gainesville, FL, USA
J.P. de Cuyper,	Vrije Univ. Brussel, Brussels, BELGIUM
C. de Loore,	Vrije Univ. Brussel, Brussels, BELGIUM
R.H. Durisen,	Indiana Univ., Bloomington, IN, USA
K. Ebina,	Osaka Univ., Osaka, JAPAN
Y. Eriguchi,	Univ. of Tokyo, Tokyo, JAPAN
T. Fukushima,	Maritime Safety Agcy., Tokyo, JAPAN
K. Fushiki,	Kyoto Univ., Kyoto, JAPAN
B. Gaffet,	Kyoto Univ., Kyoto, JAPAN
R.A. Gingold,	Monash Univ., Clayton, Victoria, AUSTRALIA
I. Hachisu,	Kyoto Univ., Kyoto, JAPAN
Y. Hagio,	Kyoto Univ., Kyoto, JAPAN
T. Hamada,	Ibaraki Univ., Mito, JAPAN
T. Hanawa,	Univ. of Tokyo, Tokyo, JAPAN
T. Hara,	Kyoto Sangyo Univ., Kyoto, JAPAN
T. Hasegawa,	Univ. of Tokyo, Tokyo, JAPAN
H.J. Haubold,	Zentralinst. Astrophys., Potsdam, GDR
S. Hayakawa,	Nagoya Univ., Nagoya, JAPAN
C. Hayashi,	Kyoto Univ., Kyoto, JAPAN
R. Hoshi,	Rikkyo Univ., Tokyo, JAPAN
S. Ichimaru,	Univ. of Tokyo, Tokyo, JAPAN
S. Ikeuchi,	Hokkaido Univ., Sapporo, JAPAN
S. Inagaki,	Kyoto Univ., Kyoto, JAPAN
T. Ishizuka,	Ibaraki Univ., Mito, JAPAN
S. Isobe,	Tokyo Astron. Obs., Mitaka, JAPAN
K. Itoh,	Rikkyo Univ., Tokyo, JAPAN
N. Itoh,	Sophia Univ., Tokyo, JAPAN
M. Izawa,	Kyoto Univ., Kyoto, JAPAN
P.C. Joss,	Massachusetts Inst. Tech., Cambridge, MA, USA
K. Kaminishi,	Kumamoto Univ., Kumamoto, JAPAN
Y. Kannari,	Tohoku Univ., Sendai, JAPAN
M. Kato,	Rikkyo Univ., Tokyo, JAPAN
S. Kato,	Kyoto Univ., Kyoto, JAPAN
M. Kiguchi,	Kyoto Univ., Kyoto, JAPAN
T. Kim,	Tokyo Astron. Obs., Mitaka, JAPAN
H. Kimura,	Purple Mt. Obs., Nanking, CHINA
R. Kippenhahn,	MPI f. Phys. & Astrophys., Garching, GFR

M. Kitamura,	Tokyo Astron. Obs., Mitaka, JAPAN
E. Kobayashi,	Osaka Kagaku Kyoiku Center, Osaka, JAPAN
H. Kodama,	Kyoto Univ., Kyoto, JAPAN
T. Komuro,	Kyoto Univ., Kyoto, JAPAN
M. Kondo,	Univ. of Tokyo, Tokyo, JAPAN
S. Kyan,	Rikkyo Univ., Tokyo, JAPAN
D.Q. Lamb,	Center for Astrophys., Cambridge, MA, USA
L.B. Lucy,	Columbia Univ., New York, NY, USA
K. Maeda,	Kyoto Univ., Kyoto, JAPAN
T. Makino,	Osaka Industry Univ., Daito, JAPAN
A.G. Massevitch,	Astronomical Council, Moscow, USSR
T. Matsuda,	Kyoto Univ., Kyoto, JAPAN
T.J. Mazurek,	State Univ. of New York, Stony Brook, NY, USA
L. Mestel,	Univ. of Sussex, Brighton, UK
L.V. Mirzoyan,	Burakan Astrophys. Obs., Burakan, Armenia, USSR
S. Miyaji,	Univ. of Tokyo, Tokyo, JAPAN
S. Miyama,	Kyoto Univ., Kyoto, JAPAN
H. Mizuno,	Kyoto Univ., Kyoto, JAPAN
T.Ch. Mouschovias,	Univ. of Illinois, Urbana, IL, USA
T. Murai,	Nagoya Univ., Nagoya, JAPAN
Y. Nakagawa,	Kyoto Univ., Kyoto, JAPAN
M. Nakamura,	Tohoku Univ., Sendai, JAPAN
T. Nakamura,	Kyoto Univ., Kyoto, JAPAN
Y. Nakamura,	Tohoku Univ., Sendai, JAPAN
T. Nakano,	Kyoto Univ., Kyoto, JAPAN
K. Nakazawa,	Kyoto Univ., Kyoto, JAPAN
H. Nariai,	Hiroshima Univ., Takehara, JAPAN
K. Nariai,	Tokyo Astron. Obs., Mitaka, JAPAN
M. Nishida,	Kyoto Univ., Kyoto, JAPAN
K. Nomoto,	NASA/GSFC, Greenbelt, MD, USA
T. Ogasawara,	Kyoto Univ., Kyoto, JAPAN
M. Oishi,	Univ. of Tokyo, Tokyo, JAPAN
A. Okazaki,	Tokyo Astron. Obs., Mitaka, JAPAN
K. Oohara,	Kyoto Univ., Kyoto, JAPAN
Y. Osaki,	Univ. of Tokyo, Tokyo, JAPAN
W. Packet,	Vrije Univ. Brussel, Brussels, BELGIUM
U. Rhee,	Univ. of Tokyo, Tokyo, JAPAN
I.W. Roxburgh,	Queen Mary College, London, UK
S.M. Ruciński,	Warsaw Univ. Obs., Warsaw, POLAND
S-A. Sørensen,	London Univ. Obs., London, UK
Y. Sabano,	Tohoku Univ., Sendai, JAPAN
S. Sakashita,	Hokkaido Univ., Sapporo, JAPAN
T. Sakurai,	Kyoto Univ., Kyoto, JAPAN
E.E. Salpeter,	Cornell Univ., Ithaca, NY, USA
H. Sato,	Kyoto Univ., Kyoto, JAPAN
K. Sato,	Kyoto Univ., Kyoto, JAPAN
E.L. Schatzman,	Obs. de Nice, Nice, FRANCE
D.N. Schramm,	Univ. of Chicago, Chicago, IL, USA
M. Sekiya,	Kyoto Univ., Kyoto, JAPAN
H. Shibahashi,	Univ. of Tokyo, Tokyo, JAPAN
K. Shibata,	Kyoto Univ., Kyoto, JAPAN

M. Simoda,	Tokyo Gakugei Univ., Koganei, JAPAN
D. Sugimoto,	Univ. of Tokyo, Tokyo, JAPAN
H. Suzuki,	Kyoto Univ., Kyoto, JAPAN
M. Suzuki,	Rikkyo Univ., Tokyo, JAPAN
F. Takahara,	Kyoto Univ., Kyoto, JAPAN
M. Takahara,	Kyoto Univ., Kyoto, JAPAN
K. Takayanagi,	Inst. Space Aeronaut. Sci., Tokyo, JAPAN
H. Takeda,	Kyoto Univ., Kyoto, JAPAN
Y. Tanaka,	Ibaraki Univ., Mito, JAPAN
Y. Tanaka,	Kyoto Univ., Kyoto, JAPAN
R.J. Tayler,	Univ. of Sussex, Brighton, UK
A. Tomimatsu,	Hiroshima Univ., Takehara, JAPAN
K. Tomita,	Hiroshima Univ., Takehara, JAPAN
W.M. Tscharnuter,	MPI f. Phys. & Astrophys., Garching, GFR
J. Tsuji,	Tokyo Astron. Obs., Mitaka, JAPAN
S. Tsuruta,	Montana State Univ., Bozeman, MO, USA
A.V. Tutukov,	Astronomical Council, Moscow, USSR
N. Ukita,	Tokyo Astron. Obs., Mitaka, JAPAN
T. Umebayashi,	Kyoto Univ., Kyoto, JAPAN
W. Unno,	Univ. of Tokyo, Tokyo, JAPAN
E.P.J. van den Heuvel,	Univ. van Amsterdam, Amsterdam, NETHERLANDS
D. Vanbeveren,	Vrije Univ. Brussel, Brussels, BELGIUM
O. Vilhu,	Univ. Helsinki, Helsinki, FINLAND
T. Wada,	Rikkyo Univ., Tokyo, JAPAN
T. Watanabe,	Res. Inst. Atmospherics, Toyokawa, JAPAN
J.C. Wheeler,	Univ. of Texas, Austin, TX, USA
F.B. Wood,	Univ. of Florida, Gainesville, FL, USA
M. Yamada,	Waseda Univ., Tokyo, JAPAN
A. Yamasaki,	Univ. of Tokyo, Tokyo, JAPAN
M. Yasutomi,	Nagoya Univ., Nagoya, JAPAN
T. Yokoyama,	Kumamoto Inst. Tech., Kumamoto, JAPAN
Y. Yoshii,	Tohoku Univ., Sendai, JAPAN
H. Yoshimura,	Univ. of Tokyo, Tokyo, JAPAN

SYMPOSIUM ON FUNDAMENTAL PROBLEMS IN THE THEORY OF STELLAR EVOLUTION — INTRODUCTORY REMARKS

Daiichiro Sugimoto
Department of Earth Science and Astronomy
College of General Education, University of Tokyo
Komaba, Meguro-ku, Tokyo 153 Japan

It is about twenty years since Professor C. Hayashi and his collaborators wrote a paper in 1962 on "Evolution of the Stars" in the Supplement series of *Progress of Theoretical Physics*. It seems proper to begin this Symposium with these words, because Friday, July 25 of this Symposium marks Professor Hayashi's sixtieth birthday and because it is hosted by the Research Institute for Fundamental Physics from which the Journal *Progress of Theoretical Physics* is issued.

In the twenty years since this paper, there has been much progress in this field of astrophysics. On the one hand observations in X-ray, ultraviolet, infrared and radio wave lengths have uncovered a variety of celestial bodies in different circumstances. Among them, we can now see a neutron star directly by means of the Einstein Observatory, and infrared and radio observations as well as ultraviolet observation have revealed physical characteristics of interstellar gas and protostars. Thus, now we can observe stars from their birth through their death. On the other hand, progress in theory has also been significant. In a sense Hayashi et al.'s paper in 1962 marked the transition to the computer age. As computers have become relatively common and increased in speed a great many theoretical models have been constructed. Here, however, we must examine for ourselves how such models have contributed to advance our understanding of nature.

The structure of astrophysics and the theory of stellar evolution, in particular, are characterized with three steps. In the first step, we have *local* physics in the sense that the physical processes taking place locally at given temperature and density must be known. In the second step the local physics is inserted into models of celestial bodies where selfgravitation plays an essential role. In the third step, the resultant global behavior of the models is compared with observations.

The second step, in particular is characteristic of astrophysics in the sense that it treats the aspect which are less familiar in ordinary laboratory physics but are inherent to celestial bodies. First

D. Sugimoto, D. Q. Lamb, and D. N. Schramm (eds.), Fundamental Problems in the Theory of Stellar Evolution, 1–3.

of all, the system is governed essentially by selfgravitation leading to a non-local nature and to spatial gradients in physical parameters, and by non-linear interactions with the range of forces equal to the size of the system. Because of this the system should be thermodynamically dissipative and open. Such a system behaves, to some degree, out of common sense. For example, the effective gravothermal heat capacity of the star is negative and the isothermal temperature distribution corresponds to a local minimum of entropy rather than local maximum in most cases. The nature of such a system is a motive force of evolution which characterizes astrophysical systems. Here we should ask how and, in particular, why they evolve, why are they both now and initially out of thermal equilibrium etc.

In order for models to be called *Theory*, they should enable us to understand not only their individual nature but also the general characteristics of such systems. For example the responses of the model to changes in physical parameters or in physical assumptions or approximations should be understood without adding further detailed numerical computations. Such generalizations are rather difficult, particularly in the second step, but we have to try to find such generalizations.

Though many computations and model constructions have been done recently, I personally think that the things in the second step still remain obscure. We are apt to say that we take into account all detailed physics as accurately as possible, but in many cases we are talking only about the local physics in the first step. Too much physics in detail sometimes causes complexities and may even obscure the essential physics acting in the second step. In many instances physics in the second step has not even acquired its citizenship. In such cases we only have a huge pile of numerical results and a description of the final results for a single special case. Only computers might know the physics involved in the second step, and we have a big response table between the input to and the output from the black box. This is something like the tale of blindmen stroking an elephant.

In order to remove such pit falls and to look into the black box, two things are important. First, we have to recognize the importance of idealized models which can be constructed by neglecting relatively unimportant factors. Second, we have to perform numerical experiments rather than construct detailed models. Such an approach is common in the field of physics and used to be common in astrophysics before the computer age. Recently, however, it is apt to be ignored in favour of detailed modeling.

Such aspects are becoming more and more important because we are now proceeding to extend our scope of studies which include the effects of rotation, magnetic fields, two and three dimensional configurations etc. These studies are too complicated to be understood from a pile of numerical results. In the early sixties when Hayashi et al.'s paper was written such aspects were primarily taken into consideration because of the shortage of computing facilities. Now computers have become much more powerful but our subjects of research have become still more complicated and difficult. In this sense our relative situation is similar to that just before the computer age.

This is one of the reasons motivating the Symposium on *Fundamental* Problems in the Theory of Stellar Evolution. At this Symposium we hope to have much discussion concerning bold assessments of known facts, interrelations between them and strategy to surmount a barrier standing in front of the coming phase in the stellar evolution theory.

THE EFFECTS OF ROTATION DURING STAR FORMATION

Peter Bodenheimer
Lick Observatory, Board of Studies in Astronomy
and Astrophysics, University of California,
Santa Cruz, California U.S.A.

ABSTRACT

Observations of molecular clouds show evidence of rotation and of fragmentation of subregions of the clouds into multiple stellar or protostellar systems. This review concentrates on the effects that rotation and pressure gradients have in a self-gravitating cloud to cause it to undergo the crucial process of fragmentation. Recent two-dimensional and three-dimensional numerical hydrodynamic calculations have made progress in determining these effects. In most cases the calculations are performed with modest spatial resolution and are limited to isothermal clouds with neglect of viscous and magnetic effects. The combined results of several calculations strongly suggest that rotating clouds that are unstable to collapse are also unstable to fragmentation.

1. INTRODUCTION

Among the fundamental problems that must be addressed by theoreticians who study star formation, the following are now being actively pursued:

1. What are the dominant physical processes that must be considered at each stage of star formation?

2. How does one predict the rate of star formation both in our galaxy and in external galaxies, and how does this rate vary with position in the galaxy and with time since the formation of the galaxy? Closely connected with this question is that of the efficiency of star formation: of the total mass of interstellar material available in a gravitationally bound cloud in the galaxy, what fraction actually is formed into stars during the lifetime of the cloud?

3. What is the mass spectrum of the stars that are formed and what are the maximum and minimum masses of stars? How do these quantities vary according to position in the galaxy and with time since the formation of the galaxy?

4. How are binary and multiple systems of stars formed? The

D. Sugimoto, D. Q. Lamb, and D. N. Schramm (eds.), Fundamental Problems in the Theory of Stellar Evolution, 5–26.

important processes that must be considered include a) encounters and captures in a cluster or association of stars and protostars, b) fragmentation of a rotating, collapsing cloud into two or more protostellar objects during the earliest stages of stellar evolution, and c) fission of a rotating object in hydrostatic equilibrium during its pre-main-sequence contraction phase. What is the relative importance of each process and how are they interrelated?

5. What processes determine whether the end product of star formation is a single star, a planetary system, a binary system, or a multiple system? What is the probability that a planetary system will be formed?

Clearly a complicated network of physical processes must be considered in an attempt to answer questions 2 through 5. Although it is clear that certain subproblems related to star formation can be approached with analytical methods, the overall problem must be attacked with large-scale numerical computations involving a 3-D spatial grid. The physical processes that must be considered include at least the following: self-gravity, gas pressure, magnetic fields, rotation, radiative transfer, turbulence and convection, formation and dissociation of grains and molecules, and molecular and grain chemistry. If star formation is taken to start in a gravitationally bound cloud, the solution will depend on various parameters needed to define this initial state, which depends on the processes by which the dense molecular cloud complexes were formed in the interstellar gas. Finally, the solution to the star formation problem must involve observational predictions. For example, the emergent infrared spectrum of a collapsing protostar can be calculated, or the profiles of the lines emitted by the CO molecule can be calculated from the models and compared to the radio observations. A considerable amount of such work has been done in the spherically symmetric case, but studies of rotating objects are just beginning (see the review by Bertout and Yorke 1978).

The effects of rotation clearly can not be studied separately from the other physical processes that are important in star formation. However, the overall problem is so complicated that no attempt has been made to solve it in full generality; rather the interactions of a few physical effects (e.g., gravity, rotation, and pressure, or rotation and magnetic fields) have been studied under restrictive assumptions and idealized initial conditions. Those recent studies, primarily numerical in nature, that emphasize the effects of rotation will be discussed in this paper. Clearly, angular momentum must be one of the dominant effects in the solution of questions 4 and 5 above, and it undoubtedly has significant indirect, and largely unexplored, effects on the solution of questions 2 and 3.

The evolution of a star, up to the time where nuclear reactions become significant, can be divided into three periods. The first, star formation, refers to the approximate density range $10^{-23} - 10^{-19}$ g cm^{-3} and concerns the processes by which a massive interstellar cloud collapses and at the same time fragments into gravitationally bound pieces of order 1 solar mass. The second period, protostellar evolution, is assumed to begin at a density corresponding to the Jeans limit for a fragment of a given mass at a temperature of 10K. Densities increase from about 10^{-19} g cm^{-3} to 10^{-2} g cm^{-3} or more for fragments of 1 solar mass, and final temperatures are ~10^6K in the interior. This extreme compression, resulting from gravitational collapse, is due to two main mechanisms. The first applies to the star formation period and to the earlier part of the protostellar period, up to densities of about 10^{-13} g cm^{-3}. The protostellar material is optically thin to infrared radiation, and as collapse proceeds the heat generated by compression is immediately radiated by ions, atoms, molecules, or grains. Thus thermal pressure is unable to halt gravitational collapse unless rotation or magnetic fields begin to play an important role. The second mechanism occurs during the later part of protostellar collapse, starting at temperatures above 1800K and densities above 10^{-8} g cm^{-3} where dissociation of molecular hydrogen results in $\Gamma_1 < 4/3$ and consequent gravitational instability (in the absence of rotation). When collapse stops in the entire mass of the protostar, the third period of evolution begins, the pre-main-sequence contraction through a sequence of quasi-equilibrium states, continuing until nuclear-burning temperatures (~10^7K) are reached in the center. The evolution through periods two and three has been calculated in a continuous fashion only for the spherically symmetric case. However, when rotation is included, the protostar must evolve through analogous periods. Many of the recent calculations involving rotation have applied to the star formation period and the early part of protostar collapse, where an isothermal collapse at approximately 10K is a reasonable assumption. This review concentrates on these results. Note, however, that there have also been significant advances regarding the effects of rotation during the pre-main-sequence contraction phase by Lucy (1977), Gingold and Monaghan (1978, 1979), and Durisen and Tohline (1980). It has also been suggested (Larson 1980) that rotation plays a significant role in the explanation of the FU Orionis phenomenon.

2. OBSERVATIONAL DATA

The time scales for the three periods of evolution just referred to are approximately 10^7 years for star formation, 10^6 years for protostellar collapse, and 4 x 10^7 years for the quasi-static contraction of 1 solar mass. Consequently, there is abundant observational material in the visual and near infrared regions of the spectrum appropriate for study of the third period, and also a considerable amount of radio data appropriate for study of the star formation period. In the intermediate period, opportunities for comparison of theory and observation are very limited, due to the short time scale, expected heavy obscuration of protostars, and difficulties of observation in the relevant (far IR) spectral region.

From the observational material available regarding rotation it is clear that (1) angular momentum must play a significant role in star formation, and (2) there is a reduction of many orders of magnitude in the specific angular momentum of spin of particular mass elements between the molecular cloud phase (period 1) and the T Tauri phase (period 3). Typical values of specific angular momenta J/M for T Tauri stars are 10^{17}cm^2 s^{-1} to at most 10^{18} cm^2 s^{-1} (Herbig 1957, Kuhi 1978) if the stars are assumed to be uniformly rotating. On the other hand, a typical diffuse interstellar cloud would be expected to have J/M in the range 10^{23} to 10^{24} if it is corotating with its orbital motion about the center of the galaxy. Although there is little direct observational evidence of rotation in H I clouds (but note Gordon 1970) it is generally argued that the galactic magnetic field will be effective in maintaining corotation.

Much more observational evidence on rotation is available for massive dark clouds and globules which have densities of 10^{-20} - 10^{-22} g cm^{-3} and presumably correspond to the star formation period. Indirect evidence suggests that the so-called Hopper-Disney clouds, which are elongated dark clouds aligned with the galactic plane, are rotating as proposed by Heiles (1976) and further analyzed by Field (1978). More direct observational evidence for rotation in a few analogous objects is provided by velocity gradients in the microwave molecular emission lines of ^{13}CO. Milman's (1977) observation of the globule B361 gives an approximate angular velocity $\Omega = 10^{-13}$ s^{-1}. Martin and Barrett (1978) have observed the two Bok globules B163 and B163 SW and find spins of $\Omega = 6 \times 10^{-14}$ s^{-1} and 1.0×10^{-13} s^{-1}, respectively. The two objects are suspected to be in orbital motion with the angular momenta of spin aligned with that of the orbit. Other globules observed by Martin and Barrett, for example B335, have no detectable rotation. Large dark clouds, such as Mon R2 with an estimated mass of 10^4 solar masses, have also been observed to rotate; ^{13}CO observations give $\Omega = 1.4 \times 10^{-14}$ s^{-1} (Loren 1977) while CS observations in the same cloud give $\Omega \gtrsim 7.4 \times 10^{-14}$ s^{-1} (Kutner and Tucker 1975). Further examples, with similar values of Ω, are given by Field (1978) and Snell (1979). The specific angular momenta of these objects are in the range 10^{22} to 10^{23} cm^2 s^{-1} although there are also objects with lower values. Thus, apparently for many clouds where star formation is taking place, reduction in J/M by up to 6 orders of magnitude must take place by the time the T Tauri phase is reached. One of the chief aims of the numerical calculations to be discussed below is the resolution of this problem. In this connection there is additional observational evidence of interest concerning the question of fragmentation. Apart from the obvious fact that many young stars are found in clusters and associations, other observations suggest that fragmentation is a dominant process in star formation: (1) A considerable number of present main sequence stars are in multiple systems with two or more periods represented. Abt and Levy (1976) show that the typical main sequence star of spectral type F3-G2 has <u>both</u> a close and a distant companion. An example of such a system is κ Peg (ADS 15821), a visual binary with a period of 11.52 years each of whose components is a spectroscopic binary having periods of 4.77 days and 5.97 days, respectively (Beardsley and King 1976). A number of other systems with multiple periods in the same range have been observed, and the phenomenon is suggestive of a multiple fragmentation process during star formation. (2) Multiple infrared sources have recently been discovered in the cores of dense molecular clouds, where star formation is suspected to be taking place. For example, Beichman, Becklin and Wynn-Williams (1979) cite numerous observations that support the suggestion that infrared sources in such clouds commonly form in groups of two or more with a characteristic separation of 0.1 parsec. (3) The double Bok globule B163, B163 SW (Martin and Barrett 1978) mentioned above is suggested to be an example of a rotating cloud that has fragmented into orbiting subcondensations. Other molecular line observations also support the picture of fragmentation associated with rotation (Ho and Barrett 1980, Crutcher _et al_. 1978).

3. TWO-DIMENSIONAL NUMERICAL CALCULATIONS OF COLLAPSING CLOUDS

Three major suggestions regarding the solution of the angular momentum problem during star formation have come forward. (1) Stars form only from interstellar material that has much less angular momentum than the average. Although this effect may be significant, it has not been shown observationally that there is sufficient such material to account for the observed rate of star formation. Even material that is not rotating to present observational limits could have very significant J/M. (2) Magnetic fields result in braking of rotation through transfer of angular momentum from the cloud to the surrounding medium as a consequence of the propagation of Alfvén waves along the twisted magnetic field lines (Mestel and Spitzer 1956, Lüst and Schlüter 1955). Although it is clear, for example from the work of Mouschovias (1980), that this effect can reduce J/M by at least 2 or 3 orders of magnitude, eventually the braking becomes ineffective since the density in the cloud increases to the point where the degree of ionization is negligible and the field decouples from the matter. In fact, one would not expect magnetic braking to account for 100% of the angular momentum reduction, since then we could not account for the angular momentum of short and moderate-period binary systems that are unlikely to have formed by capture. (3) Angular momentum is converted from the spin of a collapsing cloud into orbital motion of binary or multiple systems that form as a result of fragmentation. Since the orbital J/M in binaries ranges from 4×10^{18} cm^2 s^{-1} (3-day period) to 10^{21} cm^2 s^{-1} (10^4 year period), the formation of binaries, perhaps through multiple fragmentation stages, naturally fills in the angular momentum gap between the rotating dark clouds and the T Tauri stars. The observations of rotation and fragmentation mentioned above support this suggestion, and many of the recent 2- and 3-D calculations of rotating clouds have been directed toward study of the process.

We first summarize recent calculations of collapsing rotating clouds in two space dimensions. The standard assumptions employed are : (1) axial symmetry; (2) global and local conservation of angular momentum, that is, no physical effects such as viscosity or magnetic fields transport angular momentum; (3) an ideal gas composed primarily of molecular hydrogen; (4) isothermal collapse, with certain exceptions noted below; (5) no mass flow in either direction through the outer boundary which is generally fixed in space. As long as the collapse remains isothermal the results may be scaled to any desired value of mass M, temperature T and molecular weight μ; thus the only parameters of the calculation are the two basic dimensionless quantities α, the initial ratio of thermal energy to gravitational energy, and β, the initial ratio of rotational energy to gravitational energy. Collapse occurs if $\alpha \leq 1. - 1.43\ \beta$ (Black and Bodenheimer 1976); this condition is essentially the Jeans criterion for a rotating cloud. The initial distributions of density and specific angular momentum are also parameters, although most calculations assume uniform density and angular velocity as initial conditions. The equations solved are the standard hydrodynamic equations of continuity, motion, and (where necessary) energy, along with the Poisson equation for the gravitational potential and an equation of state. Thus the physical effects included are self-gravity, gas pressure, and rotation. Radiation transport is included in some of the non-isothermal calculations.

The first question we may ask is whether there are any comparisons between 2-D calculations and observations. The principal study now available is a comparison with a set of six Bok globules, including the rotating ones B163, B163 SW, and B361 (Villere and Black 1980). The collapse models were generated from the code of Black and Bodenheimer (1975) while the observed parameters that were fit included the ^{13}CO column density at the center of the cloud, the ^{13}CO core radius, the axis ratio in the core, the ratio of the ^{13}CO core radius to the optical radius, the width of the ^{13}CO line profile, and the rotational velocity. In five of the six cases all of these parameters were consistent with collapsing cloud models. The derived masses are ~100 solar masses and the derived ages since the beginning of collapse are 50 to 90% of the initial free fall time. By this time the collapse has resulted in a centrally condensed density distribution but not yet in extreme rotational distortion or in ring structures. Predicted central densities of 10^4 to 2×10^5 H_2 molecules/cm^3 are also consistent with observations. Another important result is that the inferred initial density at the onset of collapse is much larger than that required for gravitational collapse ($\alpha \approx 0.1$). One globule (B361) is not fit by the models chiefly because of a large observed line width. The ratios of the abundances of ^{13}CO to H_2 in each globule are also determined by the model fit, and there is a clear trend of decreasing $^{13}CO/H_2$ with increasing density in the center of the cloud, in agreement with entirely independent measurements of this quantity in molecular clouds (Wootten et al. 1978). It was later found that all six globules could be fit if the ratio of $^{13}CO/H_2$ was allowed to vary spatially within the models as determined from the observations. Further calculations have also provided fits to six Lynds clouds, three of which are rotating.

A second important question concerns the degree to which the various 2-D calculations agree with each other. A detailed comparison has been carried out for the same initial and boundary conditions by Bodenheimer and Tscharnuter (1979). The former used an explicit "fluid-in cell" method involving differencing in two space dimensions and solution on a moving Eulerian grid; the latter used an implicit code involving differencing in the (spherical) radial direction only and a Legendre-polynomial expansion in the second (Θ) dimension. The case chosen had $\alpha = .46$, $\beta = .32$. Initially, Tscharnuter's code collapsed to a considerably higher central density than did Bodenheimer's. However, this discrepancy did not affect the later evolution since only a very small amount of mass was involved. In both cases a bounce occurred at the center, and after about 5 initial free-fall times, both calculations settled down to a near-equilibrium configuration that included a mild ring feature whose maximum density was about 4 times the central density. The isothermal equilibrium configuration that was obtained was quite flattened, with a polar-to-equatorial axis ratio of 1:7 and with a final β-value of 0.25. It had been previously suggested (Biermann and Michel 1978) that such nebulae, having about 2 solar masses and radii of about 10^4 AU, would provide a suitable location for the formation of cometary nuclei, provided that the nebula were stable for a long enough time to allow the dust grains to settle into a thin layer at the equatorial plane.

A second comparison calculation was performed by Boss (1980a) who used an explicit "fluid-in-cell" code which differed in several respects from the code of Black and Bodenheimer (1975) and also employed fewer grid points (220 versus 1600). A repeat of the Bodenheimer-Tscharnuter comparison resulted in a maximum density intermediate between those obtained by Bodenheimer and Tscharnuter. The long-term equilibrium structure was also quite similar and included a ring structure with moderate density contrast. However, Boss' model showed oscillations with shorter period and higher amplitude than those of Bodenheimer and Tscharnuter. Boss ran a second comparison case with $\alpha = .55$, $\beta = .02$, the same conditions used in one of the runs performed by Black and Bodenheimer (1976). Here the cloud is definitely gravitationally unstable, no equilibrium is possible, and after collapse and flattening a pronounced, self-gravitating ring structure develops in the central region of the cloud in both calculations. At this time the distributions of angular velocity, infall velocity, and density with distance from the center calculated by Boss and by Black and Bodenheimer are in excellent agreement. In summary, the available comparisons of 2-D hydrodynamic codes show overall fair agreement.

A third question that must be addressed is the physical mechanism for the origin of ring structures and the reality of their existence. This matter is one of considerable importance since such structures turn out to be sensitive checks on the accuracy of computer codes; furthermore, they have been shown to be unstable to fragmentation when three space dimensions are considered. Of the 2-D isothermal collapse calculations, those of Larson (1972), Black and Bodenheimer (1976), Nakazawa, Hayashi, and Takahara (1976), Regev (1979), Bodenheimer and Tscharnuter (1979), and Boss (1980a) have produced rings. A wide variety of numerical techniques is represented in these calculations. On the other hand, earlier calculations by Tscharnuter (1975) as well as those of Kamiya (1977) and Norman, Wilson and Barton (1980) do not show ring formation but rather a flattened disk-like structure in the interior of the cloud. When rings are found, they approach and later pass through a stage of hydrostatic equilibrium, at which time their properties agree well with those of equilibrium isothermal rings calculated analytically by Ostriker (1964). The controversial question is, however, the mechanism for excitation of the ring mode.

It would, of course, be desirable to find an analytic argument, entirely independent of the numerical codes, that would demonstrate the reality or non-reality of the ring structure. This approach has been explored by Tohline (1980a) and extended by Boss (1980a). An analytic approach becomes possible if the collapse is assumed to be pressure-free and the gravitational potential is fixed in time. In fact, the pressure does not play a dominant role in isothermal collapse, and the interplay between gravity and centrifugal effects is primarily responsible for the flow of material perpendicular to the rotation axis. Once rotational effects have produced a highly flattened structure, pressure effects do halt the collapse flow along the rotation axis. Tohline's approach was to integrate analytically the orbits of non-interacting particles. If the initial density distribution is uniform, and if the background static gravitational potential corresponds to this distribution, the particles all orbit in the potential well with the same period, maintaining a uniform but changing density. However, a typical collapsing cloud, due to the propagation of a rarefaction wave from the surface, always develops a non-uniform density distribution. Tohline also obtained an analytic expression for the particle orbits in the equatorial plane of a cloud collapsing in the potential of a $\rho \propto (1-r^2)$ distribution. In this case, the particles in the inner part of the cloud have shorter periods than those farther out; the inner ones reach their minimum radius sooner, then they begin to move with outward velocities, plowing into particles still falling in from larger radii. In this manner a density wave is excited that is ring-like in nature and that propagates outward. Although the ring self-gravity is not included in the analytic treatment, Tohline was able to show that enough mass is contained in the off-axis density enhancement to result in a self-gravitating ring; that is, the potential minimum would move into the ring.

Tohline supplemented his analytic calculations by solving the pressure-free collapse problem with a 2-D particle code. The gravitational potential was taken to be time-varying in a manner that closely approximated one of the collapses calculated by Black and Bodenheimer (1976). Again, the results show that ring formation occurred. The analytic approach was then extended by Boss (1980a) who obtained the particle orbits, not limited to the equatorial plane, of a rotating cloud with a static potential corresponding to $\rho \propto 1/r$. The development of phase differences and the resulting ring-like density wave that propagates outward is confirmed.

With a rather strong physical argument as well as diverse

analytical and numerical calculations now supporting the existence of a ring in 2-D collapse calculations, what can be said about the calculations that do not produce rings? The absence of rings in Tscharnuter's (1975) calculation is now thought to be due to artificial numerical transport of angular momentum outwards from the center of the cloud. In this regard, small details in the difference equations representing the equation of motion in the azimuthal direction can strongly influence the results. Kamiya (1977) performed his calculation on a Lagrangian grid. Although local conservation of angular momentum must be exact in such a calculation, there are other difficulties. The zones become highly distorted (Kamiya did not rezone) so that difference representations become inaccurate, and the solution for the gravitational potential is likely to be in error. The pressure-free but non-uniform model of McNally (1976) was probably not evolved long enough in time for the ring to develop.

In the calculation of Norman et al. (1980), however, considerable attention was paid to the minimization of numerical inward diffusion of angular momentum in a Eulerian scheme, an effect they suggest is responsible for ring formation in other calculations. The collapse that they calculate (α = .52, β = .08) is followed through a continuous increase of central density by more than ten orders of magnitude. Successively smaller fractions of the cloud's mass undergo a cycle of collapse, flattening, then a halt to the collapse due to pressure effects. The end result after 1.22 initial free-fall times is a flattened disk rather than the ring structure which Black and Bodenheimer (1976) obtained after a much smaller increase in central density for similar values of α and β. The reason for the discrepancy has not been clarified. Norman et al. do obtain ring formation for an initially differentially rotating but uniform-density cloud, and they suggest that the initial distribution of angular momentum versus mass is the critical parameter. The uniformly rotating uniform-density sphere may be a singular case that produces a disk, while only slight deviations from such a distribution (including those induced by numerical effects) could result in a ring solution (Norman 1980).

The later phases of protostellar collapse, during which an adiabatic approximation can be used in place of the isothermal approximation, have not been as thoroughly explored. The adiabatic phase is of particular importance, however, since the sizes and rotation periods of clouds starting collapse at densities above 10^{-12} g cm^{-3} are likely to be comparable to the separations and orbital periods of observed binaries. Furthermore, the primitive solar nebula undoubtedly formed in this density range.

Black and Bodenheimer (1976) calculated one collapse with an angular momentum appropriate for the solar nebula. The collapse started in the isothermal phase, but in the later stages the central regions became optically thick and began to heat adiabatically. A ring formed, involving only a tiny fraction of a solar mass at the center. Cameron (1978) speculates that the ring will fragment into a binary system but that as a consequence of accretion of mass from the outer collapsing region the fragments will merge and form a protosun. Tscharnuter (1978) also calculated a collapse of a cloud of 3 solar masses whose central regions entered the adiabatic regime. A nearly hydrostatic opaque core formed with temperature and density approximately 780K and 1.3×10^{-9} g cm^{-3}, respectively. The ring mode does appear in this core, but its properties and mechanism of formation differ from those of the rings that appear in the dynamically unstable regions of the isothermal portion of the collapse. Tscharnuter (1980) is now continuing the evolution of this "solar nebula" with the inclusion of viscous transfer of angular momentum. The possible mechanisms for angular momentum transfer at this point in the evolution are discussed by Safronov and Ruzmaikina (1978) and by Cameron (1978).

Other calculations have simply assumed a fully adiabatic rotating collapse, starting from a sphere with given density and angular velocity distribution. Takahara et al. (1977) take an ideal gas with adiabatic exponent $\gamma = 5/3$. Such a configuration is, in fact, not intrinsically unstable to gravitational collapse; thus the model by assumption starts out of equilibrium, but it tends to collapse toward the available equilibrium state. For slowly rotating clouds the core oscillates about an equilibrium oblate spheroid. A shock wave on the core boundary marks the region where the outer lower-density material is still being accreted. For rapidly rotating initial clouds, the core develops an off-axis density maximum, a ring-like structure that stays in equilibrium. Analogous calculations have been performed by Boss (1980c) with adiabatic exponents 5/3 and 7/5. If the final value of β for the equilibrium core is less than 0.43, the core forms a spheroid while if $\beta>0.43$, the final model is ringlike. The ring structure in this case is probably analogous to the classical ring-mode instability for Maclaurin spheroids that starts at comparable values of β. The rings formed during isothermal collapse are formed through an entirely different process. Incidentally, the core equilibrium structures provide an excellent check on the accuracy of the numerical calculations, particularly angular momentum conservation, since they can be compared with independently calculated differentially rotating polytropes (Bodenheimer and Ostriker 1973). The agreement in Boss' work is good.

How does star formation proceed from such an equilibrium? The cores are probably unstable to fragmentation only if $\beta>0.26$ (see below); also the fragments still need $\gamma < 4/3$ if they are to collapse. The equation of state provides this condition at a temperature of about 2000K when H_2 dissociates. If the equilibrium core reaches this temperature collapse will occur. Otherwise either external compression or internal angular momentum transport is required to allow the evolution to proceed. Bodenheimer (1978) has calculated one collapse in the adiabatic regime including H_2 dissociation. A ring structure forms at the center by a process similar to that which forms rings in the isothermal collapse.

4. THREE-DIMENSIONAL NUMERICAL CALCULATIONS OF COLLAPSING CLOUDS

The critical process of fragmentation in rotating clouds, which is related to the origin of binary, multiple, and planetary systems, can be studied only with calculations that take into account non-axisymmetric effects. A number of important questions can then be addressed:

1. Under what conditions will a rotating collapsing cloud fragment?
2. Is ring formation simply a consequence of the assumption of axial symmetry in 2-D calculations or does it play a role in fragmentation as well?
3. What is the role of the Jeans length in the fragmentation process?
4. What are the properties of the fragments? Will they undergo further collapse and fragmentation?
5. How reliable are the numerical calculations?

This last question is particularly important since the calculations are restricted to rather coarse spatial grids.

The standard assumptions used in the 3-D calculations include a) symmetry about the equatorial plane, b) isothermal or adiabatic collapse, c) ideal gas equation of state, and d) no viscous or magnetic effects. Thus, local transport of angular momentum can occur only through gravitational torques. Most of the calculations involve direct solution of the equations of hydrodynamics on numerical grids containing up to 10^4 cells. Since such calculations are expensive, other workers have adopted a technique where the flow is represented by a set of "fluid elements" whose motions are followed by means of an N-body calculation. Pressure effects are represented either by smoothing over the density distribution or by introduction of a repulsive force term. The main part of the following discussion is based on the results from solutions of the hydrodynamic equations.

One of the first calculations of this type was that of Norman and Wilson (1978), performed on a 40 x 40 x 26 grid. The initial configuration was a near-equilibrium isothermal ring that resulted from one of the axisymmetric calculations of Black and Bodenheimer (1976). The density distribution in the ring was perturbed non-axisymmetrically and the evolution followed to test for stability. If perturbations of 10% amplitude were introduced, corresponding to pure modes of $m = 2$ through 6, the ring fragmented into m blobs, symmetrically located according to the density maxima of the initial perturbations. The fragmentation occurs within half a rotation period and the growth rate is most rapid for $m = 2$. Other calculations were performed starting from a super-position of such modes, with randomly chosen phases and amplitudes. The lower-order modes dominated the fragmentation, and in four of the five cases calculated the end result was expected to be a binary system, while in the final case three blobs formed, equally spaced around the ring circumference. In the binary systems, the residual spin angular momentum of a fragment was approximately 20% of its orbital angular momentum. The instability of rings to fragmentation was confirmed by Tohline (1980b) for the m = 2 case. Cook and Harlow (1978) performed similar experiments (using a 10 x 12 x 5 grid) on equilibrium polytropic rings. The calculation was performed adiabatically. If the initial perturbation (in velocity) was of mode m = 2 and had an amplitude of 1%, the end result was fragmentation into a binary. Other modes were also investigated.

The important question still remains, however, of whether rings form at all during the collapse of a cloud. Tohline (1980b), using a 34 x 34 x 16 grid, started from a uniform-density, uniformly rotating isothermal cloud into which he introduced a m = 2 density perturbation of 50% amplitude. The parameters of such a calculation include α, β, and the type, mode, and amplitude of the initial perturbation. For $\alpha = .05$, $\beta = .28$ the cloud fragmented directly into a binary. For $\alpha = 0.5$ with two different values of β the result was that pressure effects damped the initial perturbation, and a nearly axisymmetric structure with a ring developed. Although the perturbation amplitude in the ring began to grow, no significant fragmentation occurred before the ring passed through its equilibrium configuration and began to collapse axisymmetrically. Tohline (1980b) has shown that the perturbations left by the time the ring stage is reached are different in nature from those introduced directly into the ring by Norman and Wilson (1978) and that fragmentation is less likely to occur.

Cook and Harlow (1978) made analogous calculations (but with a smaller initial perturbation) of the collapse of an adiabatic cloud with $\gamma = 5/3$. Varying the initial value of α, they found that the higher-pressure case ($\alpha = .15$) resulted in damping of the perturbation while a lower value of $\alpha(=.1)$ led to growth and fragmentation. Recently Boss (1980d) has made more extensive adiabatic calculations on a finer grid, using $\gamma = 7/5$ and starting again from uniform density and uniform rotation. The perturbation imposed was of mode 2 with an amplitude of 50%. For low $\beta(0.05-0.1)$ fragmentation occurs only for $\alpha < 0.075$, while for higher $\beta(.2-.3)$ fragmentation occurs for $\alpha<0.15$. The higher-α runs settle down into near-equilibrium spheroids with final $\beta<.27$, the critical value above which they would be dynamically unstable to non-axisymmetric modes.

Returning to the isothermal case, Narita and Nakazawa (1978) calculated collapses with $\alpha = .3$ and two values of β starting from a highly non-axisymmetric cloud resembling an ellipsoid of non-uniform density and uniform angular velocity. A ring formed that eventually broke up into a binary system. A comparison calculation between two independent 3-D computer codes starting from the same initial conditions ($\alpha = .25$, $\beta = .2$, 50% perturbation of m = 2) was performed by Boss and Bodenheimer (1979). The codes both used the "fluid-in-cell" techniques but in general they were based on different coordinate systems, number of grid points, and difference methods. In both cases a binary system formed directly; the masses of the components were each 15% of that of the original cloud, their ratio of spin to orbital angular momentum was about 0.2, and their value of α was about 0.05 so that they contained many Jeans masses. Boss (1980b) has performed a number of other isothermal calculations, whose results contain two notable differences from other calculations. In particular, a calculation with $\alpha = .25$, $\beta = .2$ and no initial perturbation except for small numerical effects collapses, forms an axisymmetric ring, and then fragments into a binary. Initially axisymmetric calculations performed with the Tohline code remain axisymmetric. The second difference lies in the occurrence of single blobs in some runs. For example, from the initial condition of $\alpha = .24$, $\beta = .18$ with a centrally condensed density distribution and with no initial perturbation the collapse developed into a ring which then broke up into a single off-axis condensation. A similar result occurred for $\alpha = .63$, $\beta = .2$ with a 50% m = 2 perturbation, but the blob was not well defined. In other, more standard cases, Boss obtains the normal binary.

Boss (1981) has also performed calculations with a tidal perturbation induced by the presence of a nearby protostar, located 2 cloud radii away. For $\alpha = .25$, $\beta = .20$ and for various values of the mass of the distorting object, the collapsing cloud is distorted into a bar-like shape and it then fragments into a binary. The fragments all have spin J/M about a factor 20 lower than that of the original cloud. For $\alpha = .63$, however, the tidal forces again result in a bar-like configuration but the thermal energy is sufficient to prevent fragmentation; rather, a single dense fragment results. Another variation on the initial conditions was studied by Różyczka et al. (1980a) who imposed random subsonic variations on the velocity field of the uniform-density isothermal cloud. Fragmentation results when $\alpha = .02$ but not for $\alpha = 0.1$ and 0.5.

A comprehensive summary of the 3-D collapse of an isothermal cloud is provided by the work of Bodenheimer, Tohline, and Black (1980). A wide range of α and β is studied starting from two different types of perturbations with mode 2 and amplitude 10% and 50%. The general results show that the cloud first collapses toward a centrally condensed thin disk (Figure 1). When pressure effects become important in slowing the collapse parallel to the rotation axis, a shock forms on the edge of the disk. Only after about one initial free-fall time, when rotation has begun to stabilize the disk in the direction perpendicular to the rotation axis does fragmentation begin. Two different types were noticed. For $\alpha<0.3$ the non-axisymmetric perturbations grow during collapse, although slowly at first, and the cloud fragments directly, usually into a binary, but occasionally into four fragments. This type of fragmentation is illustrated in Figure 2; the bar-like nature of the overall structure is evident. On the other hand, for $\alpha>0.3$ the general result is damping of the initial perturbation and formation of a nearly axisymmetric ring structure, which then fragments (Figure 3). Clouds with α up to 0.6 were found to fragment, with the exception of an intermediate range around $\alpha = 0.5$, where the ring began to collapse on itself on a time scale shorter than the fragmentation time scale. In these cases the numerical code was unable to follow the fragmentation, although its occurrence is likely.

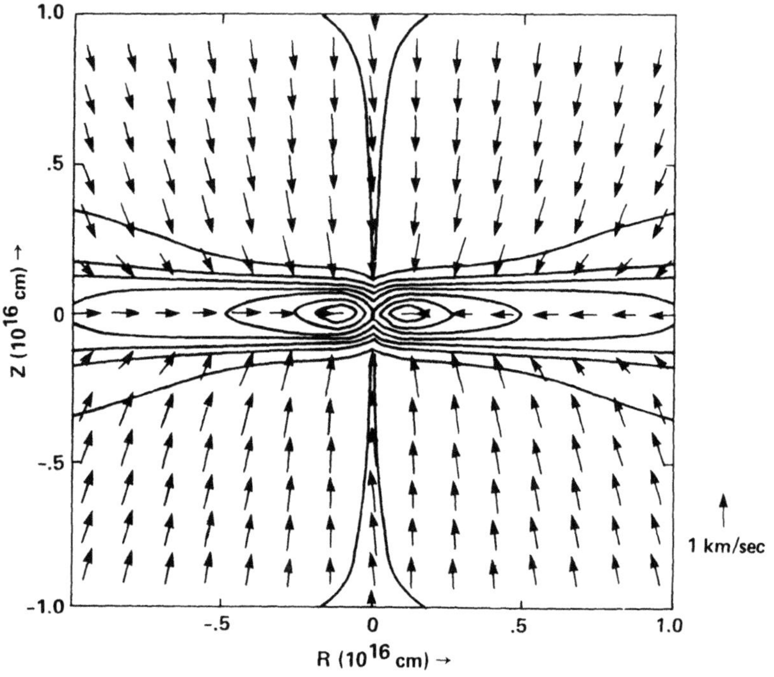

Figure 1. Cross section of the inner portion of a collapsing cloud through the rotation (z) axis after 1.21 initial free-fall times. R is the distance to the rotation axis. Equidensity contours (solid lines) with an interval of a factor of 3.2 are shown along with velocity vectors (arrows) with length proportional to speed. The calculations were done in 3 space dimensions (Bodenheimer, Tohline, and Black 1980).

A somewhat different approach is taken by Larson (1978) who uses a finite particle scheme with repulsive forces between neighboring particles to represent the thermal pressure. Dissipation due to shocks is represented by an additional viscous term which results in more transport of angular momentum than in the fluid-dynamic schemes. The initial conditions are similar to those in the other calculations with no explicit perturbation other than that given by a random initial distribution of the particles in a sphere. With $\alpha = 0.25$ and $\beta = 0.3$ a binary forms, for $\alpha = 0.35$, $\beta = 0.19$ a single dense condensation without a ring forms in the center, and with $\alpha = 0.075$, $\beta = 0.3$, the result is multiple fragmentation into 5-10 sub-condensations. The general outcome is that the number of fragments obtained is approximately equal to the number of Jeans masses contained in the original cloud.

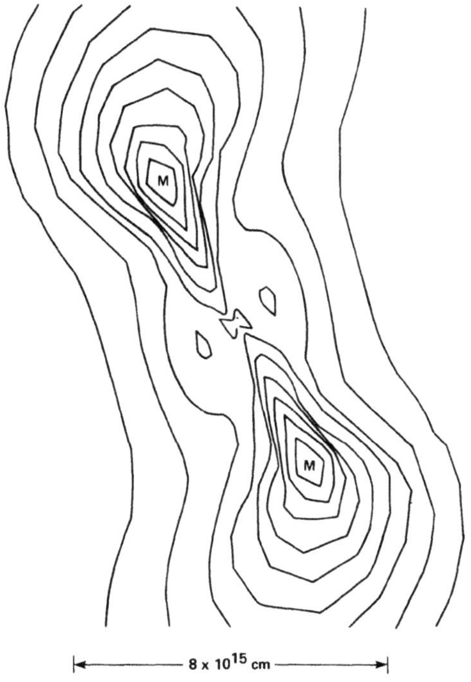

Figure 2. Equidensity contours in the equatorial plane (interval: factor 1.7) for a cloud that has undergone direct fragmentation. The maximum density is denoted by M; the center is a local minimum. (Bodenheimer, Tohline, and Black 1980).

Figure 3. Equidensity contours in the equatorial plane (interval: factor 1.7) for a cloud that has undergone a "ring-mode" fragmentation. The maximum density is denoted by M; the center is a density minimum. (Bodenheimer, Tohline, and Black 1980).

Multiple fragmentation of this type has not been observed in other calculations, partly because short-wave-length perturbations tend to be numerically and physically damped in the fluid dynamic calculations, and partly because Larson's fragments become well-developed only after several free-fall times, beyond the point where the other calculations are forced to stop for numerical reasons. Larson has deduced that a power-law mass spectrum will be the outcome. However, the finer details of these calculations must be taken with a good deal of caution because of the approximations and the small number of particles involved.

An improved numerical particle scheme that adopts some of the features of Lucy's (1977) method has been developed by Wood (1980). The smoothing length scale is defined locally which is an advantage since the development of fragments can be followed with reasonable resolution wherever they occur. The calculation starts with random 5% density fluctuations on an otherwise uniform-density uniformly rotating sphere. For $\alpha = 0.3$, $\beta = 0.23$ the cloud collapses, bounces in the z-direction, becomes unstable to a bar mode, and then fragments into a binary. For other cases, for example $\alpha = .2$, $\beta = .3$, the result was three symmetrically located condensations.

Most of the above-described calculations assumed an initial configuration that was gravitationally unstable. However, it is also of interest to mention briefly some fragmentation calculations that start from equilibria. Relevant structures include the equilibrium rings that were subjected to the analysis of Norman and Wilson (1978), the isothermal equilibrium disks (Bodenheimer and Tscharnuter 1979), and the quasi-spheroidal cores that result from adiabatic collapse (Takahara _et al._ 1977, Boss 1980c). If star formation is to proceed beyond such equilibria, some physical process must occur, and fragmentation is a strong possibility. Różyczka _et al._ (1980b) performed 3-D calculations on the isothermal equilibrium disk of Bodenheimer and Tscharnuter (1979) which has a final $\beta = 0.25$. Density perturbations of mode 2 and 4 with amplitudes of 10%, 20%, and 40% were applied. The $m = 4$ perturbations result in oscillations but no clear fragmentation, while in most cases the $m = 2$ perturbations result in fragmentation into a binary. Similar calculations were performed by the author, but the initial 50% perturbation was applied to the cloud before the onset of collapse rather than at the equilibrium phase. When the equilibrium was approached, the perturbation had damped considerably, but after about 3 initial free-fall times the central part of the cloud had clearly fragmented into a binary.

A 3-D numerical analysis (Durisen and Tohline 1980) of the dynamical stability of rapidly rotating polytropes applies to the adiabatic equilibria (Takahara _et al._ 1977 and Boss 1980c) which closely approximate polytropes of index 1.5 when $\gamma = 5/3$. Their analysis, of course, also is relevant to the possible fission of pre-main-sequence stars at a much later stage of their evolution. Starting with a density perturbation of 10% or 33% amplitude and mode 2, they follow the evolution of polytropes of n = 0.5 and n = 1.5 with the angular momentum distribution of a uniformly rotating uniform sphere, using the 3-D fluid-dynamic code of Tohline (1980b). For n = 0.5 the perturbation damped for $\beta = 0.25$ and grew for $\beta = 0.33$. For n = 1.5 it damped for $\beta = 0.08$, 0.18, and 0.27, grew for 0.33, and remained roughly constant for 0.30. Within the limits of the numerical results, the point of dynamic instability of the polytropes agrees roughly with the earlier tensor virial equation analyses which showed that the point occurs at $\beta \cong 0.26$ (e.g., Ostriker and Bodenheimer 1973). It is not yet conclusive, however, that the instability leads to fragmentation. Extensive evolution of the high-β configurations shows that an incipient binary forms at the center, but that the density contrast there does not increase noticeably during the ensuing phases, which involve the development of a spiral arm pattern in the outer regions and its evolution into an exponential disk and then into an expanding ring.

5. CONCLUSIONS

The overall results of the 2-D and 3-D hydrodynamic calculations with rotation applied to the star formation and protostar collapse periods of stellar evolution allow us to conclude the following:

1. Rotating collapsing interstellar clouds are unstable to fragmentation over a wide range of initial conditions in the isothermal phase, given even a small initial non-axisymmetric perturbation. If the cloud is unstable to collapse almost any combination of the α- and β-parameters will result in fragmentation. Flattened isothermal equilibria with β about 0.25 are also unstable.
2. The process of fragmentation differs from that suggested by Hoyle (1953) and Hunter (1962) in which fragments spontaneously appear when their mass exceeds the local Jeans mass. Perturbations can in fact damp initially due to pressure effects. After about one initial free-fall time, when the cloud collapse is slowed by pressure effects parallel to the rotation axis and primarily rotational effects perpendicular to the axis, the fragmentation begins.
3. Fragmentation can occur either directly as a consequence of the initial perturbation imposed on the cloud, or through an intermediate ring stage. The dividing line is roughly at $\alpha = 0.3$, but it also depends on the amplitude of the perturbation.
4. The dominant mode of fragmentation is the binary (m = 2) mode, although results with one, three, and four fragments have also been reported.
5. Among the results discussed above, there are some disagreements between authors on cases that were run with essentially the same initial conditions. Numerical difference techniques and numerical accuracy undoubtedly are of importance in this regard. For example, on relatively coarse numerical grids some numerical damping of perturbations inevitably occurs. Thus, some of the details of the results of the various calculations may change in the future as a consequence of improved numerical methods.
6. The properties of the fragments in the isothermal case are

such that they are unstable to collapse. The mass of a fragment in many cases is 10 to 15% of the cloud mass, the value of α is a few percent, and the ratio of spin to orbital angular momentum is very roughly 20%. The fragments form in the innermost part of the cloud which has lower J/M than the average for the cloud. This effect, combined with the conversion of spin to orbital motion, results in a reduction of spin J/M by a factor 10 to 20 from that of the initial cloud. Thus, after a series of several collapses and fragmentations the J/M as well as the fragment masses can be reduced by considerable factors. Bodenheimer (1978) showed that such a process could result in direct evolution from a massive interstellar cloud to main-sequence binary and multiple systems within the observed range of masses and orbital angular momenta.

7. During the adiabatic phase the tendency to fragment is not as universal as it is in the isothermal phase. The collapse approaches an equilibrium, and fragmentation is likely if β in the equilibrium configuration is above 0.27. Initial conditions set up at the end of the isothermal phase, which involve fragments with $\alpha<0.1$, suggest that the conditions for further fragmentation during the adiabatic phase are in general satisfied. If they are not, two possibilities are open. First, the adiabatic evolution could result in temperatures above the dissociation temperature for H_2. Fragmentation would be as likely to occur during the ensuing collapse as it is during the isothermal phase. Second, if the temperature remains too low, the equilibrium configuration can evolve only as a consequence of angular momentum transport. This final set of circumstances may provide the conditions necessary to form a single star and a surrounding rotating nebula.

REFERENCES

Abt, H.A., and Levy, S.G.:1976, Astrophys. J. Suppl. 30, pp. 273-306.
Beardsley, W.R., and King, M.W.: 1976, Publ. A.S.P. 88, pp. 200-203.
Beichman, C.A., Becklin, E.E., and Wynn-Williams, C.G.: 1979, Astrophys. J. (Letters) 232, pp. L47-L51.
Bertout, C., and Yorke, H.W.: 1978, Protostars and Planets, ed. T. Gehrels (Tucson: University of Arizona Press), pp. 648-689.
Biermann, L., and Michel, K.W.: 1978, Moon and Planets 18, pp. 447-455.
Black, D.C., and Bodenheimer, P.: 1975, Astrophys. J. 199, pp. 619-632.
Black, D.C., and Bodenheimer, P.: 1976, Astrophys. J. 206, pp. 138-149.
Bodenheimer, P.: 1978, Astrophys. J. 224, pp. 488-496.
Bodenheimer, P., and Ostriker, J.P.: 1973, Astrophys. J. 180, pp. 159-169.
Bodenheimer, P., Tohline, J.E., and Black, D.C.: 1980, Astrophys. J., in press.
Bodenheimer, P., and Tscharnuter, W.M.: 1979, Astron. Astrophys. 74, pp. 288-293.
Boss, A.P.: 1980a, Astrophys. J. 237, pp. 563-573.
Boss, A.P.: 1980b, Astrophys. J. 237, pp. 866-876.
Boss, A.P.: 1980c, preprint.
Boss, A.P.: 1980d, paper presented at IAU Colloquium No. 58: Stellar Hydrodynamics.

Boss, A.P.: 1981, to be published.
Boss, A.P., and Bodenheimer, P.: 1979, Astrophys. J. 234, pp. 289-295.
Cameron, A.G.W.: 1978, Moon and Planets 18, pp. 5-40.
Cook, T.L., and Harlow, F.H.: 1978, Astrophys. J. 225, pp. 1005-1020.
Crutcher, R.M., Hartkopf, W.I., and Giguere, P.T.: 1978, Astrophys. J. 226, pp. 839-850.
Durisen, R.H., and Tohline, J.E.: 1980, in preparation.
Field, G.B.: 1978, Protostars and Planets, ed. T. Gehrels (Tucson: University of Arizona Press), pp. 243-264.
Gingold, R.A., and Monaghan, J.J.: 1978, Mon. Not. R. Astron. Soc. 184, pp. 481-499.
Gingold, R.A., and Monaghan, J.J.: 1979, Mon. Not. R. Astron. Soc. 188, pp. 39-44.
Gordon, C.: 1970, Astron. J. 75, pp. 914-925.
Heiles, C.: 1976, Ann.Rev. Astron. Astrophys. 14, pp. 1 - 22.
Herbig, G.H.: 1957, Astrophys. J. 125, pp. 612-613.
Ho, P.T.P., and Barrett, A.H.: 1980, Astrophys. J. 237, pp. 38-54.
Hoyle, F.: 1953, Astrophys. J. 118, pp. 513-528.
Hunter, C.: 1962, Astrophys. J. 136, pp. 594-608.
Kamiya, Y.: 1977, Progr. Theor. Phys. 58, pp. 802-815.
Kuhi, L.V.: 1978, Moon and Planets 19, pp. 199-202.
Kutner, M.L., and Tucker, K.: 1975, Astrophys. J. 199, pp. 79-85.
Larson, R.B.: 1972, Mon. Not. R. Astron. Soc. 156, pp. 437-458.
Larson, R.B.: 1978, Mon. Not. R. Astron. Soc. 184, pp. 69-85.
Larson, R.B.: 1980, Mon. Not. R. Astron. Soc. 190, pp. 321-335.
Loren, R.B.: 1977, Astrophys. J. 215, pp. 129-150.
Lucy, L.B.: 1977, Astron. J. 82, pp. 1013-1024.
Lüst, R., and Schlüter, A.: 1955, Zeitschr. f. Astrophys. 38, pp. 190-211.
Martin, R.N., and Barrett, A.H.: 1978, Astrophys. J. Suppl. 36, pp. 1-51.
McNally, D.: 1976, Mém. Soc. Roy. Sci. Liège 9, pp. 487-496.
Mestel, L., and Spitzer, L., Jr.: 1956, Mon. Not. R. Astron. Soc. 116, pp. 503-514.
Milman, A.S.: 1977, Astrophys. J. 211, pp. 128-134.
Mouschovias, T. Ch.: 1980, Moon and Planets 22, pp. 31-45.
Nakazawa, K., Hayashi, C., and Takahara, M.: 1976, Progr. Theor. Phys. 56, pp. 515-530.
Narita, S., and Nakazawa, K.: 1978, Progr. Theor. Phys. 59, pp. 1018-1019.
Norman, M.L.: 1980, private communication.
Norman, M.L., and Wilson, J.R.: 1978, Astrophys. J. 224, pp. 497-511.
Norman, M.L., Wilson, J.R., and Barton, R.T.: 1980, Astrophys. J., in press
Ostriker, J.P.: 1964, Astrophys. J. 140, pp. 1067-1087.
Ostriker, J.P., and Bodenheimer, P.: 1973, Astrophys. J. 180, pp. 171-180.
Regev, O.: 1979, Santa Cruz Summer Workshop on Star Formation.
Różyczka, M., Tscharnuter, W.M., Winkler, K.H., and Yorke, H.W.: 1980b, Astron. Astrophys. 83, pp. 118-128.
Różyczka, M., Tscharnuter, W.M., and Yorke, H.W.: 1980a, Astron. Astrophys. 81, pp. 347-350.
Safronov, V.S., and Ruzmaikina, T.V.: 1978, Protostars and Planets, ed. T. Gehrels (Tucson: University of Arizona Press) pp. 545-564.
Snell, R.L.: 1979, Ph.D. dissertation, University of Texas at Austin.
Takahara, M., Nakazawa, K., Narita, S., and Hayashi, C.: 1977, Progr. Theor. Phys. 58, pp. 536-548.

Tohline, J.E.: 1980a, Astrophys. J. 236, pp. 160-171.
Tohline, J.E.: 1980b, Astrophys. J. 235, pp. 866-881.
Tscharnuter, W.M.: 1975, Astron. Astrophys. 39, pp. 207-212.
Tscharnuter, W.M.: 1978, Moon and Planets 19, pp. 229-236.
Tscharnuter, W.M.: 1980, this Symposium.
Villere, K.R., and Black, D.C.: 1980, Astrophys. J. 236, pp. 192-200.
Wood, D.: 1980, Mon. Not. R. Astron. Soc., in press.
Wootten, A., Evans, N.J., Snell, R., and Vanden Bout, P.: 1978, Astrophys. J. (Letters) 225, pp. L143-L148.

DISCUSSION

Nakamura: How have you checked the accuracy of your 2-D and 3-D numerical calculations.

Bodenheimer: A large number of tests of the accuracy of the solutions have been performed, and new tests are continually being devised. The one currently applied are
1. Comparison with analytic solutions in limiting cases, where available.
2. Intercomparison between different numerical techniques on the same physical problem. Example: Boss and Bodenheimer, Astrophys. J. 234, 289.
3. Examination of the effect of increasing the number of grid points.
4. Tests of the conservation of the distribution of angular momentum with mass, in the 2-D case (see Norman, Wilson, and Barton, Astrophys. J. August 1980).
5. Comparison with the detailed structure of equilibrium differentially rotating polytropes (Bodenheimer and Ostriker, Astrophys. J. 180, 159) in the case of adiabatic 2-D collapses, which should reach such equilibria.

Nariai: For the case of binary-type fragmentation, the loss of mass from the region outside of the outer Lagrangian points greatly reduces the J/M ratio (Nariai and Sugimoto 1976, Publ. Astron. Soc. Japan 28, 593). Is this effect included in your calculations?

Bodenheimer: This effect is included, but it has not been analyzed separately from the other effects. It would seem that this mechanism provides additional help in solving the argular momentum problem.

Mouschovias: You mentioned semi-analytical results which explain the formation of rings as due to a "density wave". Was there an implication that rotation is not important in ring formation?

Bodenheimer: Rotation is critically important in the generation of the wave. Because of rotation and a non-uniform density distribution, the particles in the inner regions overshoot their equilibrium positions, rebound, interact with particles from the outer regions that are still falling in, and so generate the wave.

Tscharnuter: According to your fragmentation scheme, the multiple stellar system should be strictly co-planar. Which processes would you consider to be responsible for the observed finite inclinations?

Bodenheimer: It is true that in some observed multiple stellar systems the short-period system and the long-period system are not co-planar. The direction of the angular momentum vectors could be influenced by (1) close encounters and interactions with other protostellar systems during the formation or early evolution of a dense cluster, or (2) magnetic fields which could affect the angular momentum of parts of a diffuse cloud before fragmentation actualy takes place.

Sugimoto: Is the number of final fragments related to the number of modes, for example, in the initial perturbation? If not, what do you expect for the number spectrum of the fragments?

Bodenheimer: In our calculations, we impose an m=2 perturbation and in most cases we get a binary as a result. However, we have also obtained 4 fragments. On general grounds I would expect that the growth rate of the lower-order modes would be faster than that of higher modes. Calculations by Norman and Wilson (Astrophys. J. 224, 497, 1978), starting from equilibrium rings, show that even if a mixture of modes (from m=2 to 6) is imposed in the initial perturbation, the final outcome is usually a binary. We still need to do more numerical calculations with better spatial resolution (to adequately represent the higher modes) to answer this question, but I expect that 2 fragments will be the usual result.

Tayler: In each stage of fragmentation only a small fraction of the mass is included in the fragments. Does this mean that you believe that star formation is very inefficient if only rotation, pressure and self-gravitation are important?

Bodenheimer: That is an important point. If only 20% of the mass of the cloud fragments at each stage (10% per fragment in a binary), there is considerable mass left over that has too much angular momentum to join the fragments. After several stages only a very small fraction of the original material would end up in stars, so the process would in fact be quite inefficient (but efficient in solving the angular momentum problem).

Unno: How many steps of fragmentation are needed for star formation in a realistic situation? Are there numerical simulations for that?

Bodenheimer: Approximately four stages of fragmentation are required for evolution from the interstellar cloud state to a main-sequence multiple system. Some approximate simulations based on the 2- and 3-D numerical solutions can be found in a paper by Bodenheimer (Astrophys. J. 224, 488).

Schatzman: What is the present situation regarding the mass spectrum problem?

Bodenheimer: From the theoretical point of view, I think that the question is completely open. The 3-D calculations that I have been discussing show that fragmentation occurs, but they make no predictions regarding the mass spectrum. Larson suggests (Mon. Not. R.A.S. 184, 69, 1978) that his fragmentation calculations result in a power-law mass spectrum. However a number of approximations are involved. Silk and Scalo, and co-workers, have investigated models involving coalescence of numerous small fragments plus accretion of the surrounding gas. However, there are problems with this model as well. For example, the origin of the original fragments is obscure: It is based on the Hoyle-Hunter fragmentation theory, which is no longer believed, and furthermore rotation is not included at all in the analysis.

THE ROLE OF MAGNETIC FIELDS IN THE FORMATION OF STARS*

Telemachos Ch. Mouschovias**
University of Illinois at Urbana-Champaign
Departments of Physics and Astronomy

ABSTRACT

We review the role of the interstellar magnetic field: (i) in the formation of interstellar clouds; (ii) in determining critical states for gravitational collapse; (iii) in affecting the collapse and fragmentation of interstellar clouds; and (iv) in resolving the "angular momentum problem" during star formation. Finally, we review the manner in which the field decouples from the matter via ambipolar diffusion; new time-dependent solutions are discussed.

1. INTRODUCTION

This review is written specifically for nonspecialists in the field of interstellar magnetohydrodynamics as it relates to star formation. The fundamental problems of star formation resolved or posed by the presence of interstellar magnetic fields, and the significant results of theoretical investigations on the role of magnetic fields in the formation of stars (and planetary systems) are described physically, with only occasional reference to the underlying mathematical formalism. We restrict our attention to the early stages of star formation. (The effects of magnetic fields on stellar structure and later stages of stellar evolution are reviewed by Mestel in this volume.) More specifically, we review some key physical processes expected to take place during (or to determine) the formation and dynamical contraction of interstellar clouds out of the mean number density of the interstellar medium ($n \sim 1\ cm^{-3}$), which is permeated by a mean magnetic field $B \sim 3$ microgauss. The relevance of the relatively diffuse stages of contraction of a cloud with a given mass stems from our current theoretical understanding that physical quantities, such as angular momentum and

*This work has been supported in part by the National Science Foundation under grant NSF AST-80-00667.
**Alfred P. Sloan Foundation fellow 1980-1982.

D. Sugimoto, D. Q. Lamb, and D. N. Schramm (eds.), Fundamental Problems in the Theory of Stellar Evolution, 27–62.

magnetic flux, present at these stages, may (i) determine to a large extent the subsequent dynamical evolution of a cloud; and (ii) reach residual (or, terminal) values during these stages. Thus, stellar rotation (referring both to single stars and multiple stellar systems) and stellar magnetic fields, at least at the onset of the Hayashi phase, may be obtainable from theoretical calculations of cloud formation and collapse.

The theoretical problems remaining unsolved on the role of magnetic fields in star formation are of such fundamental nature that analytical work is not only possible but also essential for (i) isolating the relevant physical processes; (ii) specifying the minimum set of parameters necessary for a unique determination of a solution; and (iii) obtaining a formal solution and interpreting it physically. If the qualitative nature of the conclusions is expected to change once a simplifying assumption is relaxed, such a calculation is merely an intellectual exercise. On the other hand, if relaxation of a simplifying assumption results only in a quantitative change in the conclusions, the calculation can be regarded as trustworthy, and numerical computations can be undertaken to improve quantitatively the analytical results under a more realistic set of assumptions. Some of the fundamental questions relating to magnetic fields that a theory of star formation must answer are the following.

1) Does the interstellar magnetic field play any role at all in interstellar gas dynamics in general and star formation in particular?

2) Can the field support a dense cloud against self-gravity and, if so, for how long?

3) Does it affect fragmentation and, if so, how?

4) Can the magnetic field resolve "the angular momentum problem" during star formation?

5) If it is dynamically important in the first place, does it ever become insignificant? At which stage, and by what means does this transition occur?

6) Does a cloud's magnetic energy increase or decrease upon contraction? If magnetic energy is released at some stage, in what form does it appear?

The unsettled observational questions are at least as significant as the theoretical ones. Although the evidence that a magnetic field of a few microgauss permeates the interstellar medium is firm, there is only meager observational evidence on the topology and correlation of the field strength and the gas density in diffuse H I and dense molecular clouds. The latter kind of information is essential if observations are to provide checks and input to theoretical calculations.

In §II we review the observational evidence for the interstellar magnetic field. This account is not complete in that it does not necessarily refer to the latest observations (e.g., see review by Chaisson and Vrba 1978). The emphasis is on the physics on which each observational method is based and on the difficulties in interpreting the observational results. In subsequent sections we review the progress made to date on the six fundamental issues raised above and on other matters concerning the role of magnetic fields in star formation.

2. OBSERVATIONAL EVIDENCE FOR THE INTERSTELLAR MAGNETIC FIELD

Detection of the interstellar magnetic field is based either on the fact that the field can be instrumental in the production of electromagnetic radiation or on the fact that it can modify radiation propagating through the space in which the field exists.

2.1 Synchrotron Radiation

Synchrotron radiation is produced by highly relativistic electrons gyrating in a magnetic field. It is emitted in a small solid angle about the instantaneous electron velocity, so that the line of sight must lie in the plane of the electron's orbit if the radiation is to be observed at all. The radiation from an ensemble of electrons is characterized by a power-law spectrum and by a high degree of linear polarization, with the electric field normal to the plane defined by the magnetic field and the line of sight (Ginzburg and Syrovatskii 1965; also, Bless 1968). With independent evidence for the existence of 1 GeV cosmic-ray electrons (see review by Meyer 1969), the synchrotron mechanism accounts for a major fraction of the background radio continuum emission in the Galaxy (e.g., see Spitzer 1968 or 1978 and references therein).

To deduce the magnitude of the field, one usually introduces a number of dubious assumptions, the most common of which is energy equipartition between magnetic fields and cosmic-ray protons. [At a given energy per particle, the number of cosmic-ray electrons is only about 2% that of protons (Earl 1961; Meyer and Vogt 1961).] Additional uncertainties enter in estimating the size of the emitting region. However, even if this size is known, further assumptions concerning its internal structure are necessary for estimating the strength of the magnetic field. This is so because the measured intensity of radiation at some frequency is proportional to the line integral (along the line of sight) of the product of the number density of relativistic electrons and a power (usually around 1.8) of the perpendicular (to the line of sight) component of the magnetic field. Large-scale magnetic fields ranging from 10 to 50 μgauss have been deduced (Woltjer 1965; Davis and Berge 1968). Daniel and Stephens (1970) used the fluxes of cosmic-ray electrons and synchrotron radiation observed at the earth to estimate an energy spectrum for electrons with energies $\lesssim$ 5 GeV (because the observed one has been modulated by the solar wind) and to show that this spectrum joins smoothly with the observed spectrum above 5 GeV (which does not suffer solar modulation) only if the magnetic field is in the range 6 - 9 μgauss. They assumed, however, that the region of emission was homogeneous. If the cosmic-ray density is fairly uniform, regions of strong fields are overweighted, and the mean background interstellar field may actually be weaker than the one deduced by Daniel and Stephens.

2.2 Polarization of Starlight

The observation that light from distant stars is partially polarized (Hall 1949; Hiltner 1949), and the correlation of the degree of polarization with interstellar reddening have been attributed to the dynamical alignment of elongated dust grains by the interstellar magnetic field (Davis and Greenstein 1951; Davis 1958; Miller 1962). The grains are presumed to be paramagnetic and to have a complex index of refraction. Jones and Spitzer (1967) used statistical arguments to arrive at the same conclusions.

In the absence of a magnetic field, a prolate grain in kinetic equilibrium with the surrounding gas will have equal rotational kinetic energies about each of its principal axes. The angular momentum about each principal axis is thereby proportional to the square root of the moment of inertia about that axis. A grain, therefore, tends to rotate mainly about an axis perpendicular to the axis of symmetry. In the presence of a magnetic field, dissipation of angular momentum due to magnetic torques will tend to align the axis of rotation with the direction of the field. Thus, the axis of symmetry (major axis) of the prolate grains will tend to be perpendicular to the magnetic field. [It should be emphasized, however, that alignment itself with respect to the field does not necessarily depend on the presence of dissipative torques (see Spitzer 1978, pp. 183-190).] It is essential in these considerations that the grain temperature be different from (in fact, less than) the gas temperature, so that the system will not be in thermodynamic equilibrium, which would destroy the alignment through collisions with gas atoms. The magnetic field needed to sufficiently orient the grains is of the order of 10 μgauss although a weaker field (1 μgauss) would do if the grains were ferromagnetic (Jones and Spitzer 1967). The value 10 μgauss is larger by a factor of 3 - 4 than the mean field strength obtained by reliable Faraday rotation measurements (see §2.3 below).

As starlight propagates through interstellar space, the component of the electric field perpendicular to the major axis of each grain (and, therefore, nearly parallel to the magnetic field) is less efficiently absorbed by these particles. Consequently, a map of the observed polarization vectors will also reveal the topology of the interstellar magnetic field (more precisely, of its perpendicular component) between the observed stars and the earth. The field lines, as unveiled by polarization measurements, exhibit an orderly large-scale behavior, with prominent arches over distances of a few hundred parsecs (Mathewson and Ford 1970; Davis and Berge 1968). The large-scale structure of the field correlates strongly with that of atomic hydrogen (Heiles and Jenkins 1976). Clouds and cloud complexes lie in valleys of the field lines, and arches of matter rising high above the Galactic plane coincide with the magnetic arches revealed by starlight polarization measurements.

On a smaller scale, that of individual clouds, the polarization vectors through a cloud merge smoothly with those observed in the vicinity of the cloud (Dyck and Lonsdale 1979). Polarization observations of several dark clouds by Vrba, Strom, and Strom (1976), as extended by a recent study of the R Corona Austrina dark cloud and its neighborhood by Vrba, Coyne, and Tapia (1980), support the same conclusion; the densities probed are less than about 10^3 cm^{-3}. Thus, the assumption of theoretical calculations made mainly by Mestel and his co-workers and Mouschovias and his collaborators (namely, that a cloud's magnetic field links smoothly with the field of the external medium) is beginning to find an observational foundation. The observed ordering of the field over the cloud also seems to rule out the simplifying assumption made sometimes by theorists and observers, namely, that magnetic fields in clouds are tangled up and that their effect can be represented merely by the scalar pressure $B^2/8\pi$.

In order to obtain the magnitude of the field from extinction and polarization measurements, one must know the gas temperature and density and, in addition, less certain quantities such as the shape, composition and temperature of grains. Although our understanding of the nature and evolution of interstellar grains has improved significantly (see reviews by Aannestad and Purcell 1973; Spitzer 1978), it is still premature to put much faith in field strengths deduced from polarization observations; they should only be regarded as order-of-magnitude estimates. In any case, Vrba et al. (1980) estimate a field strength of about 120 μgauss (which seems to us too high) at a density of about 350 cm^{-3}, and they find that the field scales with the gas density as $B \propto \rho^{0.38}$. Although they point out that such scaling is consistent with theoretical predictions (Mouschovias 1976b), which show an exponent 1/3 - 1/2 in the cores of dense clouds, the theoretical predictions should by no means be regarded as having been confirmed by these observations. The accurate (but somewhat conservative) conclusion is that those calculations have not been contradicted by observations yet.

2.3 Faraday Rotation

A tenuous plasma becomes optically active (or, birefringent) in the presence of a magnetic field. Faraday rotation refers to the rotation of the plane of polarization of a linearly-polarized electromagnetic wave, or to the rotation of the major axis of an elliptically-polarized wave, passing through such a medium. The angle of rotation over a distance L is given by (Spitzer 1978, p. 66)

$$\Delta\theta = \lambda^2 R_m \equiv \lambda^2 \left(0.81 \int_0^L ds\, n_e\, B \cos\phi\right) \quad \text{rad}, \tag{1}$$

where the wavelength (λ) is measured in meters, the electron density (n_e) in cm^{-3}, the magnetic field (B) in μgauss, and the distance along the line of sight (s) in parsecs. The angle between the field $\vec{B}$ and the propagation vector $\vec{k}$ is denoted by ϕ. The sign convention is that $\Delta\theta$ is positive for right-hand rotation along the direction of propagation. The <u>rotation measure</u> is denoted by R_m.

Typical rotation measures for the interstellar medium fall in the range 1 - 100 rad m^{-2}. It is therefore clear that Faraday rotation is negligible for optical wavelengths. In principle, one can use optical polarization to establish a standard and then measure $\Delta\theta$ for radio waves. Unfortunately, not many radio sources emit in the optical. To obtain R_m (see discussion by Davis and Berge 1968), one must measure $\Delta\theta$ for at least two radio wavelengths. However, because of the indistinguishability of rotation angles differing by π and because the position angle of the plane of polarization at the source is not usually known, one must measure $\Delta\theta$ at several wavelengths, plot the observed position angle as a function of λ^2, and fit a straight line through the points. In principle, several points differing by multiples of π must be plotted for each observation and that set must be selected which fits a straight line best. The slope of the line gives R_m, and its extrapolation to $\lambda^2 = 0$ gives the position angle at the source.

Once the rotation measure is obtained, one may determine the mean value of the magnetic field along the line of sight to the observed radio source only if the distance to the source and the interstellar electron density are known. To obtain the latter would have been very difficult without the discovery of pulsars. Regular signals from pulsars reaching the earth exhibit a dispersion effect (i.e., a delay in the arrival time of a signal as the wavelength increases) that can be precisely measured. This is given by

$$\Delta t = \lambda^2 \left(4.60 \times 10^{-2} \int_0^L ds\, n_e\right) \quad \text{sec}, \tag{2}$$

where λ is measured in meters, s in parsecs, and n_e in cm^{-3}. The <u>dispersion measure</u>, $D_m \equiv \int_0^L ds\, n_e$ pc cm^{-3}, is obtained by a simple measurement and constitutes a direct measure of the column density of electrons along the line of sight. If R_m and D_m are measured for the same source, one can obtain the mean value of the magnetic field along the line of sight, $\langle B_{\parallel} \rangle$, weighted by the electron density. (The contribution of the earth's ionosphere is properly subtracted.) Reversals in the direction of the field would produce cancellations in $\Delta\theta$, so that the measured $\langle B_{\parallel} \rangle$ would be smaller than its value in the general interstellar medium.

Faraday rotation measures have also been obtained and analyzed for many extragalactic radio sources (Morris and Berge 1964; Gardner and Davies 1966; Gardner, Morris and Whiteoak 1969; Wright 1973). Although these observations have the advantage that extragalactic radio sources, unlike pulsars, are distributed all over the celestial sphere, in a strict sense they only yield $<n_e B_{||}>$, rather than $<B_{||}>$ itself, since an independent determination of $<n_e>$ is not usually made.

Wright (1973) analyzed the rotation measures from 354 extragalactic radio sources, and Manchester (1974) did the same for 38 pulsars. Their results are in good agreement and indicate a large-scale magnetic field directed toward $\ell \approx 90°$ both above and below the Galactic plane. This direction of the field is in fair agreement with that determined by Appenzeller (1968) from interstellar polarization observations of stars near the south Galactic pole. He found that the mean direction of the polarization vectors was $\ell \approx 80°$. According to these workers, the local helical field, which was suggested in order to explain the starlight polarization data (Hornby 1966; Mathewson 1968; Mathewson and Nicholls 1968; Mathewson 1969), is in conflict with the Faraday rotation observations. This resolved a long-standing theoretical dilemma: a nonvanishing magnetic field in the Galactic plane, having opposite directions above and below, would imply that there exists a current sheet in the plane.

The magnitude of the field determined from Faraday rotation measurements lies in the range 1 - 3 μgauss. Superposed on the background field, both Wright and Manchester found field "irregularities" with a typical scale of a few hundred parsecs and with field strength comparable to that of the background field. The "irregularities" are reminiscent of the data of Gardner, Whiteoak and Morris (1967) on rotation measures of extragalactic radio sources, which they interpreted as the result of field lines protruding from spiral arms at least at some regions. A theoretical explanation will be provided in §4.3.

2.4 The Zeeman Effect

The splitting of the 21-cm line into three components in the presence of a weak (on laboratory standards) magnetic field allows direct observation of the interstellar field, at least in H I clouds. The frequency separation between the two shifted, or σ, components of the line depends only on the component of the magnetic field in the direction of propagation, and is given by

$$\Delta\nu \equiv \Delta\nu_{+1} - \Delta\nu_{-1} = \frac{e\ B \cos\phi}{2\ \pi\ m_e\ c} , \qquad (3)$$

where the subscripts +1 and -1 refer to values of the azimuthal quantum number m_F', and the rest of the symbols have their conventional meaning. The split $\Delta\nu$ is equal to 2.8 Hz per μgauss for propagation along the field ($\phi = 0$). Since line widths are typically measured in kHz, observations of the Zeeman effect in hydrogen are very difficult, and special techniques become necessary (e.g., see Davis and Berge 1968, pp. 762-765 for an excellent discussion and for the reason why the transverse Zeeman effect is even more difficult to detect). As in the case of Faraday rotation, only the mean component of the field along the line of sight is measured. However, fields measured through Zeeman observations are indicative of conditions in the interiors of interstellar clouds, rather than being representative of the ambient interstellar field. This is due to the expected enhancement of the field upon cloud contraction (see below). Field strengths ranging from a few to about 50 μgauss have been measured in a number of normal H I clouds (Verschuur 1971 and references therein; review and evaluation by Mouschovias 1978, Fig. 1).

Zeeman observations have also been undertaken on the 1667 MHz (Turner and Verschuur 1970; Crutcher et al. 1975), 1665 MHz (Beichman and Chaisson 1974), and 1720 MHz lines of OH (Lo et al. 1975). By and large, upper limits of a few hundred microgauss are set, and some positive detections of a few milligauss have been reported in maser regions. The paradoxes raised by several observers (e.g., Beichman and Chaisson (1974), Lo et al. (1975), and Heiles (1976)) concerning theoretical expectations and observed strengths or upper limits of the interstellar magnetic field in dense clouds do not arise if the theoretical predictions of self-consistent calculations (Mouschovias 1975b; 1976a,b) on the nonhomologous contraction and equilibria of self-gravitating clouds are adopted. These calculations determined the precise value of the exponent κ in the relation between the field strength and the gas density, $B \propto \rho^{\kappa}$. A reasonable approximation of the exact result, in a cloud core, is given by

$$(B_c/B_{bk}) = (n_c/n_{bk})^{\kappa}, \qquad 1/3 \lesssim \kappa \lesssim 1/2, \tag{4}$$

where B_{bk} and n_{bk} are the "background" values of the field and the gas number density, respectively. For example, B_{bk} = 3 μgauss and n_{bk} = 1 cm^{-3} (the approximate mean interstellar particle density). Thus, milligauss fields should be found only in regions of density $\sim 10^6$ -10^9 cm^{-3}.

It is clear that it would be extremely difficult, if at all possible, to obtain and plot much needed isopedion (i.e., equal-magnetic-field-strength) contours with the above traditional methods of measuring magnetic fields in interstellar clouds. New ways to detect magnetic fields will have to be invented.

3. BASIS FOR THE RELEVANCE OF THE INTERSTELLAR MAGNETIC FIELD

A necessary, but not sufficient, condition for the magnetic field to be important in interstellar gas dynamics (and, therefore, in star formation) is to exert forces on (or, equivalently, to impart momentum to) the bulk of the interstellar matter which are comparable in magnitude with other forces exerted on the matter (e.g., by gravitational and stellar-radiation fields, and thermal, turbulent, and cosmic-ray pressure gradients). Such forces can be estimated only indirectly. One therefore settles for a comparison of (scalar) energy densities present in interstellar space. The energy density in the magnetic field, as obtained from the observed field strength discussed in the preceding section, is comparable with the other interstellar energy densities. This by itself, however, does not establish the dynamical importance of the magnetic field. The additional condition that the decay time of the field (e.g., due to the finite electrical conductivity of the interstellar gas) be long compared with other dynamical times (or typical life times of interstellar clouds) must be satisfied. A finite electrical conductivity causes a decay (ohmic dissipation) of the field in a stationary medium at a rate given by $\partial\vec{B}/\partial t = (c^2/4\pi\sigma)\nabla^2\vec{B}$, from which it follows that the characteristic time is $\tau_{ohmic} = 4\pi\sigma L^2/c^2$. The quantity $\sigma \sim 10^7\ T^{3/2}$ (Spitzer 1962) is the electrical conductivity; c is the speed of light in vacuum; and L is the scale length over which B varies significantly --it is commonly taken as the size of the system. To obtain as small a value for τ_{ohmic} as possible, we use $T = 10°K$ and $L = 0.1$ pc, which are typical for dark clouds. Then, $\tau_{ohmic} \approx 10^{16}$ years >> age of the universe. Ohmic dissipation of the field can therefore be neglected.

There is an additional process which can render the field dynamically insignificant. Since the bulk of the interstellar matter is electrically neutral, it will be affected by magnetic forces only insofar as it is collisionally coupled to the ionized matter (ions, electrons, and charged grains), which is attached to the magnetic field. If such coupling is so efficient that no significant diffusion of the plasma (and the field) relative to the neutrals takes place within a dynamical time scale, the magnetic field is said to be "frozen in" the (neutral) matter. It varies in time only due to gas motions, in accordance with the ("flux freezing") equation $\partial\vec{B}/\partial t = \nabla \times (\vec{v} \times \vec{B})$; where $\vec{v}$ is the gas velocity. The full strength of the magnetic forces is transmitted to the neutrals in this case. On the other hand, if momentum exchange between ions and neutrals (the electrons contribute negligibly due to their small mass, and grains become important only for neutral densities $\gtrsim 10^9\ cm^{-3}$) is not very efficient, a relative drift velocity will be set up between the plasma and neutrals, and magnetic forces will not affect the bulk of the neutral matter. This process is referred to as "ambipolar diffusion" (Mestel and Spitzer 1956) because electrons and ions, tied to the field, diffuse together through the neutral matter (Spitzer 1978). It is commonly expected that such decoupling of the field from the neutral matter will occur only in the very dense molecular clouds in which the degree of ionization falls below 10^{-8}. We shall first review the dynamical effects of magnetic fields frozen in the matter, and then discuss our current understanding (and misunderstanding) of the process of ambipolar diffusion.

4. FORMATION OF INTERSTELLAR CLOUDS

Although the issue of how interstellar clouds form has received much attention during the last two decades and although significant physical problems (e.g., thermal instability, magnetic Rayleigh-Taylor instability) have been solved in the process, it is fair to state that cloud formation remains an outstanding problem in theoretical astrophysics. A clue on the nature of the mechanism responsible for the formation of interstellar clouds can be obtained if observations can determine whether clouds (atomic and molecular) are confined to spiral arms or whether they also exist in the interarm region. The physical conditions (density, temperature, strength of magnetic field) are so different in the two regions that a prospective cloud formation mechanism may be eliminated in one or the other region from the outset.

There is little doubt today that atomic-hydrogen clouds ($n \sim 20$ cm^{-3}, $T \sim 80°K$, $M \lesssim 10^3 M_\odot$) are concentrated in spiral arms, mainly in cloud complexes found in "valleys" of magnetic field lines. If they existed as such in the interarm region, they would be seen in absorption against background continuum radio sources. One cannot ignore the well-known observational evidence, however, that the Magellanic Clouds (two nearby irregular galaxies; type Irr I) contain interstellar clouds (and young stars) without any evidence for spiral structure (or for much dust). Irrespective of the nature of the mechanism responsible for the formation of H I clouds, the conclusion seems to follow that physical conditions in spiral arms are much more condusive to cloud formation than conditions elsewhere. The issue of whether molecular clouds are also concentrated in spiral arms is a very controversial one. Scoville, Solomon and Sanders (1979) conclude that no such concentration exists, while Few (1979) arrives at the opposite conclusion. The difference seems to be due to ambiguities in kinematic distances when random or systematic but noncircular (galactocentric) cloud motions are present. Blitz and Shu (1980) have recently reviewed the issue and have made the following additional point. Since interstellar dust seems to be concentrated in (and, in fact, to trace) spiral arms of external galaxies, and since it is thought that a large fraction of the interstellar dust is in molecular clouds, molecular clouds themselves must be concentrated in spiral arms. It seems certain that the last word on the subject has yet to be added. It is nevertheless the case that, given the present state of the observational evidence, if cloud formation in spiral arms is understood, the bulk of the problem will have been solved.

4.1 Jeans Instability

The oldest available mechanism which may possibly account for cloud formation is the Jeans instability (Jeans 1928; Chandrasekhar and Fermi 1953) which refers to the development of self-gravitating condensations in an infinite, uniform medium, threaded by a uniform magnetic field. Only scale lengths λ longer than the Jeans wavelength,

$\lambda_J = 1.23 \times 10^3 \ (T/6000°K)^{1/2} \ n_H^{-1/2}$ parsecs, parallel to the field lines, and longer than $\lambda_J(1 + a)^{1/2}$ perpendicular to the field lines can become unstable. The quantity a is defined by $a = 0.47 \ (B/3 \ \mu G)^2/(T/6000°K) \ n_H$, and the hydrogen number density is measured in cm^{-3}; a 10% helium abundance has been accounted for. (The temperature 6000°K is near the equilibrium temperature of a gas with $n_H \approx 1 \ cm^{-3}$, $n_e \approx 10^{-2} \ cm^{-3}$, which is heated by cosmic-ray ionization at the rate $10^{-17} \ sec^{-1}$ and cooled by e-H collisions.) The e-folding time of the instability along field lines is given by $\tau_{Jeans} = [4\pi G\rho \ (1 - h^2)]^{-1/2} = 2.3 \times 10^7 \ [n_H \ (1 - h^2)]^{-1/2}$ years, where $h \equiv \lambda_J/\lambda \leq 1$. The requirement that τ_{Jeans} be less than 3×10^7 years (otherwise stars formed by the collapse of such clouds would appear too far downstream from a galactic shock, contrary to observations --see Roberts 1969) implies that $h \leq 0.64$; i.e., $\lambda \geq 1.56 \ \lambda_J \approx 1.9$ kpc. To gather matter from such long distances into a region even as large as 100 pc (a large cloud indeed) within a time 30 million years, speeds of about 60 km s^{-1} are required. They much exceed typical free-fall velocities, and they are certainly not observed over such extended regions in the Galactic plane. Attempts to attribute cloud formation to a Jeans instability bear the additional burden of explaining the lack of central concentration (the signature of self-gravity) in H I clouds and complexes (and, possibly, in molecular cloud complexes as well [Blitz and Shu 1980]) and the fact that most normal H I clouds are not even self-gravitating, as evidenced by their relatively low densities and masses.

4.2 Thermal Instability

A thermal instability (Field 1965) is a second candidate mechanism for cloud formation. Spitzer (1951) first suggested that relatively cold and dense interstellar clouds are in pressure equilibrium with a hot and tenuous intercloud medium. Observational support for the existence of the latter was found by Heiles (1968), who obtained a density of 0.2 cm^{-3} and a velocity dispersion of 6 km s^{-1}, implying an upper limit on T of several thousand degrees. Theoretical work (Hayakawa, Nishimura and Takayanagi 1961; Field 1962; Pikelner 1967; Field, Goldsmith and Habing 1969; Spitzer and Scott 1969) established that such two thermally stable, nearly isothermal phases can exist in pressure balance. The feature of the calculations needed for the establishment of this conclusion is a heating mechanism proportional to the gas density (e.g., cosmic-ray ionization and heating by the produced secondary electrons) and a cooling mechanism proportional to the second power of the gas density (e.g., collisional excitation followed by radiative de-excitation in spectral lines at which the medium is optically thin). Thus, although some of the assumptions (e.g., that the same agent is responsible for both ionization and heating) and conclusions of the above calculations have since been superceded by observations, their main conclusion (namely, the coexistence of two stable phases of interstellar matter in pressure equilibrium) remains valid.

Matter can "condense" from the intercloud to the cloud phase if the density of the nearly isothermal intercloud gas increases beyond some critical value, thus causing a rise in pressure that cannot be maintained. The critical point marks the onset of a thermal instability (Field 1965) which proceeds almost isobarically; the denser gas cools faster and gets more compressed until the stable cloud phase is reached. This transition relieves the initial excess of pressure so that the ambient pressure is maintained at the critical value (Field et al. 1969). The isobaric nature of the condensation mode implies an upper bound on the fastest-growing wavelengths of a perturbation. It is approximately that distance within which a sound wave can establish pressure equilibrium in a time not exceeding the cooling time ($< 10^6$ yr) of the medium. Since the sound speed is $\lesssim$ 10 km s^{-1}, the wavelengths which can grow at a rate near maximum will be $\lesssim$ 10 pc. The final size of the resulting condensation is, of course, much smaller than this (< 0.1 pc), and involves only a fraction of a solar mass (see review by Mouschovias 1978, Appendix A). Wavelengths much larger than 10 pc (i) grow nearly isochorically and, therefore, cannot explain the observed cloud densities; (ii) grow at a rate slower than the magnetic Rayleigh-Taylor instability (see below), in which the magnetic field is instrumental, rather than a nuisance, in the formation of large condensations. (Field [1965] has shown that a field strength as small as 1 μgauss prevents the development of the thermal instability, except in a direction parallel to the field.) The above conclusions concerning the inability of thermal instability to form any but dwarf clouds remain valid (Mouschovias 1975b; 1978) even if one considers its development in a cooling medium, which is periodically heated by supernovae (e.g., Schwartz, McCray and Stein 1972).

4.3 Magnetic Rayleigh-Taylor (or, Parker) Instability

A light fluid can support a heavy fluid against a vertical (downward) gravitational field (assumed to be constant for simplicity) if the interface is perfectly horizontal. Deformations of the interface, however, grow, as fingers of heavy fluid protrude downward into the light fluid, thus reducing the energy of the system. Shorter wavelengths along the interface tend to grow faster than longer wavelengths. This is a classical Rayleigh-Taylor instability. It can also develop if the downward gravitational field is replaced by an upward acceleration of the heavy fluid by the lighter one. This instability may be responsible for the observed protrusions of cold neutral matter into adjacent H II regions. The nature of the instability changes if a frozen-in, horizontal magnetic field plays the role of the light fluid in supporting the gas against the gravitational field. The light fluid (the field) and the heavy fluid (the gas) now coexist in the same region of space, and analogies with the nonmagnetic case break down. The nature of the magnetic Rayleigh-Taylor instability has been worked out by Parker (1966) in the context of the interstellar medium. Is it a viable mechanism for the formation of interstellar clouds? In other words, are the unstable wavelengths of the proper size (a few hundred parsecs), and are the corresponding e-folding times short enough ($\lesssim 3 \times 10^7$ yr)? In addition, what final densities are achieved?

The pressure scale height of warm ($T \approx 6000°K$) interstellar gas in the (constant, for simplicity) vertical galactic gravitational field ($g \approx 3 \times 10^{-9}$ cm s^{-2}) is $C^2/g \approx 42$ pc; where $C = (kT/\mu m_H)^{1/2}$ is the isothermal speed of sound; k the Boltzmann constant; m_H the mass of a hydrogen atom; and μ the mean mass per particle in units of m_H ($\mu = 1.27$ to account for $n_{He}/n_H = 0.1$). This is smaller than the observed scale height by at least a factor of 3 --possibly, 4. Pressure due to the nearly horizontal frozen-in magnetic field and due to cosmic rays, which are tied to the field, provide additional support to the gas against the galactic gravitational field. Under the simplifying assumption, introduced first by Parker, that these pressures are proportional to the gas pressure (i.e., $\alpha \equiv B^2/8\pi P$ = const., and $\beta \equiv P_{CR}/P$ = const.), the scale height of the gas now becomes $H = (1 + \alpha + \beta)\, C^2/g$. With $\alpha \approx \beta \approx 1$, a value consistent with observations, one finds that $H \approx 126$ pc --not an unreasonable value. Parker showed, through a linear stability analysis, that this one-dimensional equilibrium state is unstable with respect to deformations of the field lines. The vertical gravitational field acquires a component along a deformed (non-horizontal) field line, thus causing gas to slide along the field line from a raised into a lowered portion. The unloading of gas from the raised portion leaves magnetic and cosmic-ray pressure gradients unbalanced in that region, thereby causing further inflation of the already raised portion of a field line. The component of gravity along the now more vertical field line is larger, with the result that gas can be unloaded more efficiently into the "valley" of the field line. The process will stop only when field lines have inflated enough for their tension to balance the expansive magnetic and cosmic-ray pressure gradients (Mouschovias 1974, 1975a). In this picture, the matter which accumulates in valleys of the field lines represents interstellar clouds. Or, does it?

For the instability to develop, the horizontal (along the field lines) and vertical wavelengths of a perturbation must simultaneously satisfy the following respective inequalities:

$$\lambda_y > \Lambda_y \equiv 4\pi H\{\alpha\gamma/[2(1+\alpha+\beta-\gamma)(1+\alpha+\beta)-\alpha\gamma]\}^{1/2}, \tag{5a}$$

and

$$\lambda_z > \Lambda_z(\lambda_y) \equiv \Lambda_y/[1-(\Lambda_y/\lambda_y)^2]^{1/2}. \tag{5b}$$

For the interstellar gas $\gamma \equiv d\log P/d\log\rho \approx 1$. If $\lambda_y < \Lambda_y$, the radius of curvature of a typical, deformed field line is small, hence the tension is large and it straightens the field line out --a stable regime. If $\lambda_z < \Lambda_z(\lambda_y)$, the system is stable even though λ_y may exceed its critical value Λ_y. The physical reason lies in the fact that the volume available for the field lines to expand in, and thereby decrease the magnetic energy of the system, is limited. The increase in the field strength in the valleys and the pile-up of field lines near the first undeformed field line, which forms a natural "lid" to the system below, represent an increase in magnetic energy which suppresses the instability (see Mouschovias 1975b for a detailed discussion). For a fixed $\lambda_y > \Lambda_y$, the growth rate of the perturbation increases monotonically as λ_z ($> \Lambda_z$) increases. For a fixed $\lambda_z > \Lambda_z$, the growth rate first increases and then decreases as λ_y increases. Equation (5a) shows that the horizontal critical wavelength for $\alpha \approx \beta \approx \gamma \approx 1$ is $\Lambda_y = 1.2\ \pi H \approx 477$ pc. The maximum growth rate occurs at $\lambda_y = 1.8\ \Lambda_y = 2.2\ \pi H \approx 868$ pc and $\lambda_z = \infty$, and its inverse (the e-folding time) is given by

$$\tau_{min} = 1.1\ H/C \tag{6a}$$

$$= 2.2 \times 10^7\ (T/6000°K)^{1/2}/(g/3 \times 10^{-9}\ \text{cm s}^{-2})\quad \text{yr.} \tag{6b}$$

This growth time is short enough to be relevant for cloud formation behind a spiral density shock wave. In fact, it may be smaller than the value given in equations (6a,b) because, in a strict sense, the quantity H is the scale height in the <u>initial</u> state; not its value today. It has been shown by exact determination of final equilibrium states for the Parker instability that, in the valleys of field lines, $H_{final} \approx 1.7\ H_{initial}$ (Mouschovias 1974, Fig. 2b). This implies that $\tau_{min} \approx 1.3 \times 10^7$ yr and $\lambda_y(\tau_{min}) \approx 511$ pc. A further decrease in τ_{min} can take place due to the fact that the instability is externally <u>driven</u> by a spiral density shock wave (Mouschovias 1975b, p. 73). There is yet another reason for which τ_{min} can decrease further. Giz and Shu (1980) took into consideration the actual variation of g with z, and found that the value of g which enters equation (6b) is larger than the one given above by a factor of 3. The amount of matter involved in a cylinder (along a spiral arm) of length 511 pc and diameter 250 pc (the approximate thickness of a galactic shock, as well as the width of the galactic disk in which most of the gas is found) is $8.6 \times 10^5\ M_{\odot}$. Thus the Parker instability is most suitable for the formation of large-scale condensations (or, cloud complexes), rather than individual interstellar clouds (Mouschovias 1974; Mouschovias, Shu and Woodwood 1974). The implosion by shock waves of individual clouds within these complexes can give rise to OB associations and giant H II regions, all aligned along spiral arms "like beads on a string" and separated by regular intervals of 500 - 1000 pc, in agreement with observations both in our galaxy and in external galaxies (Westerhout 1963; Kerr 1963; Morgan 1970; Hodge 1969). The nonlinear development of the magnetic Rayleigh-Taylor instability and the final equilibrium states which we have calculated also explain the large-scale intimate association between the interstellar gas and field, as revealed by the combined observations of Mathewson and Ford (1970) and Heiles and Jenkins (1976) discussed in §2.2.

In the direction $\vec{g} \times \vec{B}$ (the "third direction"), wavelengths ranging from a very small fraction of the vertical scale height H to many times H can grow with almost identical growth rates (Parker 1967). If some mechanism (other than a galactic shock) selects wavelengths $\lambda_x \sim 10$ pc in this direction, then the mass involved would be only slightly larger than $10^4\ M_\odot$. This begins to approach masses of individual clouds. Whether in fact individual clouds can form by the Parker instability will be decided only when nonlinear three-dimensional calculations are carried out. Although there is a wealth of ideas on how phase transitions and conversion of atomic to molecular hydrogen in the valleys of field lines can convert the non-gravitating clumps of gas into dense molecular cloud complexes (e.g., Field 1969; Mouschovias 1975b, 1978; Blitz and Shu 1980), no quantitative calculation has been produced yet. Initial perturbations of the field lines which have an odd symmetry about the Galactic plane are more likely to initiate the necessary phase transitions. Such perturbations allow field lines originally coinciding with the Galactic plane to deform, and they therefore can lead to a gas density (and pressure) in the Galactic plane significantly higher (a necessary condition for phase transitions) than its value in the initial (unstable) equilibrium state. Perturbations with even symmetry about the Galactic plane can lead to such phase transitions only under special circumstances (Mouschovias 1975b, pp. 55-57).

5. CONTRACTION, EQUILIBRIUM, AND COLLAPSE OF SELF-GRAVITATING CLOUDS

5.1 Criteria for Collapse

Irrespective of the nature of the mechanism responsible for cloud formation, the contraction, the available equilibrium states (if any), and the collapse of a cloud are significantly affected by a frozen-in magnetic field of initial strength comparable with the observed values (see §2). (Contraction ceases when an equilibrium state is reached. Collapse refers to indefinite contraction past the last available equilibrium state, at least as long as the cloud can be described by the same set of equations which was applicable when collapse began.) Since we are interested here in the diffuse stages of cloud collapse and star formation, $n < 10^9$ cm^{-3}, isothermality is a good approximation. Let us first recall that an isothermal, spherical, nonmagnetic cloud, which is bounded by an external pressure P_{ext} can collapse if its central density ρ_c exceeds the surface density ρ_S by at least a factor 14.3 (Bonnor 1956; Ebert 1955, 1957). The mass and radius at this critical ("Bonnor-Ebert") equilibrium state are given by $M_{BE} = 1.2\ C^4/(G^3P_{ext})^{1/2}$ and $R_{BE} = 0.41\ GM/C^2$, respectively. Thus, an isothermal, nonmagnetic cloud will have no accessible equilibrium states if its mass exceeds $M_{BE} = 2.0 \times 10^3\ (T/80°K)^2\ (\mu/1.27)^{-2}\ (P_{ext}/1600k)^{-1/2}\ M_\odot$. For molecular clouds, $\mu = 2.33$ and $T \approx 40°K$, yielding $M_{BE} \approx 1.5 \times 10^2\ M_\odot$ for the same P_{ext} as above. If thermal pressure were the only means of support of molecular clouds against self-gravity, all of them would be collapsing.

If an interstellar cloud were to form out of a medium of density ρ_i and magnetic field B_i through spherical isotropic contraction, conservation of mass ($\rho \propto R^{-3}$) and flux ($B \propto R^{-2}$) would imply that $B \propto \rho^{2/3}$. Such a model was employed for a study of the support that magnetic forces can provide against self-gravity (Mestel 1966). A spherically-symmetric density distribution, $\rho(r) = \rho_i + \rho_c \exp[-(r/r_o)^2]$, was assumed; where ρ_c (if $\gg \rho_i$) is the central density, and r_o is a radius beyond which ρ decreases rapidly to its background value, ρ_i. Thermal pressure was ignored. Then the magnetic field implied by this density distribution was calculated. Near the center of the cloud ($r \ll r_o$) the field is nearly uniform and equal to $B_i\ (\rho/\rho_i)^{2/3}$. In an intermediate region $[1 \ll r/r_o \ll (\rho_c/\rho_i)^{1/3}]$ the field is almost radial. At larger radii, $r/r_o \gg (\rho_c/\rho_i)^{1/3}$, the field becomes uniform and equal to B_i. The nearly radial field, which is solely the result of the imposed spherical contraction, causes large "pinching" forces at the equator --so much so that magnetic forces much exceed gravitational forces. Mestel argues that, if this configuration is achieved through rapid, violent contraction of the cloud, flux dissipation, reconnection and detachment of field lines will take place in the equatorial plane. He points out, however, that preferential flow of matter along field lines might prevent such configuration from being reached. The significant result, and main objective, of this calculation is a criterion for the lateral collapse of the cloud: If the total mass-to-flux ratio exceeds the critical value $(M/\Phi_B)_{crit} = 0.152\ G^{-1/2}$, the gravitational forces exceed the magnetic forces at the equator so that further contraction will ensue. (The quantity G is the universal gravitational constant.)

Criteria for the collapse of magnetic clouds which turn out to be similar with the one found by Mestel can be obtained from the Virial Theorem (Chandrasekhar and Fermi 1953; Mestel 1965; Strittmatter 1966; Spitzer 1968). The advantage of the Virial Theorem lies in the fact that it, being an integral relation, washes out the complex details of the structure of the system. That is also its greatest disadvantage because no conclusion on the internal structure of the system can be arrived at. At times, the Virial Theorem can also lead to misleading results (see discussion by Mestel 1965; Mouschovias 1975b). One would like to obtain a reliable collapse criterion for magnetic clouds analogous to the Bonnor-Ebert condition for nonmagnetic clouds; i.e., by studying exact equilibrium states.

D. A. Parker (1973, 1974) obtained equilibrium states for self-gravitating, magnetic clouds bounded by external pressure. These are true equilibria in that forces are in exact detailed balance, rather than only in an average sense. However, because of his neglect of the constraint of flux-freezing he had to adopt an ad hoc assumption concerning the form of an arbitrary function of the magnetic flux in order to close the system of the magnetohydrostatic equations. This procedure does not allow one to study a sequence of states that can evolve from one to another through continuous deformations of the field lines, or to quantify the effectiveness with which the magnetic field can prevent the gravitational collapse of a cloud.

Solution of the self-consistent equilibrium problem for isothermal, pressure-bounded, self-gravitating, magnetic clouds yielded a number of critical states for gravitational collapse under a wide variety of initial values for thermal, gravitational and magnetic energies (Mouschovias 1976a,b). Virial-theorem expressions (Spitzer 1968) were then used as interpolation/extrapolation formulae to determine, after fixing the virial constants so as to obtain agreement with the exact results, critical states for any arbitrary set of initial physical parameters (Mouschovias and Spitzer 1976). The critical mass-to-flux ratio is $(M/\Phi_B)_{crit} = 0.126\ G^{-1/2}$, which can be written in terms of the magnetic field B and the number density of protons n_0 in the cloud in the convenient form

$$M_{crit} = 5.04 \times 10^5 \frac{(B/3\ \mu G)^3}{(n_0/1\ cm^{-3})^2}\ M_\odot\ ; \qquad (7a)$$

$$= 5.04 \times 10^5\ n_0^{-(2-3\kappa)}\ M_\odot\ . \qquad (7b)$$

To obtain equation (7b), the relation $B \propto \rho^\kappa$ was used. The exact equilibrium calculations showed that $1/3 \lesssim \kappa \lesssim 1/2$ in the core (see eq. [4]). The critical mass given by equation (7a) is only about half the virial-theorem value. A similar reduction was obtained by Strittmatter (1966), but only for cold, infinitely thin clouds. The maximum external pressure which can be applied to a cloud of given mass and flux without resulting in collapse is equal to 0.60 times the value predicted by the virial theorem for a uniform, magnetic cloud. The main reason for the reduction in the critical values for mass and external pressure at a given flux is the development of a central concentration in the exact equilibrium states which produces stronger gravitational forces. Equation (7a) can be interpreted as yielding that value of B which can stabilize a cloud of given mass and number density no matter how arbitrarily large the external pressure is. For example, a molecular cloud of mass $10^4\ M_\odot$ and density $n_0 = 2 \times 10^4\ cm^{-3}$, is stabilized by a field strength of 600 μG, at any external pressure. For a finite external pressure, a value of B smaller than the one given by equation (7a) is necessary for stabilization against gravitational collapse.

5.2 Fragmentation

A necessary condition for fragmentation to occur in a contracting or collapsing cloud is a reduction of M_{crit} upon contraction. Once formed, a fragment will maintain its identity only if it can contract or collapse more rapidly than the background. An incisive discussion of the role of magnetic fields in the fragmentation process has been given by Mestel (1965, 1977). Yet, no detailed quantitative calculation has been undertaken on the subject. The basic effect of the field can be illustrated by using equations (7a,b). Spherical isotropic contraction/collapse ($\kappa = 2/3$) leaves M_{crit} unchanged; hence, it prohibits fragmentation. Our calculations have determined, however, that the contraction is both nonhomologous and nonisotropic, with $1/3 \lesssim \kappa \lesssim 1/2$ in a cloud's core. The exponent κ increases to values >> 1 at the magnetic poles of the cloud and decreases to zero, and even to negative values, in the equatorial plane toward the equator; we have argued that similar results should be obtained during collapse, even for a cooling, rather than an isothermal, cloud (Mouschovias 1976b; 1978). It follows then from equation (7b) that, as the density increases, M_{crit} decreases most rapidly in the equatorial plane. Consequently, fragments should form there first. Solar-mass blobs can separate out at densities characteristic of molecular clouds. This conclusion is in agreement with observations which show that the mean density of open clusters is comparable with that of molecular clouds. We had also argued that the extreme nonhomology introduced by the magnetic field has the consequence that low-mass stars may form first, and perhaps only, in the cores of dense clouds. Circumstantial observational evidence for that conclusion has been found recently by Vrba, Coyne and Tapia (1980).

The reason for which fragmentation has been regarded as a necessary element of any theory of star formation lies in the observation that young stars seem to form predominantly in groups. Reasonable as it may seem, it should be borne in mind that it is merely a hypothesis. One could conceive of an alternative, the "pre-existing cloudlet hypothesis". Cloudlets ($\lesssim 1\ M_\odot$) may form via a thermal instability behind a spiral density shock wave and then gathered in valleys of the field lines by the Parker instability (see §4.3). Within the available 3×10^7 yr, several of these cloudlets can coalesce (nonviolently) to give blobs of a few tens of solar masses. In the cores of cloud complexes, efficient conversion of atomic to molecular hydrogen can take place, and these cloudlets can also be shielded from ionizing cosmic-rays, and thus can reduce their magnetic flux via ambipolar diffusion (see §7). Since the Bonnor-Ebert critical mass for dark-cloud parameters is less than 9 $M_\odot$, these cloudlets may collapse, especially if imploded by shocks, and form individual stars in a small region of space. In this scenario no fragmentation is necessary, and the initial (stellar) mass function is largely determined by the mechanism responsible for the formation of cloudlets. If this scenario has anything to do with reality, the initial mass function is likely to be very different in different clusters (Mouschovias and Paleologou 1980a).

5.3 Collapse

A recent calculation (Scott and Black 1980) has followed numerically the collapse of a magnetic cloud with initial parameters near those specifying critical states determined earlier (Mouschovias 1976a, b). The strength of the calculation lies in the fact that it is fully time-dependent, and has therefore followed the increase in central density to three orders of magnitude further than our sequences of equilibrium states could go. It is not an accident, however, that the main conclusions of Scott and Black agree quantitatively with our earlier conclusions. We had taken great pains to explain the relevance of equilibrium calculations to star formation (Mouschovias 1976a, §Ic). We had shown analytically that the tension of the field lines will halt the collapse of the outlying portion of a cloud and will form extended envelopes. (This has been verified by the numerical collapse calculation.) As a consequence, the cloud flattens along field lines relatively rapidly, until pressure gradients balance the gravitational forces along field lines. Subsequently the cloud contracts only as rapidly as magnetic forces will allow it to contract laterally, with α_c (the ratio of magnetic and thermal pressure in the core) maintained near unity; hence, for an isothermal cloud, $B_c^2 \propto \rho_c$ or $\kappa = 1/2$ (Mouschovias 1976b, p. 151; 1978; 1979a). Scott and Black emphasize that the result $\kappa = 1/2$ in a flattened core is general and independent of initial conditions. This cannot be valid. For example, consider a cold self-gravitating cloud, threaded by a very strong magnetic field. Once released from its initial spherical, uniform state considered by Scott and Black, it will free-fall along field lines to an infinitesimally thin sheet. During such collapse, $\kappa \gtrsim 0$ (because the density increases without a corresponding increase of the magnetic field), even though a flattened core forms. In the case of a collapsing cloud with a non-negligible thermal pressure, the phase $\kappa \gtrsim 1/2$ is expected to be reached in about one (nonmagnetic) free fall time, because that is roughly the time required for pressure gradients to become comparable with gravitational forces due to flattening along field lines.

Both, our earlier analytical argument and the collapse calculations have shown that the tension of field lines near the equator prevents significant contraction from taking place, and thus field lines deform relatively little there. The large "pinching" forces found in Mestel's (1966) non-self-consistent model do not appear, and magnetic reconnection and detachment of the cloud's field from the background during the early phases of cloud collapse is unlikely to take place. Magnetic reconnection inside clouds, if it occurs at all, will play a role at a very advanced collapse stage, long after ambipolar diffusion has set in.

Thus far we have ignored rotation, which a cloud will inevitably acquire, no matter what its formation mechanism is, by virtue of its being in a rotating system, the Galaxy. In the following section we review the effect of frozen-in magnetic fields on rotation and vice versa.

6. THE ANGULAR MOMENTUM PROBLEM AND MAGNETIC BRAKING

A blob of interstellar matter of mass $\sim 1\ M_\odot$ at the mean density of the interstellar medium $\sim 1\ cm^{-3}$ has an angular momentum ($\vec{J}$) a few times 10^{55} g cm^2 s^{-1} by virtue of its participation in the general galactic rotation. On the other hand, a wide binary star system with members of mass $\sim 1\ M_\odot$ each, and a period of 100 yr possesses an angular momentum only a few times 10^{53} g cm^2 s^{-1}. Hence, if binary stars are to form through the collapse and fragmentation of interstellar clouds, a mechanism must exist which can transfer angular momentum efficiently from a collapsing cloud or fragment to the surrounding medium. This is "the angular momentum problem" for binary stars. It is more severe for single stars since a typical star possesses only an angular momentum of order 10^{49} g cm^2 s^{-1}.

The few ideas and calculations which aim at resolving the angular momentum problem (e.g., by putting the angular momentum of the parent cloud into the orbital motion of cluster stars) run either into observational difficulties (e.g., why aren't clusters flat?), or into theoretical problems, or both (see reviews by Mouschovias 1978, §IIc; 1979a, §I). The fastest rotating cloud is the globule B 163 SW, with an angular velocity somewhat less than 10^{-13} rad s^{-1} (Martin and Barrett 1978), and the fastest rotating massive cloud is the Mon R2 (Kutner and Tucker 1975), with $\omega \approx 6 \times 10^{-14}$ rad s^{-1} (see also review by Field 1978). If angular momentum is conserved during the contraction of a cloud, the angular velocity should increase with density as $\omega \approx 10^{-15}\ n^{2/3}$ rad s^{-1}. Thus, typical dark and molecular clouds with densities in the range 10^4 - 10^6 cm^{-3} should exhibit angular velocities in the range 4.6×10^{-13} - 10^{-11} rad s^{-1}. Clearly, this is not the case. Whatever the nature of the mechanism which resolves the angular momentum problem during star formation, it must operate efficiently during the relatively diffuse stages of cloud formation and collapse.

In a paper concerned with "ways in which the cloud can lose its magnetic energy", Mestel and Spitzer (1956) state the following: "If turbulence is negligible ... the angular momentum present leads to disk formation, and the subsequent evolution of the disk is slow enough for the field and plasma to diffuse outwards, inspite of the increased densities. The objection to this is that the strong frozen-in magnetic field will probably remove angular momentum too rapidly; the time of travel of a hydromagnetic wave across the cloud is of the same order as the time of free-fall, and so it is not obvious that a rotating disk will form" (emphasis added). The underlined, qualitative suggestion is the most promising mechanism proposed as yet for the resolution of the angular momentum problem. It is now referred to as the process of "magnetic braking" of a cloud's rotation.

Figure 1 illustrates the principle of magnetic braking. We consider for simplicity a disk-shaped cloud coinciding with the plane $z = 0$ and threaded by an initially uniform, frozen-in magnetic field

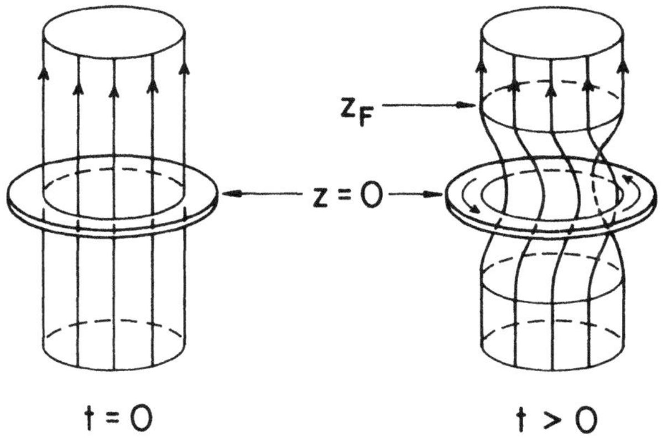

Figure 1. Illustration of the principle of magnetic braking.

(left illustration). At time $t = 0$ the cloud is imparted an arbitrary uniform angular velocity about its axis of symmetry. At a small time later, the situation is as shown in the illustration to the right in Figure 1. The frozen-in field lines (solid lines with arrows) have been twisted by the motion of the cloud. Matter in the external medium adjacent to the cloud surfaces is pulled by the field lines (assuming flux-freezing in the external medium as well) in the direction of motion of the cloud; i.e., it is imparted an angular momentum about the original axis of rotation. Exactly the same argument applied now to this rotating external matter and a layer adjacent to it, further away from the cloud, shows that a rotational disturbance propagates away from the cloud, both above and below, along the z-axis. At any one time there is a "front", shown at position z_F in the figure, beyond which matter is still undisturbed. The speed of propagation of the disturbance (which is referred to as a "torsional Alfvén wave") is the Alfvén speed in the external medium, $v_{A,ext} = B/(4\pi\rho_{ext})^{1/2}$, where ρ_{ext} is the external density. Since the external medium obviously gains angular momentum, which could only come from the original angular momentum of the cloud, the net effect is to slow down the cloud's rotation. We estimate the characteristic time within which a significant amount of angular momentum is transferred from the cloud to the external medium as follows.

A significant angular momentum transfer will take place when the wave fronts have propagated far enough from the cloud to affect a moment of inertia of the external medium comparable with that of the cloud. For a cloud (disk or cylinder) of density ρ_{cl} and half-height Z, this condition is fulfilled when the wave fronts reach a distance $Z_I = Z\ [1 + (\rho_{cl}/\rho_{ext})]$ from the equitorial plane. The time required for this to happen (and, hence, the characteristic time for magnetic braking of an aligned disk or cylinder rotator) is

$$\tau_{\|} = \int_{Z}^{Z_I} dz/v_{A,ext} = (\rho_{cl}/\rho_{ext})(Z/v_{A,ext}) \; . \tag{8}$$

It turns out that this is an exact expression for the e-folding time of magnetic braking of an aligned, rigidly rotating disk or cylinder and, if short-lived transients are ignored, an excellent approximation for a differentially rotating disk or cylinder as well, provided that ρ_{cl}/ρ_{ext} is not close to unity (Mouschovias and Paleologou 1980b). (A multiplicative factor 8/15 appears on the right-hand side of equation (8) if one considers a sphere of radius Z or an oblate spheroid of semi-minor axis Z.)

If the field lines are initially radial and perpendicular to the symmetry axis of the disk or cylinder (a "perpendicular rotator"), equality of moments of inertia is achieved when the waves (now propagating perpendicular to the axis) reach a distance from the axis of rotation given by $R_I = R\,[1 + (\rho_{cl}/\rho_{ext})]^{1/4}$, where R is the radius of the cloud. The Alfvén velocity in the external medium is now a function of radial distance r from the axis; namely, $v_A(r) = R\,v_A(R)/r$, where $v_A(R)$ is the Alfvén speed just outside the cloud surface. It therefore follows that the e-folding time for magnetic braking is

$$\tau_{\perp} = \int_{R}^{R_I} dr/v_A(r) = \frac{1}{2}\left[\left(1 + \frac{\rho_{cl}}{\rho_{ext}}\right)^{1/2} - 1\right]\frac{R}{v_A(R)} \; . \tag{9}$$

This time scale differs from the initial time scale of the exact solution (Mouschovias and Paleologou 1979) by less than a factor of 2. However, it does not reveal the oscillating nature of the approach to corotation of the cloud (or fragment) with the background --which has the important consequence of predicting retrograde rotation in some cloud fragments, and in stellar and planetary systems as a natural consequence of magnetic braking of a perpendicular rotator (see Mouschovias and Paleologou 1980a, §3). The globule B 163 SW, referred to above, exhibits such retrograde rotation. Clark (1980) claims the detection of retrograde rotation in two dense fragments.

The groundwork for calculations on magnetic braking was laid by Ebert, Hoerner and Temesváry (1960), who showed that the disturbances in the external medium obey a wave equation. Kulsrud (1971) obtained expressions, accurate to second order, for the rotational deceleration of stars with dipolar fields. These expressions were adopted by Nakano and Tademaru (1972) to study the loss of angular momentum by interstellar clouds, and also by Fleck (1976) for a similar study. Since the observational evidence in §2.2 and the theoretical arguments concerning magnetic reconnection given in §5.3 indicate that a cloud's magnetic field does not detach from that of the background, the applicability of the results of these studies to interstellar clouds is questionable. The first detailed study of magnetic braking was carried out by Gillis, Mestel and Paris (1974), who considered an aligned spherical rotator, with a field-line geometry similar to that of Mestel's (1966) cloud model described in §5.1 above. They assumed that the cloud's angular velocity was constant at all times, and they concluded that magnetic braking was never so efficient as to keep the cloud corotating with the background. A series of analytical but approximate calculations for aligned spherical and oblate rotators (self-gravity was included in the latter) found efficient magnetic braking, which can keep most H I clouds in synchronous galactocentric orbits, and which predicts low angular velocities for molecular clouds compared with expectations based on angular momentum conservation (Mouschovias 1977a, b; 1978; 1979a). A density of about 2.4×10^3 cm^{-3} was predicted above which clouds ought to begin to exhibit angular velocities appreciably higher than 10^{-15} rad s^{-1}, due to the fact that the time scales of magnetic braking and ambipolar diffusion (see §7) become comparable.

An exact calculation for a perpendicular rotator, properly accounting for the time-dependence of the cloud's angular velocity in a manner consistent with the instantaneous magnetic torques exerted on the surface, found a much higher efficiency for magnetic braking than that found for aligned rotators (Mouschovias and Paleologou 1979). The physical origin of this higher efficiency lies in the fact that the rotational waves now set in motion matter at larger and larger distances from the axis of rotation. Hence, a moment of inertia in the external medium comparable with that of the cloud is swept at a much earlier time than in the aligned rotator case --compare $\tau_{\parallel}$ and $\tau_{\perp}$ in equations (8) and (9) for the same $\rho_{cl}/\rho_{ext} >> 1$, $Z \approx R$, and $v_{A,ext} \sim v_A(R)$. It was found that enough angular momentum could be lost in less than about 10^6 yr for binaries to form. The efficiency of magnetic braking increases upon contraction because $\rho_{cl} \propto R^{-2}$ and $v_A(R) \propto R^{-a}$, where $1 \leq a \leq 2$, so that equation (9) yields $\tau_{\perp} \propto R^a$ for $\rho_{cl}/\rho_{ext} >> 1$. On the contrary, for an aligned rotator we have that $\rho_{cl} \propto R^{-2}Z^{-1}$, and $v_{A,ext} \propto R^{-2}$ if ρ_{ext} remains constant and continuity of the field across the cloud surface is assumed; therefore, equation (8) yields that $\tau_{\parallel}$ = const., independent of the stage of contraction. Binaries can form in this case in a time less than 1.4×10^7 yr from the onset of cloud contraction from a density of 1 cm^{-3}.

Gillis, Mestel and Paris (1979) subsequently relaxed their assumption of constant angular velocity of the cloud at all times, and they found an efficiency for magnetic braking similar with that of the aligned rotator described above. Both they and Mouschovias and Paleologou (1980a, b) considered the propagation of nonlinear torsional Alfvén waves within the cloud as well, and followed their numerous partial internal reflections on the cloud surfaces. These waves cause large shearing motions within the cloud which, however, are short lived. Mouschovias and Paleologou compared their solution with an exact solution for a rigid rotator. A rigid rotator can better and better approximate a cloud as contraction goes on. This is so because the Alfvén crossing time decreases upon contraction, thus tending to better fulfill the assumption implicit in a rigid-rotator model, namely, that a torque exerted on the cloud surface is communicated instantaneously to any interior point. Mestel and Paris (1979) also employed the virial theorem to study the magnetic braking of a contracting cloud while allowing the cloud's magnetic flux to vary in some arbitrarily prescribed manner, in an effort to simulate the effect of ambipolar diffusion on magnetic braking. They find that clouds with $M > M_{crit}$ contract on a time scale set by magnetic braking, rather than on the free fall time scale. Clouds with $M < M_{crit}$, on the other hand, can contract only if they lose their magnetic flux.

In summary, magnetic braking can resolve the bulk of the angular momentum problem during the diffuse stages of cloud contraction. Clouds should begin to exhibit appreciable rotation above the galactic background for densities in excess of 2.4×10^3 cm^{-3}, due to the fact that ambipolar diffusion, to which we now turn, begins to set in with a time scale comparable with that of magnetic braking for an aligned rotator. Much higher densities will need to be reached for clouds with $\vec{J} \perp \vec{B}$ before they exhibit appreciable rotation above that of their background.

7. THE MAGNETIC FLUX "PROBLEM" AND AMBIPOLAR DIFFUSION

If flux-freezing remained valid all the way up to main-sequence densities, spherical isotropic collapse would result in typical stellar fields $B_* \approx n^{2/3}$ μgauss $\sim 10^{10}$ Gauss, which are much too strong compared to observed values. Flattening along field lines reduces the exponent κ to $1/3 \lesssim \kappa \lesssim 1/2$ in a cloud's core (Mouschovias 1976b), so that a nonspherical collapse yields $B_* \approx 3 \times 10^2 - 3 \times 10^6$ Gauss, if the field remains frozen in the matter. The surface field of the Sun is near the lower limit of the above range. Consequently, observations of relatively weak stellar fields do not constitute compeling evidence for breakdown of flux-freezing during some stage of cloud collapse and star formation. Mestel (1977) also makes this point. Ambipolar diffusion, nevertheless, which was introduced in §3, can lead to a breakdown in flux-freezing at a relatively early stage in cloud collapse ($n \approx 10^4 - 2 \times 10^6$ cm^{-3}) and thus explain the entire range of periods of binary stars from 10 hr to 100 yr (Mouschovias 1977a).

There are only three significantly different solutions for ambipolar diffusion thus far. Spitzer (1968 or 1978) obtained a steady-state solution for the drift velocity of the plasma relative to the neutrals for an infinite cylinder of <u>uniform</u> density supported laterally against self-gravity by a magnetic field parallel to the axis of symmetry. (A steady-state refers to a solution found under the assumption that locally the magnetic force on the ions is balanced by the gravitational force on the neutrals.) The implied time scale for ambipolar diffusion in an H I cloud is

$$\tau_D = 5 \times 10^{13} \, x \qquad \text{yr} \, , \tag{10}$$

where the quantity $x \equiv n_i/n_H$ is the degree of ionization in the cloud. By definition, τ_D is the time required for the plasma to drift relative to the neutrals a distance r from the axis of symmetry with a drift speed equal to its value at r. This expression is therefore expected to underestimate τ_D significantly in a cloud's core (Mouschovias and Paleologou 1981).

Mouschovias (1979b) found a steady-state solution for ambipolar diffusion in a pressure-bounded cylindrical cloud which contracted to an equilibrium state from an initially uniform configuration. The characteristic time scale is now a function of distance from the axis of symmetry and, for a typical molecular cloud, is given by

$$\tau_B(r) = 1.16 \times 10^{13} \, \zeta(r) \, x(r) \qquad \text{yr} \, . \tag{11}$$

It is the time required for the plasma to drift relative to the neutrals a distance equal to the local magnetic scale height, and is thereby a measure of the efficiency of ambipolar diffusion as a function of position within a cloud. The function $\zeta(r)$ decreases monotonically from its value of ∞ on the axis of symmetry to 4×10^{-4} on the cloud boundary; the variation is very rapid for small r and gradual near the boundary (see his Table 1, column 4). Since the degree of ionization $x(r) \equiv n_i(r)/n_{H_2}(r)$ increases with r, $\tau_B(r)$ is likely to have a minimum within the cloud, near the cloud core. Stars may thus form first in a ring near the core, but not at the center or on the axis of symmetry --if, of course, such center or axis of symmetry exists in a real cloud.

Nakano (1979) devised what may be the closest approximation to an attack on the time-dependent process of ambipolar diffusion without actually solving the time-dependent fluid equations. He employed the formulation and method of solution developed by Mouschovias (1976a) for the study of sequences of equilibrium configurations of pressure-bounded, self-gravitating, magnetic interstellar clouds. A new feature was introduced which consisted of allowing Mouschovias' mass-to-flux ratio $dm(\Phi)/d\Phi$, characterizing each flux tube of a cloud, to vary slightly from one equilibrium configuration to the next in accordance with a prescription dependent on the local strength of the magnetic force. He found that equilibrium configurations are possible until the cloud's core loses a significant fraction of its magnetic flux. At that stage rapid contraction seems to ensue, but the numerical scheme stops there because it is suitable only for quasi-static evolution.

A number of other papers essentially apply Spitzer's solution to different situations. Nakano (1973; 1976; 1977) employs it for approximate studies of fragmentation after the decoupling of the field from the matter. He (1978) also studied the same problem in the compressed layer between an ionization front and a shock front, which had been suggested as the location of star formation (Elmegreen and Lada 1977). Scalo (1977) estimated the heating of a cloud due to released magnetic energy during ambipolar diffusion and concluded that a value of $\kappa \gtrsim 2/3$ would result in too much heating. Elmegreen (1979) considered the effect of charged grains on steady-state ambipolar diffusion, and concluded that flux-freezing remains valid even when the degree of ionization is much smaller than 10^{-8}. Nakano and Umebayashi (1980) find, however, that grains become important only for neutral densities $n_n \gtrsim 10^9$ cm^{-3} because only then is the density of grains (n_g) comparable with that of the ions (n_i) and much larger than the density of electrons (n_e). For $n_n \lesssim 10^9$ cm^{-3}, they find that $n_i \approx n_e >> n_g$, and the grains are not always attached to the magnetic field. A linear stability analysis of the magnetohydrodynamic equations concluded what was already known from the steady-state solutions; namely, the critical mass and the collapse time scale decrease once the field decouples from the matter, at low degrees of ionization (Langer 1978). Using Spitzer's solution we had shown that ambipolar diffusion can decouple the field from the matter at relatively low densities (10^4 - 2×10^6 cm^{-3}) in a significantly short time; thus, grain effects are not expected to become important (Mouschovias 1977a). Assuming that angular momentum is essentially conserved once ambipolar diffusion became effective, we showed that there is a one-to-one correspondence between the gas density at decoupling and the residual angular momentum in a collapsing blob which will ultimately appear as orbital angular momentum of a binary (or multiple) star system. The above range of densities for decoupling is exactly what is required to account for the entire range of periods of binary stars from 10 hr to 100 yr.

It is commonly thought that ambipolar diffusion necessarily reduces the magnetic flux and magnetic energy of a cloud. We have argued, however, that although that is possible, it is by no means the essential feature of ambipolar diffusion (Mouschovias 1978; 1979b). The essential feature of ambipolar diffusion is a redistribution of mass in at least some of the interior flux tubes of a cloud. The relatively high degree of ionization in the envelopes of self-gravitating clouds maintains the field frozen in the matter there. Ambipolar diffusion sets in when ionizing high-energy ($\gtrsim$ 100 MeV) cosmic rays are screened out of a dense core, and allows the neutrals to contract, under the influence of gravity, more rapidly than the ions, which are acted upon by the full strength of the retarding magnetic forces. Thus, the total flux of the cloud does not necessarily change, and its magnetic energy can even increase while ambipolar diffusion is in progress. It is gravitational energy, not magnetic, which is converted first into kinetic energy of neutrals and then into heat via neutral-ion collisions. The additional important consequence of this picture is that collapsing protostars may retain a larger magnetic energy than previously realized, and magnetic braking may continue past the point of initiation of ambipolar diffusion to allow the formation of single stars through only one stage of fragmentation in the parent cloud.

The first time-dependent solution for ambipolar diffusion has been obtained recently (Mouschovias and Paleologou 1981). It refers to the case in which ambipolar diffusion both redistributes mass in the flux tubes of the system as well as reduces the total flux and magnetic energy of a cloud. It applies to a layer of gas compressed relatively rapidly (e.g., by a strong shock), with a magnetic field parallel to the surfaces of the slab. The slab-cloud is in pressure balance with a hot and tenuous external medium, whose field is negligible compared to that of the cloud. The neutrals are assumed to be at rest. The drift velocity of the plasma at the cloud boundary is shown as a function of time in Figure 2 (scale on left side of the figure). Each curve is labeled by the value of the neutral density in units of cm^{-3}. It increases rapidly and reaches a maximum (filled circles) within a time τ_* equal to 130 - 300 times the ion neutral collision time (within about 6×10^{-4} - 3 yr, for neutral density in the respective range

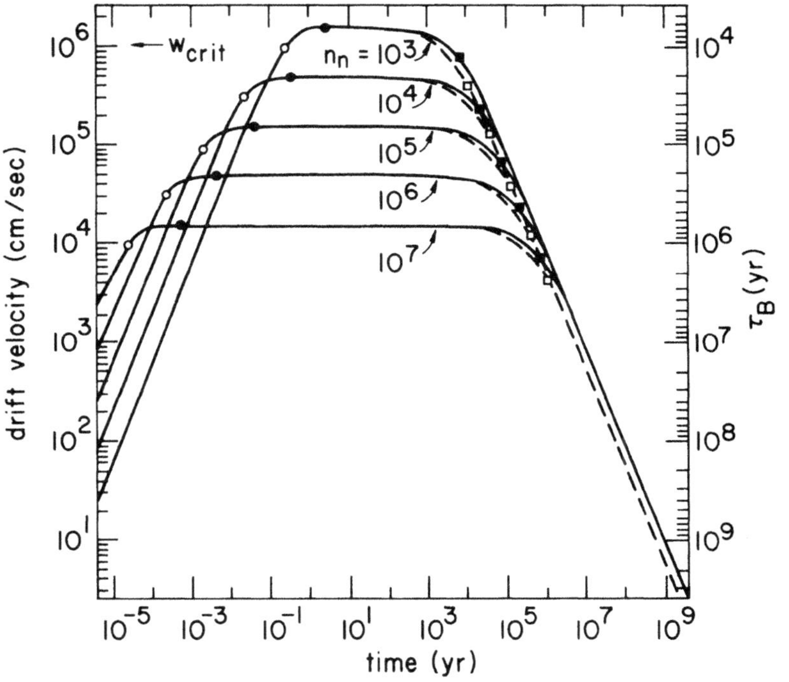

Figure 2. The drift velocity (scale to the left of the frame) and the characteristic time for time-dependent ambipolar diffusion (scale to the right of the frame) as functions of time for different values of the neutral density.

10^7 - 10^3 cm^{-3}). For this density range, the maximum drift velocity lies in the range 0.25 - 15 km s^{-1}. The corresponding characteristic time τ_B for ambipolar diffusion is in the range 6.6 × 10^5 - 6.6 × 10^3 yr (scale on the right side of the frame). Beyond the time τ_*, the driving magnetic force is almost exactly balanced by the retarding collisional force between ions and neutrals, and the asymptotic behavior of the drift velocity is t^{-1}. These results are insensitive to the rate at which ionization equilibrium tends to be re-established as plasma escapes the cloud. Solid curves are for very slow and dashed curves for very rapid re-establishment of ionization equilibrium compared with the rate at which ambipolar diffusion progresses.

The magnetic field, normalized to its initial value in the slab, is shown as a function of time in Figure 3. The labeling of the curves and the meaning of solid and dashed curves are as in Figure 2. Up to

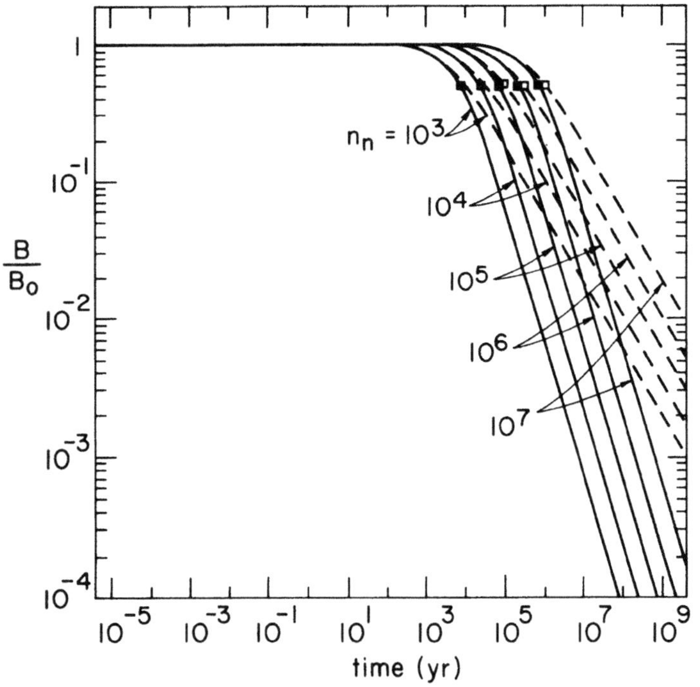

Figure 3. The magnetic field as a function of time for different values of the neutral density.

a time $\tau_{B,min}$ (indicated by filled and open squares on the two sets of curves), which is equal to the minimum value attained by the characteristic time for ambipolar diffusion, the magnetic field hardly changes. Near $\tau_{B,min}$, however, the field (and the flux) decreases to 1/2 its original value. If the original mass-to-flux ratio of the cloud (or fragment) is near the critical value for gravitational collapse (see eqs. [7a,b]), by the time $\tau_{B,min}$ the cloud (or, fragment) will begin to contract dynamically with a significant fraction of its magnetic flux trapped in (and hence its magnetic energy increasing in time), although ambipolar diffusion is in progress. A quantitative discussion of the collapse phase with ambipolar diffusion will be discussed elsewhere (Paleologou and Mouschovias 1981). The asymptotic behavior of the field is t^{-1} and $t^{-1/2}$ if re-establishment of ionization equilibrium is slow and rapid, respectively. This behavior is relevant only for clouds whose mass-to-flux ratio is well below critical, so that they can wait quiescently for at least 10^4 - 10^6 yr while their flux decreases in time. In the case of clouds with mass-to-flux ratio near critical, the asymptotic behavior of the field has only an academic significance. Dynamical contraction will set in by the time $\tau_{B,min}$.

A widespread misconception exists, that ambipolar diffusion proceeds more rapidly the smaller the degree of ionization is. The results just described show that this is not necessarily so. To illustrate the point further, we consider a slab-shaped cloud (as above) of moderate density so that self-gravity is not dominant. We then ask: (i) What is (are) the dimensionless free parameter(s) whose specification determines uniquely the solution for ambipolar diffusion? (ii) How does the efficiency of ambipolar diffusion depend on this (these) free parameter(s)? It is straightforward to show on physical grounds (see Mouschovias 1980) or rigorously (see Mouschovias and Paleologou 1981) that there is only one free parameter (ν_A) in this problem. It is the ratio of two natural time scales; namely, the time $\tau_{A,i}$ it takes an Alfvén wave (strictly, a magnetosonic wave) to traverse the thickness L of the cloud with a speed equal to the Alfvén speed in the ions, and the ion-neutral collision time τ_{in} --the latter refers to collisions of a single ion in a sea of neutrals. Thus the dimensionless parameter $\nu_A \equiv \tau_{A,i}/\tau_{in} \propto L\, n_i^{1/2}\, n_n / B$ represents the number of collisions in one Alfvén crossing time. The larger ν_A is, the more collisions an ion suffers in one Alfvén crossing time; hence, the larger the collisional drag and the smaller the rate of ambipolar diffusion. The degree of ionization is not a relevant parameter in this case. In fact, it is clear from the expression for ν_A that a given magnetic configuration can drive a given ion density through the neutrals more efficiently the smaller the neutral density (and, hence, the larger the degree of ionization) is.

The situation changes somewhat, in that x_i is a legitimate free parameter, if self-gravity is strong enough to drive neutrals through ions, which are retarded by magnetic forces (Mouschovias 1980). Yet, even in this case, x_i is only one of three free parameters. If it were for x_i alone, ambipolar diffusion would indeed tend to be more efficient as x_i decreases. The other two free parameters, however, $\bar{\nu}_A$ (the number of collisions of a neutral particle in one Alfvén crossing time) and ν_{ff} (the number of collisions of a neutral particle in one free fall time) can counter this tendency and can reduce the efficiency of ambipolar diffusion depending on precisely how and due to what physical cause x_i decreases.

8. CONCLUSION

We have reviewed the state of the art of star formation in magnetic interstellar clouds, particularly as it attempts to provide answers to the six basic questions posed in §1. Progress is being made rapidly in understanding the precise role of the magnetic field. What seems certain now, observationally and theoretically, is that the field is important, if not crucial, in star formation, particularly in resolving the thorny angular momentum problem. One should maintain an open mind, however, and renormalize one's thinking if new evidence, observational or theoretical, challenges our present conclusions or offers better alternatives.

REFERENCES

Aannestad, P. A., and Purcell, E. M. 1973, *Ann. Rev. Astron. & Astrophys.*, **11**, 309.
Appenzeller, I. 1968, *Astrophys. J.*, **151**, 907.
Beichman, C. A., and Chaisson, E. J. 1974, *Astrophys. J. Letters*, **190**, L21.
Bless, R. C. 1968, in *Stars and Stellar Systems*, Vol. **7**, *Nebulae and Interstellar Matter*, ed. B. Middlehurst and L. H. Aller (Univ. of Chicago Press, Chicago), pp. 667-684.
Blitz, L., and Shu, F. H. 1980, *Astrophys. J.*, **238**, 148.
Bonnor, W. B. 1956, *Mon. Notices Roy. Astron. Soc.*, **116**, 351.
Chaisson, E. J., and Vrba, F. J. 1978, in *Protostars and Planets*, ed. T. Gehrels (Univ. of Arizona Press, Tucson), pp. 189-208.
Chandrasekhar, S., and Fermi, E. 1953, *Astrophys. J.*, **118**, 116.
Clark, F. O. 1980, preprint.
Crutcher, R. M., Evans, N. J., Troland, T., and Heiles, C. 1975, *Astrophys. J.*, **198**, 91.
Daniel, R. R., and Stephens, S. A. 1970, *Space Sci. Rev.*, **10**, 599.
Davis, L., Jr. 1958, *Astrophys. J.*, **128**, 508.
Davis, L., Jr., and Berge, G. L. 1968, in *Stars and Stellar Systems*, Vol. **7**, *Nebulae and Interstellar Matter*, ed. B. Middlehurst and L. H. Aller (Univ. of Chicago Press, Chicago), pp. 755-770.
Davis, L., Jr., and Greenstein, J. L. 1951, *Astrophys. J.*, **114**, 206.
Dyck, H. M., and Lonsdale, C. J. 1979, *Astron. J.*, **84**, 1339.
Earl, J. A. 1961, *Phys. Rev. Letters*, **6**, 125.
Ebert, R. 1955, *Zs. f. Astrophys.*, **37**, 217.
_____. 1957, *Zs. f. Astrophys.*, **42**, 263.

Ebert, R., von Hoerner, S., and Temesvary, S. 1960, Die Entstehung von Sternen durch Kondensation diffuser Materie (Springer-Verlag, Berlin), p. 184.
Elmegreen, B. G. 1979, Astrophys. J., 232, 729.
Elmegreen, B. G., and Lada, C. J. 1977, Astrophys. J., 214, 725.
Few, R. W. 1979, Mon. Notices Roy. Astron. Soc., 187, 161.
Field, G. B. 1962, in Interstellar Matter in Galaxies, ed. L. Woltjer (W. A. Benjamin, Inc., New York), p. 183.
_____. 1965, Astrophys. J., 142, 531.
_____. 1969, Interstellar Gas Dynamics, ed. H. J. Habing (Reidel, Dordrecht), p. 51.
_____. 1978, in Protostars and Planets, ed. T. Gehrels (Univ. of Arizona Press, Tucson), p. 243.
Field, G. B., Goldsmith, D. W., and Habing, H. J. 1969, Astrophys. J. Letters, 155, L149.
Fleck, R. C. 1976, Mon. Not. Roy. Astron. Soc., 175, 335.
Gardner, F. F., and Davies, R. D. 1966, Australian J. Phys., 19, 441.
Gardner, F. F., Morris, D., and Whiteoak, J. B. 1969, Australian J. Phys., 22, 813.
Gardner, F. F., Whiteoak, J. B., and Morris, D. 1967, Nature, 214, 371.
Gillis, J., Mestel, L., and Paris, R. B. 1974, Astrophys. Space Sci., 27, 167.
_____. 1979, Mon. Not. Roy. Astron. Soc., 187, 311.
Ginzburg, V. L., and Syrovatskii, S. I. 1965, Ann. Rev. Astron. & Astrophys., 3, 297.
Giz, A., and Shu, F. H. 1980 (private communication).
Hall, J. S. 1949, Science, 109, 166.
Hayakawa, S., Nishimura, S., and Takayanagi, K. 1961, Public. Astron. Soc. Japan, 13, 184.
Heiles, C. 1968, Astrophys. J. Suppl., 15, No. 136.
_____. 1976, Ann. Rev. Astron. & Astrophys., 14, 1.
Heiles, C., and Jenkins, E. B. 1976, Astron. & Astrophys., 46, 333.
Hiltner, W. A. 1949, Science, 109, 165.
Hodge, P. W. 1969, Astrophys. J., 156, 847.
Hornby, J. M. 1966, Mon. Notices Roy. Astron. Soc., 133, 213.
Jeans, J. H. 1928, Astronomy and Cosmogony (Cambridge University Press, Cambridge).
Jones, R. V., and Spitzer, L., Jr. 1967, Astrophys. J., 147, 943.
Kerr, F. J. 1963, The Galaxy and the Magellanic Clouds, ed. F. J. Kerr (Reidel, Dordrecht), p. 81.
Kulsrud, R. M. 1971, Astrophys. J., 163, 567.
Kutner, M. L., and Tucker, K. D. 1975, Astrophys. J., 199, 79.
Langer, W. D. 1978, Astrophys. J., 225, 95.
Lo, K. Y., Walker, R. C., Burke, B. F., Moran, J. M., Johnston, K. J., and Ewing, M. S. 1975, Astrophys. J., 202, 650.
Manchester, R. N. 1974, Astrophys. J., 188, 637.
Martin, R. N., and Barrett, A. H. 1978, Astrophys. J. Suppl., 36, 1.
Mathewson, D. S. 1968, Astrophys. J. Letters, 153, L47.
_____. 1969, Proc. Astron. Soc. Australia, 1, 209.
Mathewson, D. S., and Ford, V. L. 1970, Mem. Roy. Astron. Soc., 74, 143.

Mathewson, D. S., and Nicholls, D. C. 1968, Astrophys. J. Letters, 154, L11.
Mestel, L. 1965, Quart. J. Roy. Astron. Soc., 6, 161, 265.
_____. 1966, Mon. Not. Roy. Astron. Soc., 133, 265.
_____. 1977, Star Formation, eds. T. de Jong and A. Maeder (Reidel, Boston), p. 213.
Mestel, L., and Paris, R. B. 1979, Mon. Not. Roy. Astron. Soc., 187, 337.
Mestel, L., and Spitzer, L., Jr. 1956, Mon. Notices Roy. Astron. Soc., 116, 503.
Meyer, P. 1969, Ann. Rev. Astron. & Astrophys., 7, 1.
Meyer, P., and Vogt, R. 1961, Phys. Rev. Letters, 6, 193.
Miller, C. R. 1962, Ph.D. Thesis, California Institute of Technology.
Morgan, W. W. 1970, The Spiral Structure of Our Galaxy, eds. W. Becker, and G. Contopoulos (Reidel, Dordrecht), p. 9.
Morris, D., and Berge, G. L. 1964, Astron. J., 69, 641.
Mouschovias, T. Ch. 1974, Astrophys. J., 192, 37.
_____. 1975a, Astron. & Astrophys., 40, 191.
_____. 1975b, Ph.D. Thesis, University of California, Berkeley.
_____. 1976a, Astrophys. J., 206, 753.
_____. 1976b, Astrophys. J., 207, 141.
_____. 1977a, Astrophys. J., 211, 147.
_____. 1977b, in Star Formation, eds. T. de Jong and A. Maeder (Reidel, Boston), p. 235.
_____. 1978, in Protostars and Planets, ed. T. Gehrels (Univ. of Arizona Press, Tucson), pp. 209-242.
_____. 1979a, Astrophys. J., 228, 159.
_____. 1979b, Astrophys. J., 228, 475.
_____. 1980, Astrophys. J. Letters, submitted.
Mouschovias, T. Ch., and Paleologou, E. V. 1979, Astrophys. J., 230, 204.
_____. 1980a, The Moon and the Planets, 22, 31.
_____. 1980b, Astrophys. J. 237, 877.
_____. 1981, Astrophys. J., submitted.
Mouschovias, T. Ch., Shu, F. H., and Woodward, P. R. 1974, Astron. & Astrophys., 33, 73.
Mouschovias, T. Ch., and Spitzer, L., Jr. 1976, Astrophys. J., 210, 326.
Nakano, T. 1973, Publ. Astron. Soc. Japan, 25, 91.
_____. 1976, Publ. Astron. Soc. Japan, 28, 355.
_____. 1977, Publ. Astron. Soc. Japan, 29, 197.
_____. 1978, Publ. Astron. Soc. Japan, 30, 681.
_____. 1979, Publ. Astron. Soc. Japan, 31, 697.
Nakano, T., and Tademaru, E. 1972, Astrophys. J., 173, 87.
Nakano, T., and Umebayashi, T. 1980, preprint.
Paleologou, E. V., and Mouschovias, T. Ch. 1981, in preparation.
Parker, D. A. 1973, Mon. Notices Roy. Astron. Soc., 163, 41.
_____. 1974, Mon. Notices Roy. Astron. Soc., 168, 331.
Parker, E. N. 1966, Astrophys. J., 145, 811.
_____. 1967, Astrophys. J., 149, 535.
Pikelner, S. 1967, Astr. Zh., 44, 1915.
Roberts, W. W. 1969, Astrophys. J., 158, 123.

Scalo, J. M. 1977, Astrophys. J., 213, 705.
Schwartz, J., McCray, R., and Stein, R. F. 1972, Astrophys. J., 175, 673.
Scott, E. H., and Black, D. C. 1980, Astrophys. J. 239, 166.
Scoville, N. Z., Solomon, P. M., and Sanders, D. B. 1979, in The Large Scale Characteristics of the Galaxy, ed. W. Burton (Reidel, Dordrecht), p. 277.
Spitzer, L., Jr. 1951, Problems of Cosmical Aerodynamics (Central Air Documents Office, Dayton, Ohio), p. 31.
_____. 1962, Physics of Fully Ionized Gases, 2nd ed. (Interscience, New York).
_____. 1968, Diffuse Matter in Space (Interscience, New York).
_____. 1978, Physical Processes in the Interstellar Medium (Wiley-Interscience, New York).
Spitzer, L., Jr., and Scott, E. H. 1969, Astrophys. J., 157, 161.
Strittmatter, P. A. 1966, Mon. Notices Roy. Astron. Soc., 132, 359.
Turner, B. E., and Verschuur, G. L. 1970, Astrophys. J., 162, 341.
Verschuur, G. L. 1971, Astrophys. J., 165, 651.
Vrba, F. J., Coyne, G. V., and Tapia, S. 1980, preprint.
Vrba, F. J., Strom, S. E., and Strom, K. M. 1976, Astron. J., 81, 958.
Westerhout, G. 1963, The Galaxy and the Magellanic Clouds, ed. F. J. Kerr (Reidel, Dordrecht), p. 78.
Woltjer, L. 1965, in Stars and Stellar Systems, Vol. 5, Galactic Structure, eds. A. Blaauw and M. Schmidt (Univ. of Chicago Press, Chicago), pp. 531-587.
Wright, W. E. 1973, Ph.D. Thesis, California Institute of Technology.

DISCUSSION

Bodenheimer: How many orders of magnitude in specific angular momentum can be lost through magnetic braking given sufficient time? Can you summarize your results?

Mouschovias: Our results depend neither on the origin nor on the magnitude of the initial angular momentum of the cloud--that's the advantage of solving a problem in dimensionless form. Exactly by how many orders of magnitude the initial angular momentum will be reduced, depends on the density n_{dec} at which ambipolar diffusion will decouple the field from the neutral matter relatively efficiently. I have shown that enough angular momentum is lost to account for the entire range of periods of binary stars from 10 hours to 100 years (through a single fragmentation process). To form the Sun-Jupiter "binary", n_{dec} must be $\sim 10^9$ cm^{-3}.

Bodenheimer: Can you elaborate on your statement that the degree of ionization has nothing to do with the ambipolar diffusion rate?

Mouschovias: My statement, in the form you quoted, it referred to clouds in which gravity is not dominant. In such objects, the dimensionless free parameter is the ratio of the Alfvén crossing time in the ions and the collision time of an ion in a sea of neutrals; this varies as $n_i^{1/2} n_n$.

Physically, it means that each ion has a harder time diffusing through the neutrals as the neutral density n_n increases. So, the efficiency of ambipolar diffusion decreases as the degree of ionization decreases. In self-gravitating clouds, the degree of ionization is relevant in the "classical" sense but, still, it is only one of three free parameters; its decrease does not necessarily mean more efficient ambipolar diffusion.

Bodenheimer: I do not entirely agree with your statement that the angular momentum problem has already been solved at the molecular cloud stage. The observations I quoted earlier show that J/M is at least 10^{23} in massive clouds. Could you clarify your statement?

Mouschovias: That was not my statement. I said that the bulk of the angular momentum problem has been resolved. In other words, the same mechanism which removed so much angular momentum from the "rapidly" rotating clouds (which, as you mentioned, rotate slowly compared to the angular velocities implied by conservation of angular momentum from an initial density of 1 cm^{-3} and $\omega \sim 10^{-15}$ sec^{-1}), will have an even easier time removing the necessary additional, relatively small amount of angular momentum. To re-iterate, I meant to convey the message that, as you stated in your talk, there is an angular momentum problem for dense clouds. But the angular momentum problem is much more severe if one considers the earlier, more diffuse, stages of such dense clouds. Still, magnetic braking can resolve even this more severe angular momentum problem.

Sugimoto: What is the physical situation which corresponds to $\kappa = 1/3$?

Mouschovias: First, let me recall that $\kappa = 1/2$ corresponds to rapid establishment of near hydrostatic equilibrium between gravity and pressure gradients along field lines. Virtually "instantaneous" re-adjustment along field lines can take place, and the contraction of a cloud then proceeds only as rapidly as magnetic forces allow the cloud to contract perpendicular to the field lines -- while near-hydrostatic equilibrium is maintained along field lines. A value $\kappa < 1/3$ means, of course, a smaller increase of the magnetic field strength for a given increase in the gas density. This could be the result of the development of a central condensation in which the magnetic force is partly determined by field lines that are frozen in the envelope as well as in the external (intercloud) medium. There is no theoretical lower limit on κ; e.g., if $\underline{B}$ is very strong, only motion along field lines will occur, and $\kappa \approx 0$ for such motion.

Nariai: Is not the energy density of the gravitational field of the galaxy comparable to or larger than the energy density of the magnetic field?

Mouschovias: The galactic gravitational field is indeed important. In fact, it is responsible for the Parker instability, which, when triggered by a galactic shock, may account for the formation of cloud complexes, OB associations, and giant H II regions along spiral arms separated by regular intervals of about one kiloparsec, like "beads on a string" (see Mouschovias, Shu, and Woodward 1974, Astron. & Astorophys., 33, 73).

Nariai: In the case where most of the magnetic lines of force lie in the plane of rotation, wouldn't you expect transfer of mass among neighboring clouds connected by the magnetic field, which may reduce the timescale of the change in $m(\Phi)$?

Mouschovias: No, I would not. The mean separation of interstellar clouds along the same field lines is much too large (at least several hundred parsecs and maybe larger) and matter velocities much too small ($v \leq v_{Alfvén} \sim 10$ km/sec) for exchange of mass to be relevant. In addition, field lines "buckle" in the space between clouds and extend high above the galactic plane. This geometry makes the kind of mass exchange which you are suggesting very unlikely.

Schatzman: I suggest that you consider mass loss from the cloud, since such mass loss can carry away a large amount of angular momentum.

Mouschovias: Mass loss due to what? And over what time scale? If somebody estimates significant mass loss (e.g., due to compressional hydromagnetic waves), then we'll surely have to consider it. However, I am finding that magnetic braking by itself can resolve the angular momentum problem during the early, diffuse stages of star formation.

Nakano: What configuration did you take for the cloud?

Mouschovias: As shownin Figure 4 below, the magnetic fields vector of a cylindrical (or disk) cloud, rotating about its axis of symmetry, is initially perpendicular to the axis of rotation. Under our assumptions, the results are independent of the length of the cylinder (or, the thickness of the disk).

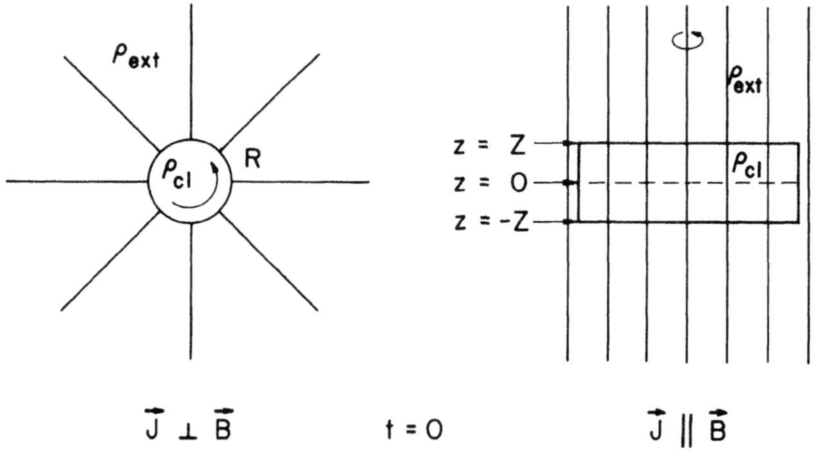

Figure 4. The geometries employed by Mouschovias and Paleologou in their studies of magnetic braking of perpendicular (left) and aligned (right) disk or cylinder rotators. Solid lines are field lines, shown at time t = 0.

van den Heuvel: Would not field line reconnection rapidly destroy a spiral-shaped magnetic field pattern?

Mouschovias: We have considered the issue of reconnection and concluded that, if it happens at all, it happens far from the cloud -- at least 20 cloud radii away. In fact, we estimated the energy of cosmic rays which may be produced in situ in the intercloud medium by such reconnection (see 1979, Ap. J., 230, 204).

Nariai: Have you solved the equations for i, B, and E as well as the equation of motion, or have you solved only the equation of motion using the frozen condition? I would not say that a problem of this type is solved rigorously unless i, B, and E are given as functions of time.

Mouschovias: For the case $\vec{J} \parallel \vec{B}$, we impose the condition that both components of $\vec{B}$ are continuous across the cloud surface, and indeed we solve the problem rigorously. The current density is always given by $\nabla \times \vec{B} = (4\pi/c)\vec{j}$ and the electric field by $\vec{E} = -(\vec{v}/c) \times \vec{B}$, since the magnetic field is strictly frozen in the matter. We also solved this same problem by relaxing the condition of continuity of the azimuthal component of $\vec{B}$ across the cloud surface (i.e., by considering a rigidly rotating cloud), and showed that the behavior of the cloud (as far as its angular velocity is concerned), except for short-lived transient effects, is virtually identical in the two cases. We also found that the rigid-body approximation becomes better and better as the ratio ρ_{cl}/ρ_{ext} increases (see 1980, Ap.J., 237, 877). For these reasons, we assumed rigid-body rotation of the cloud in the case $\vec{J} \perp \vec{B}$, which concerns us here. One should always distinguish between assumptions which, if relaxed, alter the qualitative results and assumptions which, if relaxed, only affect somewhat the quantitative nature of the conclusions.

QUASISTATIC CONTRACTION OF MAGNETIC PROTOSTARS DUE TO MAGNETIC FLUX LEAKAGE

Takenori Nakano
Department of Physics, Kyoto University

A protostar with sufficient magnetic flux can be in quasistatic equilibrium. The structure at the time t of such an axisymmetric protostar immersed in a medium of finite pressure is determined when the mass $m(\phi, t)$ in each axisymmetric magnetic tube with the flux ϕ is given. A new distribution $m(\phi, t+\delta t)$ is found by calculating the amount of matter which crosses the surface of the magnetic tube ϕ in a time δt due to plasma drift (ambipolar diffusion), and then the structure at $t+\delta t$ is determined. By repeating this procedure we can follow the quasistatic contraction of a protostar. Here we take into account the effect of charged grains in addition to ions upon the plasma drift.

We have investigated the contraction of a protostar of $50M_{\odot}$. The protostar is fairly flat initially. The high-density central part contracts faster than the outer part of the disk. The increase of the density accelerates the contraction further. Finally the central part contracts rapidly leaving the outer part nearly unchanged. When the magnetic flux of the central part becomes insufficient to maintain equilibrium, it begins to contract dynamically. Thus, only a part of the protostar becomes a star. For the model investigated several $M_{\odot}$ contracts finally, which is a few times the Jeans' critical mass at the initial state of the protostar. The friction of grains increases the contraction time by about 30 percent in this model.

DISCUSSION

Bodenheimer: Could you compare the magnetic field strengths that you need for your equilibrium models with observed fields?

Nakano: The upper limits to the magnetic field strengths have been obtained for many molecular clouds, and they are consistent with the strength I have adopted in my model.

Massevitch: In comparing theory and observation, are there not enough observational results or are the observed magnetic fields smaller than required by theory?

D. Sugimoto, D. Q. Lamb, and D. N. Schramm (eds.), Fundamental Problems in the Theory of Stellar Evolution, 63–64.

Nakano: The upper limits to magnetic field strengths that I have mentioned were obtained by attributing the widths of molecular lines to the Zeeman effect, and are fairly large-much larger than required by theory. More direct methods of dtermining B are desirable.

Mouschovias: Please allow me to answer the previous two questions on the status of observations of magnetic fields in dense molecular clouds. The direct Zeeman measurements are performed on the 1720 MHz line of OH; these refer to masers ($\sim 10^{16}$cm), with the result that the "split" in the line may also be due to two masers having some relative velocity. If the OH results are due to the Zeeman effect, they imply that B ~ a few milligauss. This, however, may not represent the mean field in the molecular cloud itself. Observations of the latter field are only indirect (mainly through optical polarization observations). Although their results are consistent with my theoretical prediction $B \propto \rho^{1/2}$, one must await more direct methods of observing B before making definitive statements.

The linewidth argument given by Dr. Nakano is actually a generous overestimate of the upper limit on the magnetic field strength.

Vilhu: I have a very simple and general question in mind. Your computations of magnetic braking in protostars do not have any outflow of matter, like a stellar wind. Is such outflow unimportant?

Nakano: Because I am considering a non-rotating cloud, I do not have a magnetic braking problem. Further, I am investigating the quasistatic contraction of the cloud up to the stage when the central part begins to contract dynamically. Therefore, there is not yet a protostar which could produce a stellar wind.

ROLE OF GRAINS IN THE DRIFT OF PLASMA AND MAGNETIC FIELD IN DENSE INTERSTELLAR CLOUDS

Toyoharu Umebayashi and Takenori Nakano
Department of Physics, Kyoto University, Kyoto

The magnetic flux through an interstellar cloud or through a part of the cloud must decrease considerably by the drift of plasma and magnetic field in order that stars form in the cloud. Because most grains are negatively charged in dense clouds (Elmegreen 1979; Umebayashi and Nakano 1980), they retard magnetic flux leakage in addition to ions (Elmegreen 1979). We investigate this effect for different situations and obtain the following results (for details, see Nakano and Umebayashi 1980):

1. For nearly spherical clouds of mass $\gtrsim 10^3 M_\odot$ sustained by magnetic force the friction of grains is efficient at $n_H \gtrsim 10^5\,cm^{-3}$, and the magnetic flux leakage time t_B is greater than a few million years at any density. Grains drift as fast as ions and electrons.
2. For spherical clouds of smaller mass and disk-shaped clouds, t_B becomes much smaller and the drift of grains is much slower than ions and electrons at $n_H \gtrsim 10^6\,cm^{-3}$.
3. Thus magnetic flux leakage occurs mainly in the condensations described in 2, and the abundance of heavy elements in stars deviates little from that of the parent clouds because of small grain drift.

REFERENCES

Elmegreen, B.G. 1979, Astrophys. J., 232, 729.
Nakano, T. and Umebayashi, T. 1980, submitted to Publ. Astron. Soc. Japan.
Umebayashi, T. and Nakano, T. 1980, Publ. Astron. Soc. Japan, 32, No.3.

DISCUSSION

Unno: How do you estimate the charge of a grain?
Umebayashi: In a dense cloud shielded from the interstellar ultraviolet radiation, the photoelectric effect is not efficient at all. Then the charge state of a grain is determined by the collision of ions and electrons with the grain surface. We have calculated the sticking probability, and have found that it is between 0.3 and 1.0 and almost 1.0 for electrons and ions, respectively. Using these values, we have estimated the charge state of a grain in a steady state condition.

D. Sugimoto, D. Q. Lamb, and D. N. Schramm (eds.), Fundamental Problmes in the Theory of Stellar Evolution, 65.

ON THE CORRESPONDENCE OF OH/H_2O MASER SOURCES TO THE STAGE OF PROTO-STAR FORMATION

Masa-aki Kondo
Dept. of Earth Science and Astronomy, College of General Education, Univ. of Tokyo

VLBI observations have revealed that OH/H_2O maser sources are the aggregations of blobes with the sizes of a few A.U., spreading over the distance of several 10^{17} cm. From the standpoint of masing mechanism, especially for the hydroxyl case, we will argue that the blobes correspond to proto-stars in the stage of forming an opaque core, and the aggregations correspond to proto-clusters.

In the radiative pumping case, the size of maser sources is required to be larger than 10^{16} cm, because they must be tenuous to avoid thermalization. In order to shorten the size, we consider the mechanism of the population inversion in the collision-dominant state, especially in the OH case. Since the collisional cross-section of OH is largest in the ground Λ-doublets, these levels are likely to be firstly thermalized. Then, we consider the chemically open and non-steady state, with regard to the dominant reaction of $H_2O + H \rightleftarrows OH + H_2$, which is realized at the accretion shock boundary of the opaque core. In such a state, the population of OH is affected by the way how the produced OH molecules enter into the rotational and hyper-fine structure levels: In the above cited reaction, the produced OH molecules are situated in high rotational states, and those in the term of $^2\Pi_{3/2}$ selectively enter into the upper levels of the four Λ-doublet levels in each rotational level. Subsequently, the OH molecules in such states are collisionally cascading down through the upper Λ-doublet levels of lower rotational states. Hence, we obtain the population inversion at $J = 3/2$ and $5/2$, which takes on the characteristic of the type I OH-maser sources.

In order to obtain the brightness temperature of $10^{12\sim 14}$ K, we need the width of 10^9 cm, in the state where $N_H/10^{14} \sim 6\times10^{-3}\cdot T - 12$ and $N_H > 10^{14}$: Here N_H and T denote the total hydrogen number density (cm^{-3}) and kinetic temperature, respectively. the such state is realized in the accretion shock of the opaque core with the radius of 1 A.U. Concerning the H_2O maser sources, the collisional mechanism in the chemically open and non-steady state is also applicable. However, a detail consideration is necessary for answering the question what stage do H_2O maser sources by the above mechanism correspond to.

D. Sugimoto, D. Q. Lamb, and D. N. Schramm (eds.), Fundamental Problems in the Theory of Stellar Evolution, 66–67.

DISCUSSION

Bodenheimer: During the main accretion phase of a spherically symmetric protostar, the theoretical calculations show that the accretion shock is located only 10^{12} cm from the center or less. Is this small scale consistent with your model?

Kondo: The accretion shock in my model is concerned with the boundary of the opaque core (which some people call the first core). This location depends on the opacity of dust grains. The distance from the center is nearly equal to 1 A.U., under the usual conditions for dust grains, as shown by my calculations (1978, the Moon and the Planets, 19, 245), and those of Larson (1969, M.N.R.A.S., 145, 271).

Mouschovias: You made a point too quickly for me to understand. Do your necessary chemical reactions proceed fast enough compared to the dynamical time scale for contraction of your object, which has $n \sim 10^{16}$ cm^{-3}, to produce an effect?

Kondo: Yes, the reactions of hydroxyl radicals proceed on a time scale of about $10^{-11} n_H$ sec, where n_H is the total hydrogen number density. When $n_H > 10^{14}$ cm^{-3}, this time scale is shorter than the free-fall time scale of $3 \times 10^{14} n_H^{-1/2}$ sec.

ON THE STAR FORMATION IN EARLY STAGE OF GALACTIC EVOLUTION

Y. Yoshii and Y. Sabano
Astronomical Institute, Tôhoku University, Sendai, Japan

Evolution and fragmentation of a gas cloud are investigated for the primordial chemical composition which is the same as the products of the Big Bang. A pure-hydrogen gas cloud collapses isothermally at 500-1000 K when a low fraction of molecular hydrogen works as a coolant, and breaks into small subcondensations with mass less than 10 $M_{\odot}$ due to thermal instability associated with molecular dissociation. On the other hand a pure-hydrogen gas cloud which contains no molecular hydrogen collapses isothermally at 6000-8000 K in a thermally stable condition, and enters the region where thermal energy exceeds radiation energy when thermal equilibrium between matter and radiation is achieved in the cloud. Consideration of energetics in the subsequent stage of the cloud evolution leads to the mass range of 0.1-20 $M_{\odot}$ for the stable nuclear-burning protostars of the first generation. The thermal behavior of a gas cloud in the regime of z (the ratio of heavy element abundance to solar one) less than 10^{-4} is essentially similar to that in the case of no heavy element, and the heavy element cooling brings about thermal instability in a wide range of parameters in the regime of z greater than 10^{-3}. Linear perturbation analysis gives growth time of the instability much shorter than the free-fall time, and suggests the efficient excitation of density fluctuation driven by thermal instability. Thus the possibility of the initial mass function relatively enhanced in massive star at early times is denied, and the slow rate of metal enrichment in the interstellar medium is suggested.

DISCUSSION

<u>Mouschovias</u>: The wavelength of the thermal instability has an upper bound set by the product of the speed of sound and the cooling time. Isn't it the case that, for the temperature and density you are using, this characteristic size is too small to involve very large masses? And wouldn't you have to collect matter for distances exceeding a kiloparsec along field lines if you include even a weak magnetic field? (Field showed in 1965 that the thermal instability is suppressed, even by a relatively weak field, in the two directions perpendicular to the field.) And wouldn't

D. Sugimoto, D. Q. Lamb, and D. N. Schramm (eds.), Fundamental Problems in the Theory of Stellar Evolution, 68–69.

this require much too long a time to take place?

Yoshii: The mass contained within the characteristic length of the thermal instability amounts to 10 $M_{\odot}$, which is sufficiently large to be bound gravitationally. There are many uncertainties concerning the early evolution of the Galaxy and we intend to clarify the thermal state of the gas in the early stages, assuming no magnetic field. However, it may be, of course, that magnetic pressure has a important influence on star formation in the early stages of galactic evolution.

Joss: In your scenario, some stars from the earliest stellar generations must still be in existence. The exact numbers will depend upon the relevant birthrate function. Is it possible to reconcile your results with the stringent observational upper limits on the present number of stars in the galaxy with very low heavy element abundances?

Yoshii: The earliest stars may mostly be contained in the halo component, for which the observational constraints are, unfortunately, poor at present. As a matter of fact, extremely metal-poor stars have not been detected yet, but we think such objects will possibly be observed in future. However, it should also be remarked that metal-free stars could accrete metal-rich interstellar matter during the galactic evolution, and the metal abundance of the stellar surface could change considerably.

ON THE STRUCTURE AND EVOLUTION OF MASSIVE INTERSTELLAR CLOUDS

H. Kimura and Liu Cai-pin
Purple Mountain Observatory, Academia Sinica,
Nanking, China

The evolution of massive clouds is discussed with emphasis on inhomogeneous and heterogeneous nature of the system. We start from the two-phase model of interstellar medium and choose initial conditions of the clouds as being in the vicinity of critical state for gravitational instability. The equilibrium and stability problem for clouds is formulated in terms of polytropic models. Combining this with the thermal and chemical balance problem, we find that, even before the commencement of gravitational collapse, the fragmentation due to thermal instability can occur. The system begins to contract at a rate much slower than free fall, and thereafter, its evolution is governed mainly by collisions between fragments. The collision dynamics of the N-fragment system should be examined carefully by numerical experiments. The general trend to be expected is that the system tends to develop into a core-halo structure, accompanying by sporadic local events of star formation. The core or a clustering of massive fragments near the center could grow up into a favourable formation site of star clusters.

DISCUSSION

Hasegawa: What is the typical size of the fragments? In what range of mass and density do thermal instabilities occur?

Kimura: The mean size of fragments can be roughly estimated by diameter $d = (\rho_0/\rho_1)^{1/3} c_s \tau_c$, where c_s denotes the sound speed and τ_c the timescale of CO cooling. The degree of compression, (ρ_1/ρ_0), may be ~5. Adopting typical values of $c_s \lesssim 1$ km/s and $\tau_c = 10^5 \sim 10^6$ years, we have d = 0.07~0.7 pc. With regard to your second question, it seems better to show the minimum values of local density $\underline{n}$ and effective column density N_H from the surface that are required; the answer is $N_H \gtrsim 10^{21}$ cm^{-2} and $n \gtrsim 10^2$ cm^{-3}, if the instability is due to CO cooling.

Hasegawa: Recent analyses of the 21-cm line of HI show the existence of cold gas fragments with a total particle density of ~100 cm^{-3} and a typical size of 20 pc in diameter. The formation of molecular clouds may occur via such a stage.

D. Sugimoto, D. Q. Lamb, and D. N. Schramm (eds.), Fundamental Problems in the Theory of Stellar Evolution, 70–71.

Kimura: A scenario that we would like to propose is shown below. It seems to me that the comment by Hasegawa refers to clouds of level (C) as components of (B). In my opinion, these clouds are formed not through fragmentation but by coalescence due to inelastic cloud-cloud collisions. We have discussed the possible occurrence of fragmentation in such clouds [level (C)], and the uniqueness of the heterogeneous system thus obtained.

A senario for the earliest stage of star formation

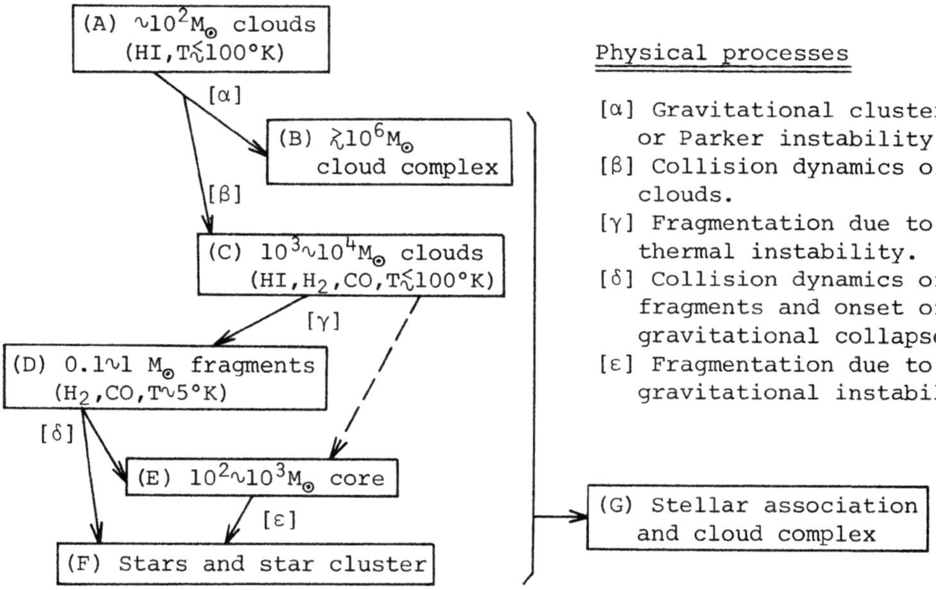

CHAIN-REACTING THERMAL INSTABILITY AND ITS IMPLICATION ON STAR-FORMATION IN INTERSTELLAR CO CLOUDS

Y. Sabano* and Y. Sofue**
* Astronomical Institute, Tôhoku University, Sendai, Japan
**Department of Astrophysics, Nagoya University, Nagoya, Japan

A new type of non-linear, chain reacting instability is presented that a sequential condensation occurs in a thermally unstable interstellar CO cloud, triggered by a local density perturbation. Non-linear growth of the instability was followed numerically with a one-dimensional and slab symmetric hydrodynamical code. The result shows that a local, small density enhancement onto initial density of n_0=510 cm^{-3} around x=0 (x: spatial coordinate) grows to a maximum density of n_{max}=1300 cm^{-3} in a time scale of 0.6×10^6y, and at the same time the gas in neighbouring region at x=0.1-0.2 pc shifts into a low-density phase of n_{min}=200 cm^{-3}. Since we assume a constant heating per particle by photons and cosmic rays, the gas in low density region is relatively more heated so that the pressure therein becomes higher than the background pressure. Then the low density gas pushes its neighbouring gas foward increasing x to lead to a second density enhancement at x=0.2 pc. In this way many further condensations are produced in a sequential manner. The spacings and sizes of condensations are uniquely determined by the characteristic parameters, the growth time of instability and the sound velocity of the background gas alone. The present calculation gives the spacing, 0.24 pc, and the mass of each fragment, $0.2M_\odot$. This mechanism could be related to sequential formation of less-massive stars in interstellar CO clouds, if the condensations evolve further into star formation sites.

DISCUSSION

Mouschovias: Your graphs showed a pressure of a few $\times 10^3$ K·cm^{-3} at the instability region and a density of $\sim 10^3$ cm^{-3}. This implies a temperature of T~5 K. Isn't this too cool for a molecular cloud and, if such a phase exists, shouldn't it be observable in absorption?

Sabano: The temperature in CO molecular clouds is observed as 20 K $\gtrsim$ T $\gtrsim$ 7 K, which may correspond to the condensations in our computation. The absorption of the CO line is indeed detected in the direction of the galactic nucleus and implies T $\lesssim$ 10 K. So the temperature of our condensations is not so unusual, although our simplified analysis might overestimate the cooling efficiency by CO molecules.

D. Sugimoto, D. Q. Lamb, and D. N. Schramm (eds.), Fundamental Problems in the Theory of Stellar Evolution, 72.

GLOBAL STRUCTURE OF INTERSTELLAR MEDIUM AND STAR FORMATION RATE

S. Ikeuchi
Department of Physics, Hokkaido University, Sapporo 060, Japan

Assuming that the interstellar medium (ISM) is composed of the hot ionized medium (HIM), the warm ionized medium (WIM) and the cold neutral medium (CNM), we examine the interchange processes among them by supernova remnants. These are the evaporation of CNM, the shock heating of WIM and the cold shell formation at the shock front. Calculating the time variation of each component, the timescale and evolutionary characteristics till attaining a steady state are deduced. Generally speaking, the final steady state is classified to two types.

One is that a considerable fraction, f_h ($0.6\sim1.0$), is occupied by a hot ($\gtrsim10^{5.3}$ K) and rarefied ($\leq10^{-2}$ cm^{-3}) gas. Even if the supernova explosions are so rare as 10^{-3} SNe y^{-1}, this steady state is maintained because the supernova remnants can expand greater than 200 pc and heat up the hot gas. The cloud abundance is so low as $\lesssim10^{-4}$ clouds pc^{-3} that the star formation is inhibited.

The other type of ISM is so-called two-phase model. The interstellar space is almost occupied by WIM and the cloud abundance is as high as $\gtrsim10^{-2}$ clouds pc^{-3}. The ejected energy from supernovae is not used to heat up the ambient medium, but is emitted away by radiation at the dense shell.

In relation to the star formation at the shocked region of a molecular cloud, we discuss the sequential formation of stars. As its observational evidence, we consider the superbubbles, which are observed in Cygnus region, Gum nebula and Eridanus-Orion region, and the supershells, which seem to associate with O, B associations. These huge loops extend with the diameters $500\sim1\times10^3$ pc. We study the expansion laws and their structures under the assumption that a chain of supernova explosions at every 10^6 years have occurred. In the case of Cygnus superbubble, about 30 supernovae are necessary for its exptension and X-ray emission. These phenomena are expected when many massive stars are formed at once as an OB association, and OB associations are formed sequentially.

D. Sugimoto, D. Q. Lamb, and D. N. Schramm (eds.), Fundamental Problems in the Theory of Stellar Evolution, 73.

THE FORMATION OF BINARY STARS

L. B. Lucy
Department of Astronomy
Columbia University
New York, New York 10027

INTRODUCTION

The speed and storage capacity of present-day computers have stimulated the development of numerical techniques allowing the investigation of three-dimensional gas dynamical problems of astronomical interest. Of the problems that can be attacked with these techniques, that of the formation of binaries has historically attracted the greatest interest. Several investigators have therefore tackled this problem, and their efforts will be discussed here insofar as they relate to the formation of close binaries. Earlier work, including the classical investigations of Kelvin, Poincaré, Jeans, and Cartan, have been reviewed by Chandrasekhar (1969) and Tassoul (1978).

Two mechanisms have dominated discussions of the origin of close binaries, and both can be at least crudely investigated with available computers and techniques. The first of these is fission, a term denoting the bifurcation of a rotating protostar during the quasi-static (i.e., Kelvin-Helmholtz) phase of its contraction to the ZAMS. The second is fragmentation, a term denoting the creation of a double or multiple system by the break up of a rotating protostar during, or immediately following, a phase of dynamical collapse. The possibility that a close binary might form by fragmentation following the dynamical collapse precipitated by the onset of dissociation and ionization of hydrogen in a protostar's deep interior was pointed out by Larson (1972) and has been illustrated quantitatively by Bodenheimer (1978).

The attraction of investigating these mechanisms with 3-D gas dynamic codes are: (1) that we can check earlier conjectures that they do indeed result in binary formation; and (2) that, if these conjectures are confirmed, the properties of the resulting model binaries can be compared with those of observed systems.

OBSERVATIONAL EVIDENCE

The large body of heterogeneous data that astronomers have compiled on binaries is far from an ideal sample if one wishes to

D. Sugimoto, D. Q. Lamb, and D. N. Schramm (eds.), Fundamental Problems in the Theory of Stellar Evolution, 75–83.

discover clues to formation mechanisms: The sample is greatly affected by a variety of selection effects, and many catalogued binaries have undergone major evolutionary changes since formation. In addition, observational errors are by no means inconsequential - typically, for each type of binary, we have a handful of systems with well-determined elements and then a vast number whose elements are poorly-determined. Nevertheless, some progress has been made in isolating seemingly reliable formation clues.

With regard to long-period binaries ($P \gtrsim 100$ dys), the most reliable work is that of Abt and co-workers - see Abt (1979) - which is largely based on observational programs specifically designed to lessen or eliminate the above mentioned problems. This work has not revealed any formation clues of high information content: the distributions of these binaries' observable characteristics are smooth and featureless. This surely implies that their formation mechanism neither forgets nor is insensitive to initial conditions; consequently, since the spectrum of initial conditions is not likely soon to be known or predicted, decisive confirmation of a formation theory will not quickly be forthcoming. Thus, although we may well be convinced that hierarchical fragmentation (Heintz 1978; Bodenheimer 1978) is essentially correct, we are not likely soon to strengthen the observational basis for this conviction.

In contrast to the long-period binaries, those of short period ($P \lesssim 25$ dys) do appear to offer formation clues of high information content, and these seem to imply that such binaries are created by a mechanism that at least partially forgets initial conditions. If so, we can reasonably hope to demonstrate decisive observational confirmation for a theory of the formation of close binaries.

The strongest evidence for the forgetting of initial conditions is the spike at $q=1$ in the distribution of mass ratios (Lucy and Ricco 1979). Taken together with the work of Garmany and Conti (1980) on O-type systems, this result seems to imply that a formation mechanism operates over the entire range of stellar masses that, in its ideal form, creates close binaries ($P \lesssim 25$ dys) with identical components. That not all such detached binaries have $q=1$ can be interpreted as a lingering memory of initial conditions.

Further probable formation clues for close binaries come from the period - spectrum diagram for spectroscopic binaries. Figure 1 shows such a plot for SB2's with $q > 0.8$ catalogued by Batten et al. (1978) - contact binaries and eruptive variables are excluded. Also shown is the locus of zero-age contact binaries (ZACB) with $q=1$ calculated using the models of Morris and Demarque (1966) for the upper main sequence and extrapolating to later spectral types using observational data for the Sun and YY Gem.

The dashed lines in Fig. 1 indicate possible formation clues. Two of these show that the lower envelope to the observed systems

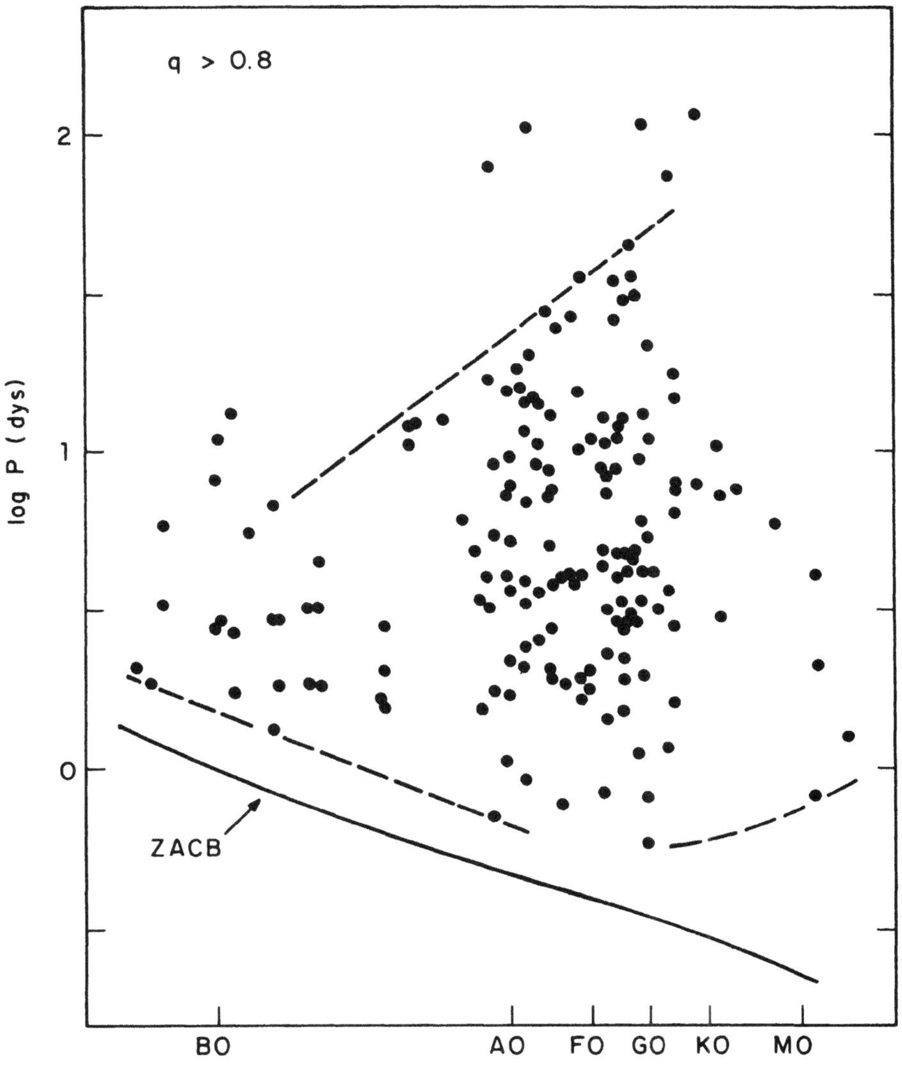

Fig. 1: Period-spectrum diagram for spectroscopic binaries with q>0.8.

departs markedly from the ZACB locus. Because we can mentally construct binaries that would populate this gap and be readily discoverable, the gap cannot be attributed to selection effects. For upper main sequence binaries, the gap probably indicates that binary formation stops well before massive protostars complete their contractions to the ZAMS. For spectral types later than about Go, the increasingly pronounced gap is probably a consequence of post-formation orbital evolution in consequence of magnetic braking (cf. Huang 1966; Mestel 1968), with the more massive of such binaries ending up as W UMa and short-period β Lyrae systems.

The uppermost dashed line in Fig. 1 calls attention to a sharp drop in the number density of SB2's. This is probably a real and therefore significant effect, but an attempt (Lucy and Acierno, unpublished) to construct arguments against all possible selection effects has not yet proved successful.

Also worth noting is the fact that galactic clusters are not markedly deficient in SB's, when one allows for the incompleteness of searches (Batten 1973). Given the low escape velocity from clusters, this implies that short-period binaries seldom result from the disruption of close multiple systems.

FISSION

Three-dimension numerical calculations of binary formation by fission using the finite-size particle technique (Lucy 1977) have been carried out by Lucy (1977) and by Gingold and Monaghan (1978, 1979). In both these investigations, the ideal problem is substantially modified because of the difficulty in achieving adequate spatial resolution and in treating a problem with two widely different time scales. Nevertheless, despite their different compromises, these investigators agree in finding that fission can lead to the formation of a binary and that the resulting system has small mass ratio, $q \sim 0.3$.

Although improved calculations would be worthwhile, these initial numerical experiments, together with recent observational results, strongly suggest that fission is not the mechanism responsible for the bulk of close binaries. Firstly, in the absence of a post-formation mass-exchange instability on a dynamical time scale (Lucy and Ricco 1979), fission seems not to provide an explanation for the many close binaries with components of comparable mass. Secondly, and perhaps more decisively, it now seems clear observationally (Cohen and Kuhi 1979) that the Kelvin-Helmholtz contraction phase starts at rather small radii, thus severely limiting the orbital periods of binaries formed by fission. For example, taking $5R_{\odot}$ as the radius at which a $3\mathfrak{M}_{\odot}$ protostar first appears on the H-R diagram and assuming that this immediately fissions with conservation of mass and angular momentum into a binary

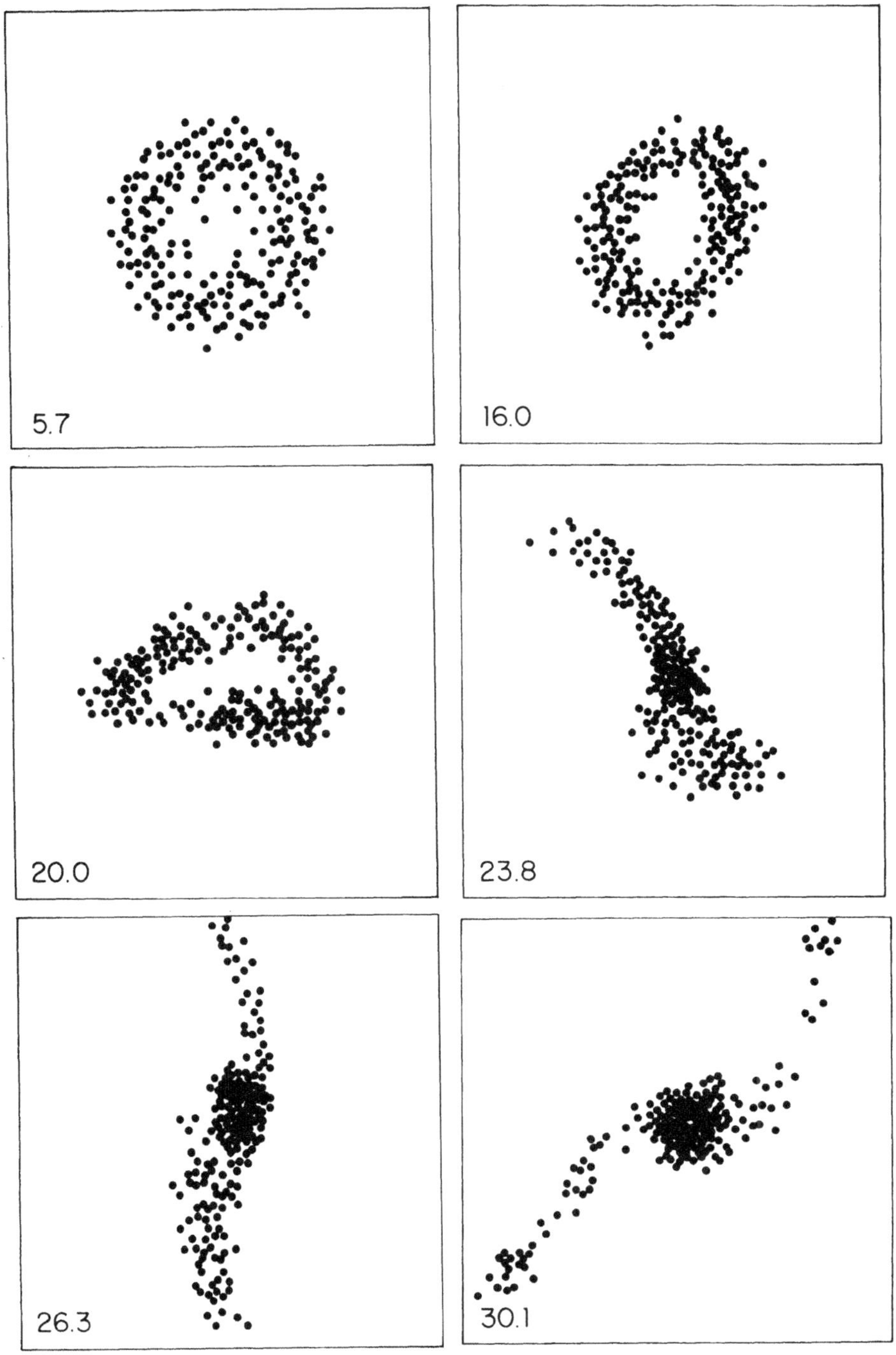

Fig. 2: Fragmentation of toroidal protostar having t_T=0.14 initially.

with q=1, we obtain P ~ 1.5 dys. Thus the bulk of short-period binaries in this mass range have too much angular momentum to attribute their formation to fission.

FRAGMENTATION

For this mechanism, there are two cases to consider, depending on whether or not dynamical collapse gives, immediately before fragmentation, a toroidal or a spheroidal configuration. A toroidal structure was first found by Larson (1972), but this result was not unanimously confirmed by later workers. The position now seems to be that most investigators agree that initial conditions exist from which toroidal structures do indeed result. Nevertheless, since not all initial conditions yield such configurations, the fragmentation of spheroidal protostars also needs to be investigated.

Norman and Wilson (1978) and Cook and Harlow (1978) were the first to carry out three-dimensional calculations of the fragmentation of toroidal protostars. Their results certainly confirm Larson's (1972) conjecture that such a protostar will break up into a binary or multiple system. But since these authors start with imposed perturbations that fix the mode of fragmentation, they cannot be regarded as having securely determined the exact outcome of fragmentation. Accordingly, the finite-size particle scheme has also been used to investigate the fragmentation of toroidal protostars (Lucy 1980), with numerical noise being relied upon to provide seed amplitudes for unstable modes. These results show that the dominant mode of fragmentation depends on t_T, the ratio of thermal to gravitational energy of the initial model, with fragmentation into many components occurring if t_T is small. As t_T is increased, one might anticipate finding a significant range for which an m=2 mode dominates, with a q=1 binary as the end result. That this is not the case is seen by comparing the previously published results (Lucy 1980) with those shown in Figure 2, which extends the earlier sequences to higher t_T. We see that the dominant mode now becomes a global distortion of the ring and that the subsequent evolution yields a rapidly rotating star with two companions of small mass.

The finite-size particle scheme has also been used to investigate the fragmentation of protostars that end their dynamical collapse phases with spheroidal structures. Figure 3 shows a typical evolutionary sequence. The initial model is a highly flattened ($a/c \simeq 6$) spheroidal configuration that is symmetric about its invariable plane, uniformly rotating and in virial equilibrium. As in the toroidal sequences, subsequent changes are isentropic ($\Gamma = 5/3$) and a small bulk viscosity term is included in the equation of motion. We see that the model evolves into a bar ($\tau = 16.2$), but any hope that this might subsequently bifurcate is thwarted by gravitational torques that transfer angular momentum from the bar to exterior debris. Thus the end result is a rapidly rotating axisymmetric single star star with an equatorial disc of debris.

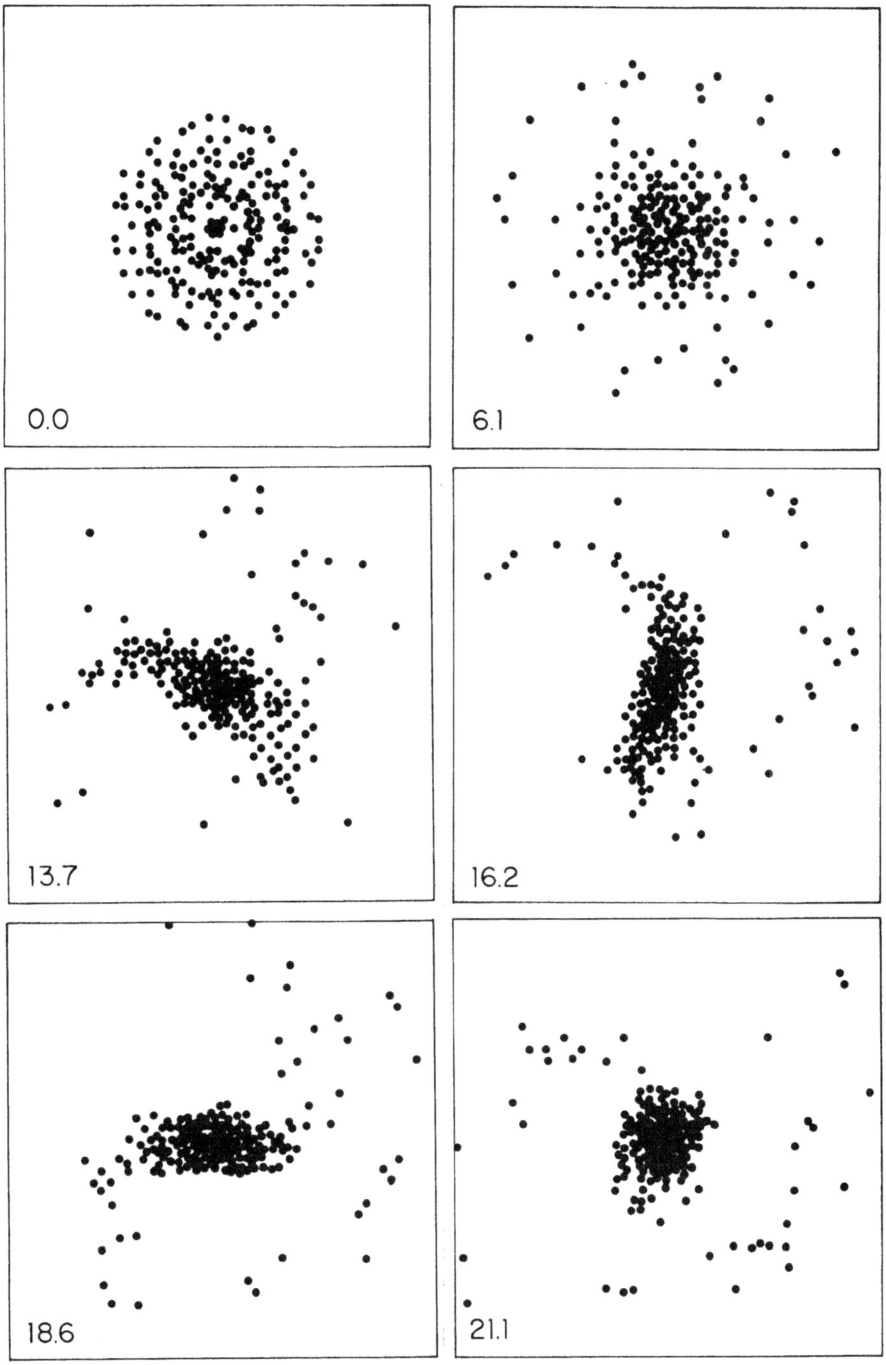

Fig. 3: Fragmentation of spheroidal protostar having t_T=0.12 initially.

DISCUSSION

The various 3-D calculations reviewed here have not yet provided a definitive explanation for the existence of any class of binary star. In particular, an understanding of the formation of q=1 systems has proved elusive. The belief (Lucy and Ricco 1979) arising from the work of Cook and Harlow (1978) and Norman and Wilson (1978) that such systems form directly by fragmentation is not confirmed since q=1 binaries do not appear when the initial models are not seeded with an m=2 perturbation of large amplitude. Nevertheless, these early 3-D calculations have suggested two conjectures for the formation of q=1 systems:

One possibility (Lucy 1980) is that, as a result of the dynamical collapse starting from a pressure-supported configuration, fragmentation proceeds from essentailly noise-free initial conditions and therefore yields a multiple system with identical components. The subsequent reduction to a binary through collisions will then fairly often produce a q=1 system if collisions occur with little mass loss.

A serious flaw in this conjecture, however, is the assumption that the $\overline{\Gamma} < 4/3$ dynamical collapse is preceeded by a phase of mechanical equilibrium (B. G. Elmegreen, private communication). As shown by Gaustad (1963), only protostars with $\mathfrak{M} \lesssim 0.3\,\mathfrak{M}_\odot$ have their free-fall collapses halted when dust opacity inhibits direct energy loss by radiation.

A second conjecture envisions a binary with nearly identical components forming in three stages: 1) fragmentation of a toroidal protostar into a close multiple system; 2) reduction of the multiple system to a binary via collisions; 3) a secular increase of the binary's mass ratio by accretion from an exterior disc formed from late-infalling gas. The first two stages are directly illustrated in the cited fragmentation calculations; the third stage is the gaseous analogue of the process invoked by Hayashi et al. (1977) to explain the growth of planets.

This work has been supported by the National Science Foundation under grant AST 79-13179. A grant of computer time at the Goddard Institute of Space Studies, NASA, is also acknowledged.

REFERENCES

Abt, H. A. 1979, A.J., *84*, 1591.
Batten, A. H. 1973, Binary and Multiple Systems of Stars (Oxford: Pergamon Press).
Batten, A. H., Fletcher, J. M., and Mann, P. J. 1978, P.D.A.O, *15*, 121.
Bodenheimer, P. 1978, Ap. J., *224*, 488.
Chandrasekhar, S. 1969, Ellipsoidal Figures of Equilibrium (New Haven: Yale University Press).

Cohen, M., and Kuhi, L. V. 1979, Ap. J. (Letters), L105.
Cook, T. L., and Harlow, F. H. 1978, Ap. J., *225*, 1005.
Garmany, C. D., and Conti, P. S. 1980, IAU Symposium No. 88, p. 163.
Gaustad, J. E., 1963, Ap. J., *138*, 1050.
Gingold, R. A., and Monaghan, J. J. 1978, M.N.R.A.S., *184*, 481.
Gingold, R. A., and Monaghan, J. J. 1979, M.N.R.A.S., *188*, 39.
Hayashi, C., Nakazawa, K., and Adachi, I. 1977, P.A.S.J., *29*, 163.
Heintz, W. D. 1978, Double Stars (Dordrecht: Reidel).
Huang, S.-S. 1966, Ann. d'Ap., *29*, 3.
Larson, R. B. 1972, M.N.R.A.S., *156*, 437.
Lucy, L. B. 1977, A.J., *82*, 1013.
Lucy, L. B. 1980, IAU Symposium No. 88, p. 7.
Lucy, L. B., and Ricco, E. 1979, A.J., *84*, 401.
Mestel, L. 1968, M.N.R.A.S., *138*, 359.
Morris, S. C., and Demarque, P. 1966, Z.f. Ap., *64*, 238.
Norman, M. L., and Wilson, J. R. 1978, Ap. J., *224*, 497.
Tassoul, J-L. 1978, Theory of Rotating Stars (Princeton: Princeton University Press).

DISCUSSION

Mouschovias: You mentioned at the beginning of your talk that observations do not give us a clue about the formation mechanism of binary stars. Wouldn't you consider Abt and Levy's (1976) result, that there is a single maximum in the period distribution for the 88 available systems, as at least evidence that a single mechanism may be responsible for the formation of all binaries with periods in the range 10 hours to 100 years -- as also suggested by theory (see, Ap. J., 211, 147)?

Lucy: When you bring together data for binaries with periods from 10 hours to 100 years into a single distribution, uncertain correction factors for different selection effects have to be applied. Such data is therefore not decisive in deciding whether or not a single mechanism is operative over such an enormous range of periods.

Tutukov: Even very close binaries have non-zero eccentricities. What is the reason for this?

Lucy: Binaries formed by fragmentation do in general have substantial eccentricities.

Tutukov: Unevolved double-line spectroscopic binaries with separation $a \lesssim 10\ R_\odot$ and mass $M \gtrsim 1.5\ M_\odot$ are absent (see Tutukov, Yungelson 1979 in Proc. IAU Symp. No.88), while the density in units of $\ell n\, a$ of double-line spectroscopic binaries with $a/R_\odot \lesssim 15\ M_\odot/M$ ($M \lesssim 1.5\ M_\odot$) is about 30 times lower than the same density for wider systems. These peculiarities seem very important for the problem of close-binary formation and, possibly, for their pre-nuclear evolution.

PRE-MAIN-SEQUENCE STELLAR EVOLUTION*

G.S. Bisnovatyi-Kogan
Space Research Institute, Moscow, USSR

* This paper was read by A.G. Massevitch

ABSTRACT

The problems of the stellar evolution to the main sequence are reviewed, taking into account the effects of mass loss, rotation and binarity. Properties of T Tauri stars are discussed which are connected with the recent observations of these stars in ultraviolet and X-ray regions. FU Ori phenomen is considered briefly.

* * * * * *

The foundations of the theory of pre-main sequence evolution have been installed by C. Hayashi in his famous paper, published in 1961. Since then the concepts of "Hayashi track", "Hayashi limit" have been constantly used by the astrophysicists, working on the problems connected with stars.

Here I want to make a review of the mordern state of the theory and to discuss the observational properties of pre-main sequence stars of T Tauri type. These stars have been investigated not only in optical but also in UV and X-ray regions from the satellites. The interpretation of the observations can be made, if we assume that T Tauri stars have a hot corona and chromosphere.

I. EVOLUTIONARY CALCULATIONS

1. Evolutionary tracks without rotation and mass loss

The main features of the quasi-hydrostatic pre-main sequence evolution have been established by Hayashi (1961) (see also Hayashi, Hoshi and Sugimoto, 1962 and Hayashi, 1965). On the early stages of contraction the luminosity L is high, the temperature T inside the star is low, so the noncomplete ionization and high opacity result in the formation of a totaly convective, contracting star. During the contraction L decreases, T increases and for $M \gtrsim 0.2\ M_\odot$

D. Sugimoto, D. Q. Lamb, and D. N. Schramm (eds.), Fundamental Problems in the Theory of Stellar Evolution, 85–97.

the radiative core is formed before the star comes to the main sequence (see fig.1). The relative mass of the radiative core at the beginning of the hydrogen burning increases with increasing the stellar mass. Burning of deuterium leads to increasing the life time of the

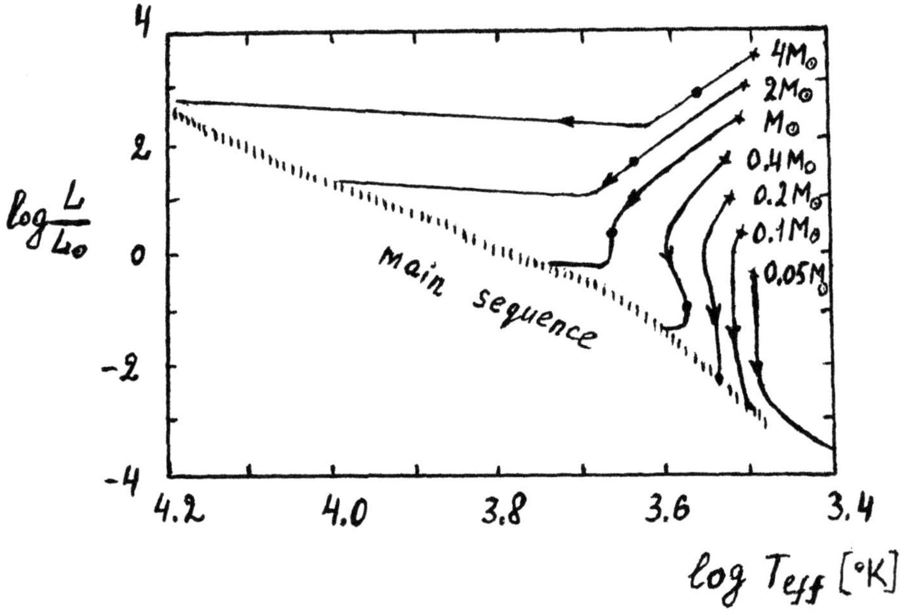

Fig. 1. The evolutionary tracks of pre-main-sequence stars in the stages of quasihydrostatic contraction according to Hayashi (1961, 1965). The crosses denote the initial models, the circles indicate the ends of the wholly convective stages or the zero-age main-sequence stages.

star on the pre-main-sequence stage (see calculations of Mazzitelli and Moretti, 1980).

Considerable progress has been made in recent years in the field of the pre-main sequence evolution of massive stars with $M > (3\div5)\ M_\odot$ (Appenzeller and Tscharnuter, 1974; Larson, 1977). The radiative cores are formed in these stars immediatly after the dynamical collapse at the beginning of the contracting phase. The rate of the core contraction is greater than the rate of the accretion of the matter from the extended envelope. So, the star comes to the point of the hydrogen burning still surrounded by the extended cold envelope. After the beginning of the hydrogen burning the outer envelope may be blown out by radiation and the star completely changes its appearance: on the place of the cold infrared star appears a hot main sequence star. This process may be related to the phenomenon FU Ori (Larson, 1977).

2. Taking into account the mass loss

Kuhi (1964,1966) has shown, that the spectra of T Tauri stars indicate on the mass loss. The rate of the mass loss is evaluated as $(0.3\div6)\ 10^{-7}\ M_\odot/yr$ and may influence the evolution. The calculations with a simplified version of mass loss

$$\dot{M} = -\alpha \frac{R^3}{M} \quad (1)$$

have been made by Ezer and Cameron (1969). The parameter α is equal to $3\cdot10^{-14}\ M_\odot/yr$ for the solar wind and α is between $10^{-11}\div10^{-8}$ for T.Tauri stars. The calculations have been made for $\alpha = 10^{-10}$, 10^{-9} and $3\cdot10^{-9}$. The characteristic results for initial masses $M = 2.93$ and $2.31\ M_\odot$ are shown in Fig. 2. Ezer and Cameron (1971) have found that equal age time lines for different mass loss rates lie very close to one another in the HR diagram. So, the large spread of stellar points on HR diagrams of young clusters cannot be explained as a result of variations in mass loss rate. The authors suppose that the spread in the times of formation of the stars in a few million years can explain the spread in the HR diagram.

3. Evolution of rotating pre-main-sequence stars

The angular momentum of the protostars plays an impor-

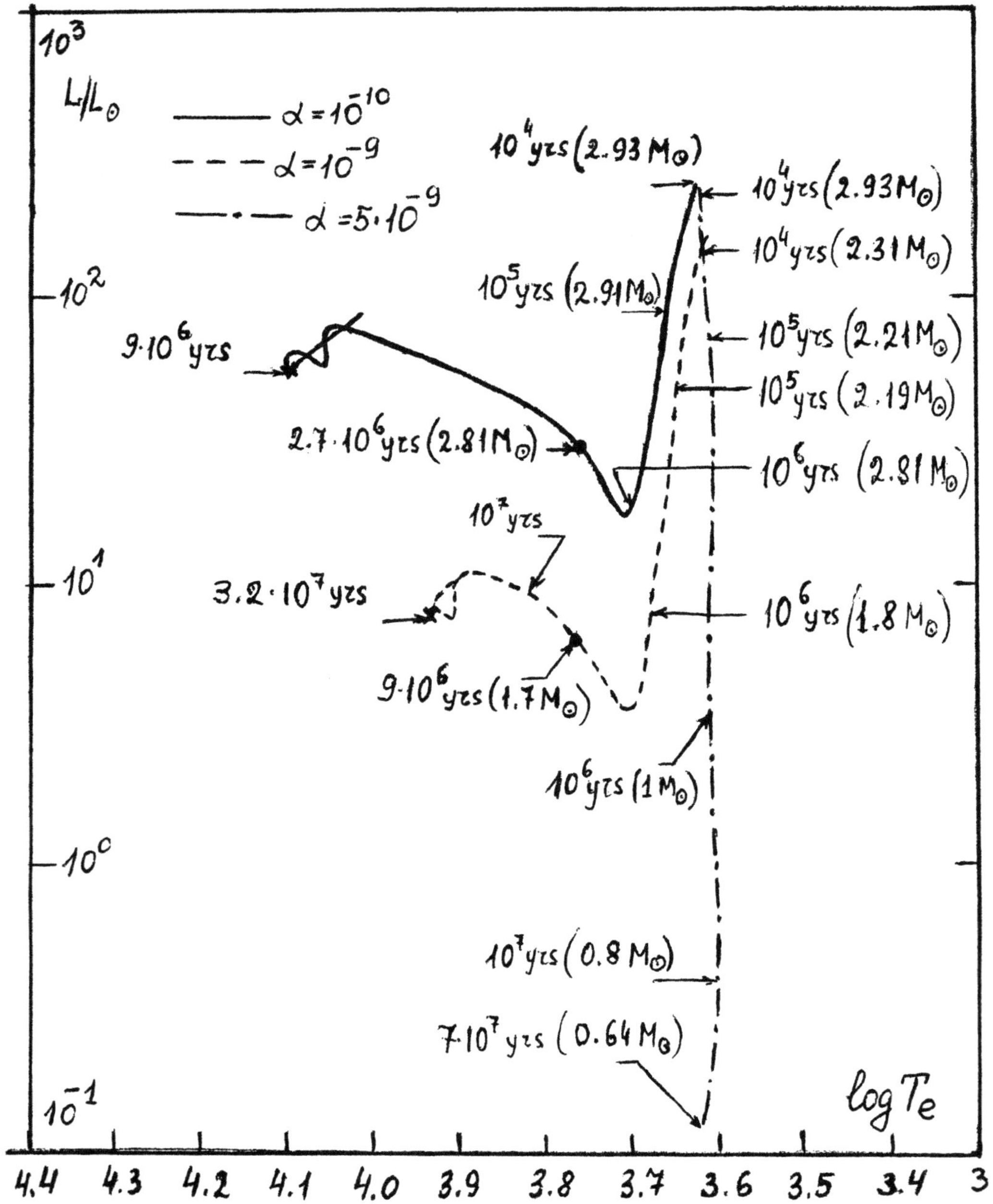

Fig.2. The evolutionary tracks of the deterium burning contracting stars with initial masses 2.93 $M_\odot$ and 2.31 $M_\odot$ and different mass loss rates following Ezer and Cameron (1971). The circles indicate the end of the outer convective zone, crosses - the main sequence. Ages and masses are given on the figure.

tant role in their evolution, leading to additional mass loss and determining the formation of binaries. The observational detection of the rotation in pre-main sequence stars is rather difficult because of very broad emission linea in T Tauri stars. Herbig (1957) found the value $\langle v \sin i \rangle = 20 \div 65$ km/s, observing four T Tauri stars. The indications on the rapid rotation of T Tauri stars are present in the work of Wilson (1975).

The investigations of the evolution of rapidly rotating stars need complex computations and have been made using some simplifications by Bodenheimer and Ostriker (1970) and Moss (1973). It occurs however, that for totally convective stars which approximate the pre-main sequence stars over considerable interval of their evolution, the evolutionary calculations may be done exactly for arbitrary rotation. This calculations have been made in the papers of Bisnovatyi-Kogan et al. (1977,1979). The main body of the convective star with constant entropy is calculated, using the self-consistent field method in the variant of Blinnikov (1975). The fitting of the envelope to the core is made separately for different points between the pole and equator which permits to find the distribution of the temperature over the star. The results of calculations for $M = 1\ M_\odot$ and $0.5\ M_\odot$ are given in Fig.3. The evolution time up to the main sequence almost doubled compared to the evolution of nonrotating stars. The method of evolutionary calculation gives good accuracy up to the point where the radiative core is not greater than 25% of the stellar mass. The evolution was calculated with the constant angular momentum of the star. During the evolution, the star becomes more flattened. The change of the shape of the star during the evolution is shown in Fig.4. The pre-main-sequence evolution of low-mass stars $M = 0.07-0.16\ M_\odot$ has been calculated in the papers of Fedorova and Blinnikov (1978); Fedorova (1979). It was shown there, that the minimum mass of a main-sequence star increases from $0.08\ M_\odot$ without rotation up to $0.1\ M_\odot$ for rigid rotation and up to $0.16\ M_\odot$ for differential rotation.

So, the changes in the premain sequence evolution, including moderate mass loss and rotation have a quantitative character. If the limiting rotation reaches in the early stages of contraction then the equatorial mass shedding begins. This process is probably connected with the formation of binary stars and planetary systems.

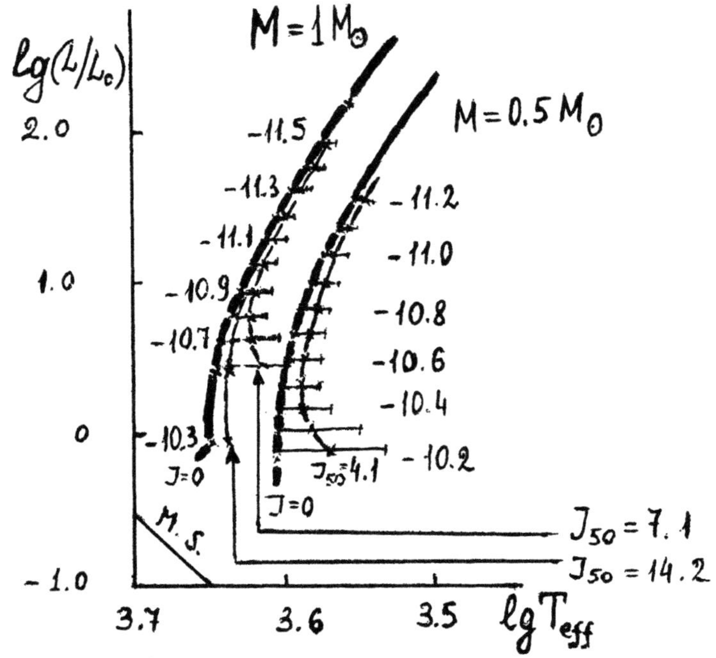

Fig.3. Evolutionary tracks of the contracting stars with masses $0.5M_{\odot}$ and $1M_{\odot}$ at different values of the angular momentum J_{50} (in units 10^{50} g cm^2/sec) following the paper of Bisnovatyi-Kogan et al.(1979). The thick lines indicate the results of the calculations by Henyey method for non-rotating stars, crosses - the models calculated by Bisnovatyi-Kogan et al.(1979). The numbers ($\lg\rho_0$ at $\lg T=3.3$ characterize the entropy of the stellar core. The horizontal lines show the spread of the effective temperature along the surface of the star. The main sequence line is indicated below left for X=0.70,Z=0.02. The "entropy" is connected with the age by following relations,where only photospherical luminosity was taken into account.

$\lg\rho_0$	−11.3	−11.0	−10.7	−10.4	−10.2
t(yr)(J =0)	0	$8\ 10^3$	$6\ 10^4$	$3.7\ 10^5$	$1.1\ 10^6$
t(yr)(J_{50}=4)	0	$7.5\ 10^3$	$5.5\ 10^4$	$3\ 10^5$	$8.1\ 10^5$
$\lg\rho_0$	−11.5	−11.2	−10.9	−10.6	−10.3
t(yr)(J =0)	0	$1.5\ 10^4$	$1.1\ 10^5$	$6.5\ 10^5$	$3.6\ 10^6$
t(yr)(J_{50}=7.1)	0	$1.5\ 10^4$	$1.1\ 10^5$	$6.4\ 10^5$	$3.3\ 10^6$
t (yr)(J_{50}=14.2)	0	$1.4\ 10^4$	$9.7\ 10^4$	$3.1\ 10^5$	$5.4\ 10^5$

4. Binaries among the contracting stars

Finding binaries among the contracting stars is also a difficult observational problem because of the broad lines. Only few pairs containing T Tauri stars have been found (Gahm, 1977). Evidently the real number of binaries among these stars is comparable with their number among the main-sequence stars and is about 50% (Martynov,1979).

It is tempting to connect some unexplained phenomena in pre-main-sequence stars with their hidden binarity (Bisnovatyi-Kogan and Lamzin, 1977a). Let us consider the star V1057 Cyg which suddenly increased its luminosity in 1970. It is known that before the flash the T Tauri type star had been observed in that place (Herbig, 1977). One meets difficulties trying to explain the flash in V1057 Cyg as well as the earlier flash in FU Ori, as a phenomenon on T Tauri stars, The masses of T Tauri stars are essentially less than the evaluated masses of FU Ori, V1057 Cyg after the flash (Petrov, 1977). The difficulties are removed, if to suppose, that the star V1057 Cyg is in the binary system, containing a T Tauri star and another young contracting star of an essentially greater mass. This star undergoes a transition from the hydrodynamical contraction to the state of radiative star close to the main sequence, giving the observed flash, similar to Larson (1977). In this case the T Tauri star still remains in this system and it is worth searching for it in the observations.

II. PROPERTIES OF T TAURI STARS

5. Observational features of T Tauri stars

Optical observations show that T Tauri-type stars have a cool photosphere with $T_{eff} = (3\div5)\cdot10^3$ K and strong UV excesses (Kuhi, 1974). The interpretation of line profiles leads to the conclusion of a mass outflow from these stars with the velocities 150-300 km/sec (Kuhi, 1964,1965; see also § 2). It is important that the observations of line profiles show deceleration of the outflowing gas (Kuan, 1975). Let us note that the observed mass-outflow velocities are much greater than the characteristic sound velocities, corresponding to the photosphere $V_{s\varepsilon} \lesssim 10$ km/sec.

Simultaneous observations of the star DF Tauri in the region of the line H_α and in UBV show the coherence in their changes (Zaytseva and Lynty, 1976). It was obtained that the variations in H_α emission approximately repeat those in ultraviolet continuum (U-filter) with the time

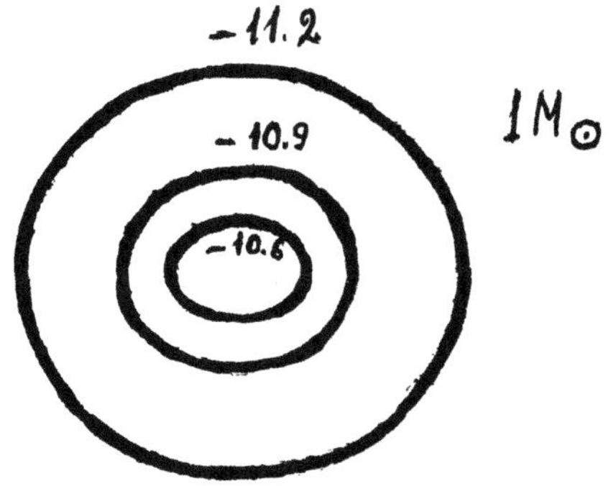

Fig.4. The form of the surface of the contracting star with 1 $M_\odot$, J_{50}= 14.2 for different parameters lg ρ_0, connected with the entropy and age of the star (see Fig. 3), following Bisnovatyi-Kogan et al. (1979).

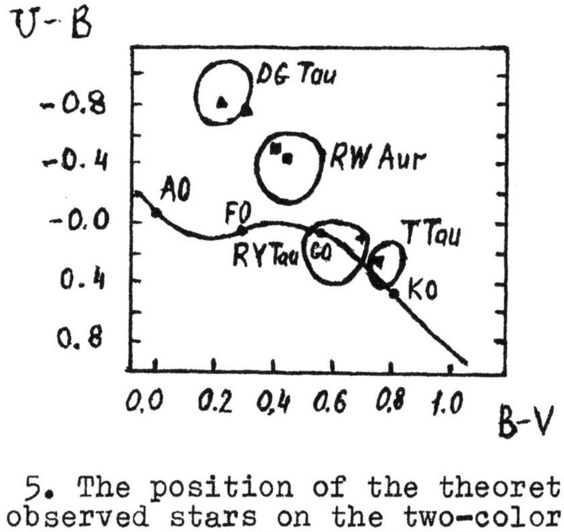

Fig. 5. The position of the theoretical models and observed stars on the two-color diagram from the paper of Bisnovatyi-Kogan and Lamzin (1977). The ovals mark the regions, inside which move the mentioned stars. The line is the main sequence. The same marks correspond to the models with the same stellar radius and different T_c and n_c.

delay ~ 70 min. There are rapid variations in the line profiles Δt = 10-20 min (Ismailov, 1973; Kolotilov and Zaytseva, 1975) which have been interpreted, as the existance of rather cold ($T \sim 10^4$ K) bulks of the matter moving in different directions. These observations have been explained in the model of a hot outflowing corona, proposed by Bisnovatyi-Kogan and Lamzin (1977b). The model predicted a hot chromospherical layer with $T \sim (5 \div 9) \cdot 10^4$ K for T Tauri stars.

The observations in X-ray and ultraviolet regions permited the chromospheric and coronal emission from these stars to be detected. The indications of the X-ray emission from T Tauri stars have been obtained in the rocket observations of the Orion nebulae (Zwijnenberg, 1976) and ANS satellite (den Boggende et al., 1978). Direct observations of T Tauri stars in Orion nebulae have been done by the Einstein (HEAO-B) Observatory (Ku and Chanan, 1979).

Observations of T Tauri and related stars by the IAU satellite in the region 1150-3100Å lead to the discovery of the chromospheric lines of the ions CIV, SiIV et al., which correspond to the temperature $(5 \div 10) \cdot 10^4$ K (Gahm et al., 1979; Appenzeller et al., 1980). The chromospherical luminosity in this region is equal to 0.3 $L_\odot$ for the star RU Lupi whose optical luminosity is about 2 $L_\odot$ (Gahm et al., 1979).

6. The model of the outflowing corona

All these observations of T Tauri stars are well explained in the model of the hot outflowing corona, proposed by Bisnovatyi-Kogan and Lamzin (1977), based on the optical data. Strong convection in T Tauri stars leads to the formation of a mechanical flux of energy which heats the layers above the photosphere, thus forming a hot corona. The mechanism of the mechanical energy transformation into heat is evidently similar to the mechanism of heating of the solar corona and essentially connected with the magnetic field (Syrovatskii, 1966; Rosner et al., 1978). The characteristic parameters of the corona in T Tauri stars are: $T_c \sim 2 \cdot 10^6$ K, $n_c \sim 10^{11}$ cm^{-3} (at the base of the corona). The characteristic velocity of the matter outflow from the corona is ~ 150 km/sec which is of the order of the observed outflow velocities in T Tauri stars. The thermal emission of the corona may be compatible with/or even greater than, the photospherical optical luminosity. The main flux is in the soft X-ray region, but the ultraviolet emission of the corona may be important for the explanation of strong excesses of these stars in "U" filter.

The origin of line emission of the hydrogen and other elements which show the matter outflow, is connected in the model of Bisnovatyi-Kogan and Lamzin (1977) with the relatively cold bulbs of the matter, which are formed in the outflowing hot corona due to the development of thermal instability. These bulks reemit the hard corona radiation in the hydrogen and other optical lines. Such a picture explains the observed time delay in the luminosity rising of the flux in contimuum (first) and H_α line (later) radiation (Zaitseva and Lynty, 1976). This time delay is connected in this model with the time of the development of the thermal instability. The changes in the coronal energy flux may explain the shift of the position of different T Tauri stars in the UBV diagram (Fig. 5). The main prediction of this model is the strong X-ray flux of the coronal origin which may even exceed the optical flux. In such a way the observed X-ray flux from the Orion nebulae may be interpreted as a sum of the emission of T Tauri stars (den Boggende, 1978; Mewe, 1979). It also follows from this model that X-ray emission of the corona leads to the heating of the stellar surface and to the formation of a hot chromospherical layer on its surface with $T = (5 \div 10) \cdot 10^4$ K.

Another model of a T Tauri star, which is based on the accretion instead of the mass outflow, was considered by Ulrich (1976). However observations of absorption variabilities (Zajtseva and Kolotilov, 1974) contradict the accretion model (Lamzin, 1980).

7. The hot chromosphere

Most of the mechanical energy flux goes out into the corona and dissipates there transforming into heat. The hot corona heats the star because of its thermal conductivity and the absorption of its hard thermal radiation. The thermal conductivity is the main source of heating the hot intermediate layer with $T = 5 \cdot 10^4 - 10^6$ K on the Sun (Shmeleva and Syrovatskii, 1973). The situation is different for T Tauri stars where the thermal radiation of the corona is much stronger. In these stars the absorption of the X-ray flux from the corona is the main source of heating the intermediate layer. This was shown by Bisnovatyi-Kogan and Lamzin (1980) for the T Tauri-type star RU Lupi as an example.

Application of the thermal conductivity model to explain the radiation of the hot chromosphere ($T \approx 8 \cdot 10^4$ K) leads to a very high density at the base of the corona ($\sim 10^{12}$ cm^{-3}) which in turn leads to an unrealistically large coronal luminosity ($>100\ L_\odot$). Considering the absorption of the coronal energy flux as a mechanism of chromosphere heating it is possible to calculate parameters of the chromosphere and corona, basing on the observed chromospheric energy flux $\sim 0.3\ L_\odot$ (Gahm et al., 1979). The hot chromospheric layer must absorb part of the coronal X-ray flux, falling on the star, so that its average depth for the absorption should be equal to unity. For the known temperature and luminosity of the layer it is possible to calculate the density and thickmess of this layer which are equal to $n_{ch} \simeq 3 \cdot 10^{12}\ cm^{-3}$, and $h_{ch} = 10$ km, respectively.

Using hydrodynamical equations to describe the outflowing isothermal corona the main coronal parameters have also been calculated for the given coronal temperature $T_c \simeq 2 \cdot 10^6$ K (Bisnovatyi-Kogan and Lamzinm 1980). It was predicted, that the X-ray luminosity of this star may be of the order of its optical luminosity. The expected fluxes near the Earth are equal to $(1 \div 2) \cdot 10^{-12}$ erg/cm^2 sec in the region 0.2-0.28 KeV, and $(1 \div 5) \cdot 10^{-11}$ erg/cm^2 sec in the regions 0.25-0.5 KeV and 0.5-1.6 KeV taking into account the absorption. This expected fluxes are sufficiently high for detection by the existing X-ray satellites. The two-layer chromosphere must exist on T Tauri stars containing a "cold" layer $T \sim 10^4$ K heated by the dissipation and a "hot" layer ($T \simeq (5 \div 10) \cdot 10^4$ K) heated by the X-ray coronal emission.

8. Some speculations

The existence of mighty chromospheres and coronas on T Tauri type stars as well as on the other stars of late spectral types discovered by "Einstein" arouses a question of the theory of the formation of a strong mechanical energy flux on the stars with connective envelopes. The most optimistic estimates show, that this flux may be of the order of and even greater than the photospherical energy flux. Let us stress the fact that this situation does not contradict the second thermodynamical law. Here it is necessary that in the place where the mechanical energy flux is born, the temperature T_m should be considerably greater than the photospherical temperature. The efficiency of the transformation of the thermal energy into the mechanical form by the star as a thermal machine, does not exceed the value $\eta \lesssim (T_m - T_{ph})/T_m$ which may be more than 0.5.

If the coronal emission of T Tauri type stars is equal to or greater than, the photosphere luminosity it may solve in part the contradiction appearing in the determination of the age of young stellar clusters. The age determined by the turning point of the massive stars from the main sequence occurs to be less than the age determined as a time of contraction of low-mass stars. These ages are $(4\div8)\cdot10^8$ and $2\cdot10^9$ years for Hyades; 10^6 and $2\cdot10^7$ years for NGC 2264, respectively (Kraft and Greenstein, 1969). When the total bolometric luminosity is several times greater than the photospherical one, the evolutionary calculations must take it into account, and it would weaken or even remove this contradiction (Bisnovatyi-Kogan et al., 1979).

REFERENCES

Appenzeller I., Chavarria C., Krautter J., Mundt R., Wolf B., 1980, preprint.
Appenzeller I. and Tscharnuter W., Astron. Ap., 1974, 30, 423.
Blinnikov S.I., 1975, Astron. Zh. USSR, 52, 243.
Bisnovatyi-Kogan G.S., Blinnikov S.I., Fedorova A.V., 1977, in the book "Early stages of stellar evolution", p. 40, ed. I.G. Kolesnik, Kiev, Naukova dumka (in Russian).
Bisnovatyi-Kogan G.S., Blinnikov S.I., Kostyuk N.D. and Fedorova A.V., 1979, Astron. Zh. USSR, 56, 770.
Bisnovatyi-Kogan G.S. and Lamzin S.A., 1977a, in the book "Early stages of stellar evolution", p. 107, ed. I.G. Kolesnik, Kiev, Naukova dumka (in Russian).
Bisnovatyi-Kogan G.S. and Lamzin S.A., 1977b, Astron. Zh. 54, 1268.
Bisnovatyi-Kogan G.S. and Lamzin S.A., 1980, Pis'ma Astron. Zh., 6, 34.
Bodenheimer P. and Ostriker J., 1970, Astrophys. J., 161, 1101.
den Boggende A., Mewe R., Gorenschild E., Heise J. and Grindley J., 1978, Astron. Astrophys. 62, 1.
Ezer D. and Cameron A.G.W., 1971, Astrophys. Space Sci., 10, 52.
Fedorova A.V. and Blinnikov S.I., 1978, Nauch. Inf. Astron. Council Acad. Sci. USSR, N 42, p. 75.
Fedorova A.V., 1979, Nauch. Inf. Astron. Council Acad. Sci. USSR, N 46, p. 3.
Gahm G.F., 1977, in Proc. of Symposium "Flare stars" in Byurakau Obs. Oct. 1976, p. 117, ed. L.V. Mirzoyan, Erevan.
Gahm G.F., Freedge K., Liseau R., Dravins D., 1979, Astron. Astrophys. 73, L4.
Hayashi C., 1961, Pub. Astr. Soc. Japan, 13, 450.
Hayashi C., Hoshi R. and Sugimoto D., 1962, Prog. Theor. Phys. Suppl. N 22.
Hayashi C., 1965, Ann. Rev. Astron. Ap., 4, 171.
Herbig G.H., 1957, Astrophys. J., 125, 612.
Herbig G.H., 1977, Astrophys. J. 217, 693.
Ismailov Z., 1973, Astron. Tsirk., N 763 p.
Kolotilov E.A. and Zajtseva G.V., 1975, Variable stars, 20, 153.
Kraft R.P. and Greenstein J.L., 1969, in Low luminosity stars, ed. S.S. Kumar, Gordon and Breach, New York, p.65.
Ku W.H.-M. and Chanan G.A., 1979, Astrophys. J. Let., 234, L59.

Kuan P., 1975, Astrophys. J., 202, 425.
Kuhi L.V., 1964, Astrophys. J., 140, 1409.
Kuhi L.V., 1966, Astrophys. J., 143, 991.
Kuhi L.V., 1974, Astron. Astrophys. Suppl. 15, 47.
Lamzin S.A., 1980, Astron. Tsirk. N 1101.
Larson R., 1977, in Star Formation, ed. T. de Jong and A. Maeder, P. 249.
Martynov D.Ya., 1979, Kurs obsehei astrofiziki (in Russian), Nauka, Moscow.
Mazzitelli I. and Moretti M., 1980, Astrophys. J., 235, 955.
Mewe R., 1979, Space Sci. Rev. 24, 101.
Moss B.L., 1973, Month. Not. RAS 161, 225.
Petrov P.P., 1977, in the book "Early stages of stellar evolution", p.66, ed. I.G. Kolesnik, Naukova dumka (in Russian).
Rosner R., Tucker W.H. and Vaiana G.S., 1978, Astrophys. J. 220, 643.
Shmeleva O.V. and Syrovatskii S.I., 1973, Solar Phys. 33, 341.
Syrovatskii S.I., 1966, Astron. Zh. 43, 4340.
Ulrich R.K., 1976, Astrophys. J., 210, 377.
Wilson L.A., 1975, Astrophys. J., 197, 365.
Zajtseva G.V. and Lyuty V.M., 1976, Variable stars, 20, 266.
Zajtseva G.V. and Kolotilov E.A., 1974, Astrophysika, 10, 365.
Zwijnenberg E., 1976, Ph. Thesis, Huygens Lab., Leiden.

DISCUSSION

Massevitch: I should like to summarize briefly several problems left to be solved in pre-main sequence evolution. They are the followings. 1) Precise computations of evolutionary tracks are needed with mass loss, rotation, and magnetic fields taken into account. 2) The effect of rotation on the pre-main sequence evolution of massive stars (the results mentioned in the review are valid only for low mass stars) must be investigated. 3) The "starting model" should be taken directly from the theory of star formation. 4) Discrepancies in the ages obtained for stars in young open clusters must be explained. 5) A theory is needed for the chromospheres and coronae of T Tauri stars. 6) The H-R diagram needs to be interpreted in regions where "very young" and "very old" stars overlap.

AGE SPREAD OF THE ORION NEBULAR STARS

Syuzo Isobe and Goro Sasaki

Tokyo Astronomical Observatory, University of Tokyo,
2-21-1, Osawa, Mitaka, Tokyo, Japan, 181.

The Orion Nebula is a recent star formation place on a sequence of star formation in the Orion Association Ia to Id and further to the Orion Molecular cloud as shown by Kutner, Tucker, Chin, and Thaddeus (1977). From the photoelectric observations, Penston (1973, 1975) obtained the age of the Nebula younger than 3×10^6 years.

The photographic observations of the Orion Nebular stars with an image splitter have been carried out in 1970 and 1971. The magnitude of faint stars is calibrated by the 2nd images of the bright stars in the same field, and the mean error of magnitude in each plate is estimated to be about 0.3 magnitude. Then, the diagram of R magnitude to R - I color is obtained for the stars in the region within 15' from the Trapezium stars. The numbers of observed stars for two different color plates are about 1000.

From the modified H - R diagram, the age of stars spread over 10^4 years to 3×10^7 years which is same as that the Orion Association Ia. Here, we conclude that stars in the Orion Nebular region were continuously formed during the whole life of the Orion Molecular cloud.

D. Sugimoto, D. Q. Lamb, and D. N. Schramm (eds.), Fundamental Problems in the Theory of Stellar Evolution, 99.

MONITOR OBSERVATIONS OF THE ORION SiO MASER

Nobuharu Ukita
Department of Astronomy, University of Tokyo

Monitor observations of the SiO v=1, J=2-1 transition were made toward the Kleinman-Low infrared nebula. Observations were made from September 1976 to February 1980 with the 6 m mm-wave telescope of the Tokyo Astronomical Observatory. Data were taken with an acousto-optical spectrometer with a frequency resolution of about 50 kHz(Δv=0.17 km s^{-1}).

The appearance of the Orion SiO maser profile is characterized by two strong emission peaks. We have found that velocity structures in the double emission feature tend to shift toward the central velocity. The inward shifts in velocity structures recur at a typical interval of about 2 years. The changes in radial velocity of individual peaks can be followed over 2 years. The rates of velocity shifts were found to be $0.9 \sim 2.7 \times 10^{-3}$cm s^{-2}, were different for each recurrent component. These results are discussed in terms of a decelerating expanding envelope of a pulsating star. The velocity shifts of the maser emission peaks can be explained by (i) expanding shells of dense material expelled recurrently from a central star undergoing a deceleration, or by (ii) shock waves propagating out through an envelope with decreasing velocity. In either case, the size of the region where maser amplification can occur is estimated to be about 1×10^{14} cm, and the maser emitting region is confined to a fraction of this region.

The central star seems to be less luminous than about 1 to 2 x 10^4 $L_\odot$(Genzel et al. 1979). Assuming that the mass of the central star is 1 $M_\odot$, we estimate the radius of the central star to be about 700 $R_\odot$ from the period density formula. A candidate star would be a protostar in the early stage of star formation immediately after flare-up(Narita, Nakano, and Hayashi, 1970). It is likely that the protostellar core would become unstable against radial pulsation in the sequence of evolution and that the collapse of the envelope is stopped and reversed by gas ejection due to the pulsation.

References;

Genzel, R., Moran, J.M., Lane, A.P., Predmore, C.R., Ho, P.T., Hansen, S.S., and Reid, M.J. 1979, Astrophys. J. Letters, **231**. L73.

Narita, S., Nakano, T., and Hayashi, C. 1970, Prog. Theor. Phys. **43**, 942.

D. Sugimoto, D. Q. Lamb, and D. N. Schramm (eds.), Fundamental Problems in the Theory of Stellar Evolution, 100.

AN OBSERVATIONAL APPROACH TO THE EARLY STAGES OF STELLAR EVOLUTION

V.A. Ambartsumian and L.V. Mirzoyan
Byurakan Astrophysical Observatory, Armenia, USSR

During last three decades an observational approach was being applied at the Byurakan Astrophysical Observatory to the problems of the evolution of astronomical bodies and systems. In contradiction to the views developed by many theoreticians who postulate as an initial state of each body or system a gaseous nebula of some kind and suppose that the processes of condensation are dominant in the Universe this approach makes use of the observed predominance of expansions, ejections and explosions.

The observational approach has led in the past to the prediction of expansion of some stellar associations confirmed later by the analysis of observations. It became clear that the stellar associations should be regarded as very young systems where the star formation processes is still continuing in our days. The new approach has brought to the discovery of many phenomena in the galaxies and presented the problem of the evolution of galaxies in a new light.

Instead of trying to derive the origin of stars from condensation of nebular matter the new approach considers as the phenomenon of the primary importance the formation of nebulae as a consequence of the activity of dense bodies (nebulae surrounding the novae, planetary nebulae, supernovae remnants, cometary nebulae and the diffuse nebulae in O-B associations).

The new approach in application to the early stages of stellar evolution is discussed. The T Tauri stage is considered as a phase following the more dense protostellar state. Developing the idea, expressed by Haro, the flare stars are regarded as the next phase of evolution. Each star of a small mass passes through these two evolutionary phases. The phenomena of fuors (FU Ori-type brightenings) can be considered as an expression of the same tendency (the transformation of dense matter into rarefied state). The observational data relating to the common origin of giant and dwarf stars in stellar associations can provide the clue to an understanding of the evolution of high luminosity stars.

D. Sugimoto, D. Q. Lamb, and D. N. Schramm (eds.), Fundamental Problems in the Theory of Stellar Evolution, 101–102.

DISCUSSION

Mouschovias: I think that all of us would agree with you that stars die and return their matter to the interstellar medium, and that this matter will eventually find itself in one cloud or another. However, did you also claim that a massive cloud or cloud complex ($M \sim 10^5\ M_\odot$) results from the death or mass loss of local stars?

Mirzoyan: We consider mass loss as an important property of young stars (for example, O-B stars, T Tauri stars etc.) which are in their early stages of evolution. Therefore, it is natural to assume that the diffuse matter coexisting with these young stars is a result partly of the mass loss from them. It is equally natural to regard the remaining part of the mass of the diffuse matter of the nebulae (perhaps the greater part) as being ejected from the initial dense protostar at the time of its disintegration into separate stars. Thus the hypothesis of dense protostars provides a common origin for the stars and for the diffuse matter.

FRAGMENTATION OF COLLAPSING GAS CLOUDS

S-A. SØRENSEN, S. NARITA* and D. McNALLY
University College London, London, U.K.

In simulations of the collapse of gas clouds it is usual to start from a uniform sphere the evolution of which is determined by the two energy ratios = Thermal/Gravitational and = Rotational/Gravitational. In numerical calculations the two parameters are unfortunately supplemented by a number of unknown factors related to the exact numerical treatment of the collapse. It is therefore crucial to compare the results of several different methods before any judgement is made concerning the correct evolutionary track of the cloud.

The first comparative study was made by Bodenheimer and Tscharnuter (1979) of an axisymmetric cloud with = 0.46 and = 0.32 which would reach a state of equilibrium. We have here recalculated this model using two different numerical codes:

	(N)	(S)
Coordinates	Spherical polar	Cylindrical
Difference schemes	First order (FLIC#)	Second order (AFBD)
Selfgravity	SOR	ADI

#) Angular momentum distribution corrected by Lagrangian test particles.

We find our results to be in good agreement with the B-T calculations except that no rings are formed in the later stages of the collapse. Both schemes conserve angular momentum satisfactorily but while the (N) code produces very faint transient rings these are absent from the result of the (S) code.

Bodenheimer, P. and Tscharnuter, W.M.: 1979, Astron & Astrophys. 74

*On leave of absence from Doshisha University, Kyoto, Japan

DISCUSSION

Bodenheimer: As far as the fragmentation problem is concerned, I don't think it is particularly important whether a ring forms or not. In the

D. Sugimoto, D. Q. Lamb, and D. N. Schramm (eds.), Fundamental Problems in the Theory of Stellar Evolution, 103–104.

particular case you considered, we (Bodenheimer and Tscharnuter) didn't find that the ring was a dominant feature in the calculation. Equilibrium isothermal disks have been shown to be unstable to fragmentation.

Mike Norman has come to the conclusion that the uniformly rotating uniform sphere is a singular initial condition which probably produces a disk solution rather than a ring solution. However, any slight deviation from that angular momentum distribution results in a ring solution. Thus even very small numerical effects could induce the ring solution. What happens when you apply your calculation to a cloud that is unstable to collapse, rather than to one that reaches an equilibrium?

Sørensen: In these cases I still obtain no ring. The response is increased slightly, but the gradients are also larger. The boundary conditions are, however, unusual and could affect the ring formation.

Roxburgh: Have different codes been used for the same problem and obtained the same results? If not, is this not a cause for concern and why should we believe anyone's results?

Sørensen: No, exactly the same results have not been obtained, but you would not expect that. With more time available, I could point out all the similarites in the results from different codes. The major features are, in fact, quite similar. What we must do is to take the intersection of the different simulations and bilieve in those; but it is just as important to note the differences and to try to determine why they arise.

ACCUMULATION OF A RAPIDLY ROTATING PROTOSTAR AND THE FORMATION OF AN ASSOCIATED NEBULA AS A RESULT OF ANGULAR MOMENTUM TRANSPORT BY TURBULENT FRICTION

W M Tscharnuter
MPI für Astrophysik, Garching b. München, W-Germany

The influence of angular momentum transport by turbulent friction on the structure of a rotating protostar has been investigated. Turbulence is characterized by a simple viscosity parameter $\eta = \xi \cdot c_s \cdot l$, where c_s denotes the local speed of sound, l the typical length scale of the largest eddies (thickness of the nebula) and ξ the "efficiency" parameter (= 1/10 in our model).

In modelling the solar nebula we started out from a 3 $M_\odot$-cloud of density 10^{-20} g/cm^3 for which the ratios thermal/gravitational and rotational/gravitational energy are about 1 and 1.2 10^{-4}, respectively, i.e. the cloud is assumed to be marginally unstable according to the Jeans criterion and to rotate with an angular velocity inferred from the galactic rotation. During about 3 10^4 yr after the formation of a stellar core containing initially only a few 10^{-3} $M_\odot$ the central condensation has accreted 0.5 $M_\odot$. At the same time an almost stationary disk-like nebula has taken shape whose densities and temperatures range from 10^{-11} to 10^{-13} g/cm^3 (the surface densities are 10-30 g/cm^2) and from 10^3 to 10^2K, respectively, within a distance of 5 10^{12} up to 2 10^{14} cm. Its rotation is very nearly Keplerian. We have thus covered the collapse and accretion phases up to the point where the further evolution is dominated by a quasi-stationary accretion flow due to turbulent friction. From our numerical model we may estimate this evolutionary timescale to be of the order of 10^7 yr.

DISCUSSION

Bodenheimer: Could you comment on the stability of the core? Is it likely to fission?

Tscharnuter: Unfortunately, at present it is not possible to study the internal structure of a rapidly rotating protstar consistently with the accretion flow. Estimates for the ratio $|E_{tot}/E_{grav}|$ yield numbers in the range 0.27 - 0.28, so that fission cannot be excluded.

Unno: In your formula for turbulent friction, why do you use the sound

D. Sugimoto, D. Q. Lamb, and D. N. Schramm (eds.), Fundamental Problems in the Theory of Stellar Evolution, 105–106

speed instead of the characteristic speed in the system?

Tscharnuter: Since the driving mechanism for turbulence is unknown, the speed of sound, as an upper limit for the turbulent velocity, is at least reasonable. Of course, typical velocities generated by differential rotation, as long as they are smaller than the speed of sound, ought to be considered too.

FRAGMENTATION OF ISOTHERMAL GAS CLOUDS

R.A. Gingold and J.J. Monaghan
Department of Mathematics, Monash University, Australia

Research supported by the Australia Japan Foundation and the Ian Potter Foundation and ARGC grant B 77/15346

An analysis of the collapse and fission of the isothermal gas cloud examined by Boss and Bodenheimer (BB, 1979) has been made using SPH (Lucy 1977, Gingold and Monaghan 1977, 1978). While SPH is not as effective as finite difference methods for problems with spherical symmetry, it has the advantage in fragmentation problems that the description of the fragments is independent of their position and orientation. Furthermore, the SPH algorithm has been tested by applying it to non axisymmetric problems for which accurate solutions are known as well as spherically and axially symmetric problems.

The initially non-axisymmetric rotating cloud is found to undergo a fission as in the BB calculation. Up until $t \sim 2.4 \times 10^4$ years, the central and maximum densities of the two calculations are in good agreement. From $t \sim 2.4 \times 10^4$ years until $\sim 2.8 \times 10^4$ years, when the BB calculation was discontinued because of its poor azimuthal resolution, the two calculations differ considerably. At the end of this time the SPH fragments are rotating with a $\beta \sim 30$ times the BB value. Furthermore, the SPH fragments have become elongated whereas the BB fragments are roughly spherical. The rapid rise in the BB maximum density is not found in the SPH calculation, possibly because of the abovementioned differences. Because the SPH calculation has a better resolution of the fragments it can be continued for a much longer time. We find the fragments become subject to strong tidal forces and eventually coalesce to form a bar surrounded by a halo. The conclusion of BB that the final state is a binary must therefore be doubted. The difference between the BB and SPH calculations may be due to the fact that initially the mass which forms each fragment in the BB calculation only occupies about 4 cells in azimuth, and at the time the two results begin to diverge from each other, the fragments in the BB calculation occupy ~ 2 cells in azimuth. The BB calculation therefore has an initially low and continually degraded azimuthal resolution. In contrast the SPH resolution increases by a factor of 3 during the collapse.

References
Boss,A.P. and Bodenheimer,P.: 1979, Astrophys. J., 234, pp.289-295.
Gingold,R. and Monaghan,J.: 1977, Mon. Not. Roy. Astr. Soc.181,pp.375-389.
Gingold,R. and Monaghan,J.: 1978, Mon. Not. Roy. Astr. Soc.184,pp.481-499.
Lucy,L.B.: 1977, Astron. J., 82, pp.1013-1024.

D. Sugimoto, D. Q. Lamb, and D. N. Schramm (eds.), Fundamental Problems in the Theory of Stellar Evolution, 107–108.

DISCUSSION

Bodenheimer: Could numerical diffusion of angular momentum play a role in the calculation, and is it a possible cause of the discrepancy between your results and ours?

Gingold: Since the SPH code is a Lagrangian formulation, it does not suffer from the angular momentum problem that arises in Eulerian codes. If the system is axisymmetric, then we conserve local angular momentum exactly. Tests on polytropes with moderate rotation show that the small deviations from axisymmetry that arise from the use of a finite number of paticles do not significantly change the distribution of local angular momentum. The total angular momentum is always conserved exactly.

Sørensen: You mentioned the resolution problems facing the hydrodynamic calculations. The smoothing of your particles will, however, also limit your resolution. Could you give a value for this limit?

Gingold: The resolution is of the order of half the smoothing length h shown in the diagrams (which will be published elsewhere). At the stage when the fragments first clearly form ($t \sim 2.4 \times 10^4$ yr), the fragments subtend about 6×3 smoothing lengths. At this stage, our fragments would subtend only one azimuthal cell in the finite difference schemes.

Sørensen: Bodenheimer increased his resolution by doubling the number of grid points. Have you tried to reduce your resolution too see whether or not your result approaches Bodenheimers?

Gingold: We have also used only 400 particles and a resolution twice as coarse as the present resolution. The results were similar. However, this resolution may still be better than that achieved by the finite difference codes. To degrade the SPH resolution further may cause the initial density distribution to differ significantly from that used by Boss and Bodenheimer. Also, various ways of generating the initial density distribution have been tried and all behave similarly.

NUMERICAL STUDIES OF THE FISSION HYPOTHESIS FOR ROTATING POLYTROPES

R. H. Durisen* and J. E. Tohline**
*Indiana University
**Los Alamos Scientific Laboratory

The fission hypothesis suggests that close binary stars form due to global nonaxisymmetric instabilities in rotating, quasistatically contracting stars. We study this hypothesis by using an explicit, donor cell, finite difference 3-D hydrodynamic code with self-gravitation to follow dynamic two-armed instabilities in rapidly rotating polytropes. Typical grids are 32 x 16 x 16 in cylindrical coordinates (ϖ, ϕ, z) and assume reflection symmetry about the equator plane and rotation axis. Initial conditions are obtained by applying density perturbations $\delta\rho = a\rho \cos 2\phi$ with a = 0.10 or 0.33 to axisymmetric equilibrium models with the same angular momentum distribution but with various values of polytropic index n and of $\beta \equiv T/|W|$ where T = total rotational kinetic energy and W = total gravitational energy. We find dynamic growth of perturbations when $\beta \gtrsim 0.30$ for both n = ½ and 3/2. To within the limitations of our methods, this agrees well with the classical dynamic stability limit of $\beta \simeq 0.274$ for the bar modes of the Maclaurin spheroids. An unstable case with n = 3/2 and $\beta = 0.33$ is evolved for about ten initial central rotation periods. The part of the star inside corotation develops into a stable bone-shaped or dumbbell-shaped structure after about three pattern rotations. At about the same time, material outside corotation is ejected in the form of two trailing spiral arms. These arms wrap due to differential rotation, merge into a detached disk, and eventually narrow into a radially expanding ring with slight cos 2ϕ density enhancements. The ring contains 16% of the mass but more than half the angular momentum. The central bone-shaped object is an analog of the Riemann S-type ellipsoids. Fluid circulates dynamically and stably from one knob of the bone to the other. In this sense, the object is probably better described as a triaxial star than as a contact binary. Similar behavior is exhibited by an extensive n = ½ and $\beta = 0.33$ evolution. This work was supported by U.S. National Science Foundation Grant AST-7821449.

D. Sugimoto, D. Q. Lamb, and D. N. Schramm (eds.), Fundamental Problems in the Theory of Stellar Evolution, 109–110.

DISCUSSION

Tscharnuter: How much mass and angular momentum is contained within the "bone-shaped" region?

Durisen: The final central bone-shaped object contains 84% of the mass but less than half of the angular momentum, and has a $T/|W|$ of about 0.19.

Kippenhahn: Your calculations show damping which you called "numerical" damping, and they show growth which you called "physical" growth. What are the time scales of these two phenomena compared to each other?

Durisen: For $T/|W| \lesssim 0.30$, the amplitude of the $\cos(2\phi)$ density perturbation decreases from 10% to about 1% in one to two pattern rotations. For $T/|W| \gtrsim 0.30$, the growth is roughly exponential in the linear regime with an e-folding time of one to two pattern rotations. The similarity of these time scales is probably responsible for the fact that our dynamic stability limit is 10% higher in $T/|W|$ than the classical Maclaurin spheroid value. We are planning to repeat our calculations with different azimuthal resolutions and with second-order instead of first-order transport. We also plan experiments on various other idealized problems to determine numerical diffusion coefficients and their dependence on the parameters of the numerical scheme. Comparison calculations using other 3-D codes are also underway. In this way, we hope to clarify which effects in our calculations are numerical and which are physical.

Gingold: Using the SPH particle code we also found that when the polytrope was of index 1.5 spiral arms formed, soaking up angular momentum. For an index of 0.5, however, we got a fission. Have you performed calculations with a lower polytropic index?

Durisen: Yes. We have carried out an evolutionary calculation with $n = 1/2$ and $T/|W| = 0.33$, and we have followed it for about three pattern rotations. Again, a central bone-shaped object forms, and material is ejected in the form of spiral arms. However, because of time-step size problems, we have not yet evolved this case beyond the phase of active spiral arm ejection. It is possible that the central bone might fragment at a later evolutionary stage.

Sørensen: How flat is your cloud at different stages?

Durisen: The starting axisymmetric equilibrium model with $n = 3/2$ and $T/|W| = 0.33$ has an equatorial to polar ratio of about six. During the intermediate, highly nonaxisymmetric phase of spiral arm ejection it is difficult to characterize the flattening uniquely because it depends on which meridional section is taken. Roughly speaking, the overall flattening from the tip of the arms to center increases due to expansion of the arms and contraction of the central bone. At the end of the evolution, it makes more sense to describe the ring and the bone separately. In computational units, the expanding ring has an inner radius of 14, a radial width of about 6, and a halfthickness in z of about 4. The central bone-shaped object has a semimajor axis of about 5.5 in computational units. We have not yet carefully scrutinized meridional sections of the bone, but it is centainly much less flattened than the starting model.

PRE-MAIN-SEQUENCE EVOLUTION OF CLOSE BINARIES WITH MASS TRANSFER

Kwan-Yu Chen
Department of Astronomy, University of Florida

The scenario begins with two spherical masses of Roche radii, as given by Paczynski (1971)[1], at a separation of centers, A. The mass flows from the initially more massive star *a* in rapid rotation to its companion *b*. The separation changes in accordance with the conservation of orbital angular momentum. The corresponding Roche radius of *b* is considered to be its radius. The luminosities of the stars are given in the approximate expressions following Schatzman (1963)[2] for convective or radiative equilibrium. The luminosity of *b* determines the time for a given mass transfer with the use of the virial theorem. This time step and the virial theorem, then, determine the change of the radius of *a*. The angular speed of *a* is maintained such that the centrifugal and gravitational accelerations are equal at the equator. If the total angular momentum of the system is conserved, the angular momentum of *b* can be computed. Initially, *b* is considered to rotate in synchronism with the orbital revolution. The mass transfer stops when the centrifugal and gravitational accelerations become equal at the equator of *b*, or when the change in potential energy becomes zero. In the table, S_f is the ratio of final rotational and orbital angular speeds; t is the time interval in 10^3y; initial M_a is 3 solar masses; M_{bi} = 0.7 for part 1, and A_i = 32.2 solar radii for part 2.

1)	A_i	A_f	S_{af}	S_{bf}	t	ΔM	log L_a		log L_b		log T_a		log T_b	
	86.0	83.8	2.82	3.40	0.34	.012	2.66	2.59	1.37	1.37	3.61	3.61	3.44	3.44
	53.7	52.4	2.82	3.40	1.09	.012	2.36	2.29	1.07	1.07	3.63	3.64	3.46	3.47
	17.2	16.8	2.82	3.40	18.6	.012	1.62	1.55	0.33	0.33	3.70	3.70	3.53	3.53
	15.0	14.7	2.83	3.40	24.4	.011	1.54	1.55	0.24	0.25	3.71	3.73	3.53	3.54
	10.7	10.7	2.92	3.42	19.9	.004	1.32	1.86	0.03	0.03	3.72	3.88	3.55	3.55
2)	M_{bi}													
	2.0	34.0	17.3	2.31	45.8	1.15	1.90	0.94	1.55	1.98	3.67	3.72	3.63	3.68
	1.85	32.9	19.2	2.48	56.5	1.21	1.91	0.95	1.49	1.94	3.67	3.74	3.61	3.67
	1.5	29.0	4.21	3.15	21.4	.180	1.94	1.65	1.33	1.36	3.67	3.68	3.59	3.61
	0.7	31.4	2.82	3.40	3.89	.012	2.03	1.96	0.74	0.74	3.66	3.67	3.49	3.49

[1] Paczynski, B. 1971. Ann. Rev. Astron. Astrophy. 9, p. 183.
[2] Schatzman, E. 1963. In "Star Evolution", ed. L. Gratton, p. 177.

D. Sugimoto, D. Q. Lamb, and D. N. Schramm (eds.), Fundamental Problems in the Theory of Stellar Evolution, 111.

FORMATION OF THE PLANETS

Chushiro Hayashi
Dept. of Physics, Kyoto University, Kyoto, Japan

A timetable for an evolutionary sequence of processes, which begins with the formation of the solar nebula being nearly in equilibrium and ends with the planetary formation, is presented. Basic features of the processes and grounds for the estimation of time-scales are explained for each of the processes.

1. INTRODUCTION

On the origin of the solar system, a wide variety of theories have so far been presented since the days of Kant and Laplace in the 18th century. Theories have become more and more realistic with time and, at present, we are at a stage of very scientific investigation.

There still remain many fundamental problems to be solved. For example, it is not yet clear to what degree the magnetic force is important in determing the structure of the pre-planetary nebula. The magnetic force will be effective in a collapsing stage of the nebula and also in the outer regions of the nebula after it has settled into equilibrium, i.e., in stages and in regions where the gas is ionized above a certain level by cosmic rays and the solar UV radiation. However, in most of the regions of the nebula where the planets are formed, the ionization degree of the gas is lower than, say, 10^{-15} and the magnetic effect may be neglected compared to the solar gravity and the gas pressure.

Besides the magnetic effect, a variety of solar nebula models have so far been proposed. Here I choose recent three models and indicate their basic differences. First, in Cameron's model (Cameron 1973, DeCampli and Cameron 1979), the mass of the nebula is as large as $2M_\odot$ (including the Sun) and the surface density of the disk is as high as 10^5 g/cm^2. Then, the disk fragments into gaseous protoplanets with mass as large as that of the present Jupiter (see Section 4 for the condition of fragmentation).

D. Sugimoto, D. Q. Lamb, and D. N. Schramm (eds.), Fundamental Problems in the Theory of Stellar Evolution, 113–128.

On the other hand, in models of Safronov (1969) and us (Kusaka et al. 1970, Hayashi et al. 1977 and others) the nebular mass is in the range, 0.01-0.04$M_\odot$, and the gas disk is stable against fragmentation. The fragmentation occurs only in a thin dust layer which is formed later as a result of sedimentation of dust grains towards the mid-plane of the gas disk. A basic defference between Safronov and us lies in that we have fully considered the effect of the nebular gas on planetary formation, such as on the growth time of the planets and on the formation of the giant planets.

It is not possible to review here all the topics on planetary formation. Then, I choose to talk mainly about the results of our Kyoto group obtained in these ten years, which are summarized in the following timetable (Table 1) where an evolutionary sequence of many important processes are listed.

Table 1. Main events and time intervals.

Interval	Events and processes
10^4y ---	Collapse of a rotating molecular cloud of about 1 $M_\odot$.
━━━━	Formation of a pre-planetary gas disk being in equilibrium.
10^4y ---	Growth and sedimentation of grains to form a dust layer.
━━━━	Fragmentation of the dust layer into planetesimals ($\sim 10^{18}$ g).
10^5y ---	Accumulation of planetesimals to protoplanets ($\sim 10^{25}$ g).
────	Formation of protoplanets with surrounding gases.
10^6y ---	Radial migration and trap of planetesimals in the Hill sphere.
────	Formation of the Earth surrounded by a hot H_2 atmosphere.
10^7y ---	Growth of a Jupiter's core to 10 times the Earth mass.
━━━━	Formation of Jupiter with collapsing of its atmosphere.
10^{7-8}y -	Escape of the nebular gas and the Earth's H_2 atmosphere.
━━━━	Complete gas escape, formation of present Earth atmosphere.
4×10^9y -	Planetary perturbation and other processes in the absence of gas.
	Present.

There are four processes or stages of special importance in the evolution, as denoted by the thick lines in Table 1, where very drastic changes occur in the physical condition of the solar nebula. These are (1) the collapse of a cloud and the formation of a gas disk which is rotating (in nearly Keplerian motion) around the proto-Sun and is nearly in a state of thermal and dynamical equilibrium under the influence of the solar gravity and radiation, (2) the fragmentation of a thin dust layer due to gravitational instability, (3) the collapse of surrounding H_2 atmospheres onto the protoplanets which have grown above a certain limit (about 10M_E, where M_E being the Earth's mass) and (4) the disappearance of the gas component of the disk due to the solar wind and UV radiation which are expected to be very strong in the T Tauri stage of the Sun.

In stages lying between the above four stages, the change of physical conditions is more continuous and gradual and we may follow the evolution with more precise calculations. Explanation of the processes listed in Table 1 will be given in each of the following sections.

2. STRUCTURE OF A GAS DISK

At the end of the collapse, the solar nebula will be heated to relatively high temperatures (say, 2×10^3K in the Earth region) by shock waves and will also be in turbulent motion. This turbulence as well as the magnetic field, if it is strong enough, may have a role of transferring angular momentum outwards. The turbulence will decay in a time of several Keplerian periods and, further, in a time of the order of 10^3 yrs the disk is flattened (i.e., the Helmholtz-Kelvin contraction) and the gas cools to reach a thermally steady state where the heating of the nebular gas by the solar radiation balances with the radiative cooling.

Let us now consider the structure of the equilibrium disk in helio-centric cylindrical coordinates where r and z denote the distance from the Sun and the height from the equatorial plane of the disk, respectively. Under the assumption that the radial displacement of all the dust materials (which form the planets later) during the time from this stage to the present was minimum, Kusaka, Nakano and Hayashi (1970) determined the distribution of the surface density $\rho_s(r)$. For example, they distributed the present Earth's mass uniformly in the region, r=0.85-1.23 AU, and multiplied this by a factor, 300 =(H+He)/(Mg+Si+Fe). Then, from considerations of thermal and dynamical equilibrium, they constructed a disk model where the gas density ρ and the temperature are determined as functions of r and z.

According to this model, the total mass of the gas disk is in the range 0.02-0.04$M_\odot$ and the structure has such features as shown in Table 2, where $z_o(r)$ denotes the half-thickness of the nebula and $\eta(r)$ $(=1-\Omega(g)/\Omega)$ is the deviation of the gas angular velocity $\Omega(g)$ from the Keplerian angular velocity, $\Omega=(GM_\odot/r^3)^{1/2}$, which is due to the existence of a pressure gradient in the r-direction.

Table 2. Structure in regions of the Earth and Jupiter.

r(AU)	z_o(AU)	$T_{z=o}$(K)	$\rho_{z=o}$(g/cm^3)	η
1.0	0.04	230	6×10^{-9}	0.002
5.2	0.33	100	2×10^{-10}	0.004

The radial distribution of the surface gas density is approximated by the form r^{-k} with a constant k lying in the range 1.5-2.0. Adopting, for example, the case of k=2, we plot the surface density of dust mass in Fig. 1, where also the ranges of the present planetary perturbation are indicated. Keeping in mind Bode's law, we may imagine from Fig. 2 how the giant planets were formed in the outer regions and also how the gap in the asteroid region resulted from Jupiter's perturbation.

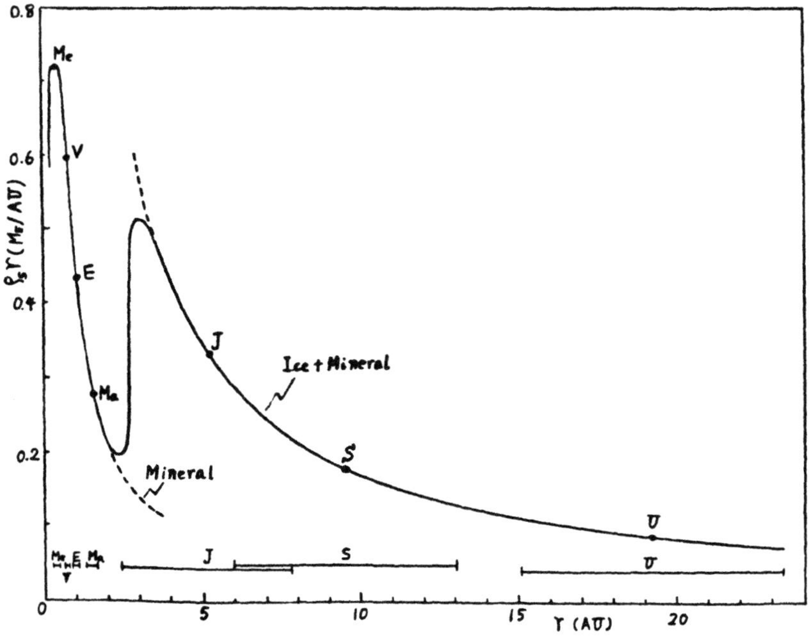

Fig. 1. Distribution of dust mass and ranges of planetary perturbation. The ordinate is the surface density ρ_s of dust mass multiplied by the distance r. The ranges of planetary perturbation, indicated by the segments in the lower region of the diagram, correspond to $a \pm (7h+ea)$ where a, e and h are the semi-major axis, the eccentricity and the Hill radius of the present planets.

3. GROWTH AND SEDIMENTATION OF DUST GRAINS

Old results of Kusaka et al. (1970) have been greatly improved by recent calculations of Nakagawa, Nakazawa and Hayashi (1980), where the disk are divided into 10 layers in the z-direction and the time variation of the mass spectrum of grains in each layer is calculated. The details will be talked by Nakagawa in this session. The result indicates that a thin dust layer, composed mainly of cm-size grains, is formed near the mid-plane of the gas disk in about 5×10^3 yrs. This layer begins to fragment into a number of planetesimals as shown in the following section.

4. FRAGMENTATION OF THE DUST LAYER

Gravitational instability of a thin disk (a homogeneous mixture of dust grains and gas molecules) for a ring-mode variation in the form, $\exp(i\omega t+ikr)$, was studied by Safronov (1969), Hayashi (1972) and Goldreich and Ward (1973). The dispersion relation and the condition for instability are given by

$$\omega^2 = \Omega^2 + c^2k^2 - 2\pi G\rho_s k < 0 \quad \text{(unstable)}, \tag{1}$$

where ρ_s is the surface density of the dust layer and c is the sound velocity given by

$$c^2 = \gamma p/\rho = \gamma(n_m+n_d)kT/(\rho_m+\rho_d) \simeq \gamma n_m kT/\rho_d \; . \tag{2}$$

Here, the subscripts m and d denote gas molecules and dust grains, respectively.

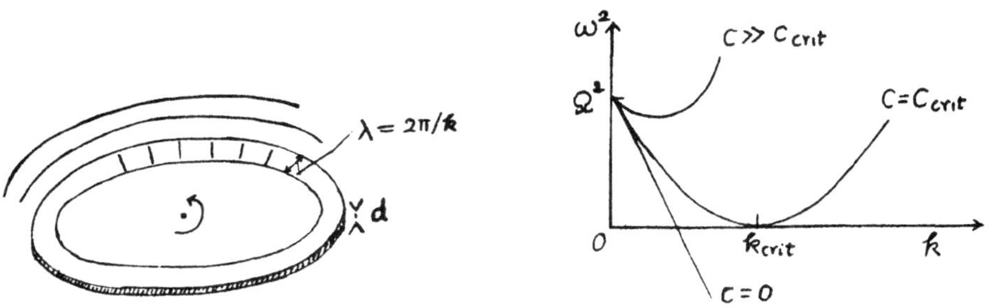

Fig. 2. Ring-mode instability and dispersion relation.

As the sedimentation of grains proceeds, the sound velocity decreases gradually and when it becomes smaller than the critical value, $c_{crit}=\pi G\rho_s/\Omega$, the layer fragments into a large number of rings (Fig. 2) and it is expected that, nearly at the same time, a ring fragments into a number of beads. At a critical stage, where $c=c_{crit}$ and $k=k_{crit}=\Omega^2/\pi G\rho_s$, and at r=1 AU with $\rho_s=20\,g/cm^2$, we have for the size and the mass of a fragment, $\lambda=2\pi/k=7\times10^8$ cm and $m=\rho_s\lambda^2=1\times10^{19}$g, respectively. The thickness and the density of the layer at this stage are about 200 cm and 0.1 g/cm^3, respectively. The total number of the fragments in the whole disk is about 10^{12}.

The above dispersion relation was obtained for the case of one-component fluid. The case of two-component fluid (i.e., gas fluid and dust fluid with mutual frictional interaction) was studied by Hayashi (unpublished). The result indicates that the fragmentation begins at a stage earlier than the above-mentioned critical stage, i.e., at a stage when the density of the dust layer becomes nearly equal to the Roche density ($=2\times10^{-6}g/cm^3$ at r=1 AU) and the layer thickness is about 10^7 cm. At this stage, sedimentation is faster than fragmentation. Then, the final size and mass of a fragment will not be greatly different from the above-mentioned values.

The fragment, formed in the above process, still contains a small amount of gas molecules and shrinks gravitationally to a so-called planetesimal of 10 km size, which is made of sands of silicates containing also ices if r is greater than about 2.7 AU.

5. ACCUMULATION OF PLANETESIMALS TO THE PLANETS

Our final aim for the accumulation process will be to find the time variation of a distribution function f(m; a, e, i; t) (where m is the mass, and a, e and i are the semi-major axis, the eccentricity and inclination, respectively, of a Keplerian orbit around the Sun) for an ensemble of solid bodies with size ranging from planetesimals to the present planets and also to find the time variations of the density and velocity of the gas present in the solar nebula. This is to solve a problem of long-term stochastic irreversible processes occuring in a coupled system of the gas and solid bodies, both of which are under the strong influence of the solar gravity. It will be very difficult, at present, to find a complete solution but we have to know, at least, the relative velocities of colliding planetesimals and also the rate of their radial migration, both of which determine the rate of their accumulation.

Hayashi, Nakazawa and Adachi (1977) studied this problem under the simplifications that (1) all the solid bodies, except for one massive protoplanet under consideration, have the same mass but this mass increases with time and (2) all the gas is in circular motion with velocity as mentioned in Section 2. The orbital motion of solid bodies is treated as composed of regular motion (with e=i=0) and random motion which is proportional to e and i. If we set aside the accumulation process for a moment, the change of orbital motion is determined by three effects: gravitational scattering, gas drag effect and planetary perturbation, as will be explained in the following.

5.1. Gravitational scattering (a large number of two-body encounters).

We use the result of Chandrasekhar on the problem of stellar dynamics by modifying it in order to apply to the case of Keplerian particles under consideration, where the mean random velocity v is given by

$$v^2 \simeq (e^2 + i^2)\, v_k^2. \tag{3}$$

Here, v_k is the Keplerian velocity in circular motion. Obviously, the mean change in Δt due to scattering is zero, i.e., $\langle\Delta a\rangle=\langle\Delta e\rangle=\langle\Delta i\rangle=0$, but the mean square change in Δt is finite and the diffusion coefficients in the three-dimensional (a, e, i) space are given by

$$\frac{<(\Delta a)^2>}{a^2 \Delta t} \simeq \frac{<(\Delta e)^2>}{\Delta t} \simeq \frac{<(\Delta i)^2>}{\Delta t} \simeq \frac{e^2 + i^2}{t_c} , \tag{4}$$

where t_c is the collision time which is related to the cross section σ for two-body scattering, the mass m and the surface number density n_s of solid bodies:

$$t_c = \frac{1}{nv\sigma} , \quad \sigma = \pi\left(\frac{2Gm}{v^2}\right)^2 \ln\left(\frac{v^2 d}{2Gm}\right) , \quad n = \frac{n_s}{ai} . \tag{5}$$

Here d is the mean separation of solid bodies. By means of Eq.(5), the above diffusion coefficient is written as

$$\frac{e^2 + i^2}{t_c} = \frac{1}{\tau_c} \frac{1}{i(e^2+i^2)^{1/2}} , \quad \tau_c = 9 \times 10^{63} (\Omega a^2 n_s m)^{-1} \text{ (c.g.s.)}. \tag{6}$$

It is to be noticed that diffusion is very rapid if e and i are small.

5.2. Gas drag effect.

The gas drag effect which changes the orbital elements a, e and i was calculated by Adachi, Hayashi and Nakazawa (1976) and Weidenschilling (1977). The difference in the angular velocity between the gas and a solid body, as mentioned in Section 2, gives rise to a gradual change in a, e and i. The drag force acting on a body is given by

$$F \simeq \pi r_b^2 \rho v^2 , \quad v \simeq (e + i + \eta) v_K , \tag{7}$$

where r_b is the radius of a solid body, v is the relative velocity between the gas and the body, and η is a small quantity as shown in Table 2. Calculations with perturbation method give

$$\frac{1}{a}\frac{da}{dt} \simeq -\frac{2}{\tau_g} (e + i + \eta)(\eta + e^2) , \tag{8}$$

$$\frac{1}{e}\frac{de}{dt} \simeq \frac{1}{i}\frac{di}{dt} \simeq -\frac{e + i + \eta}{\tau_g} , \quad \tau_g = 7 \frac{m^{1/3}}{\rho v_k} \text{ (c.g.s.)} . \tag{9}$$

It is to be noticed that the rate of decrease of a (i.e., the rate of radial migration) is much smaller than those of e and i and that all the rates are sensitive to the values of e, i and η.

5.3. Effect of planetary perturbation.

This effect is important only in the neighborhood of a protoplanet (i.e., if the distance from a planet is smaller than about seven times the Hill radius). The excitation of random motion of a solid body due to this effect was estimated by Hayashi et al. (1977) by means of numerical calculation of the orbits in the frame of the Restricted Three-Body Problem.

5.4. Interplay of the effects of scattering and gas drag.

From Eqs.(4), (6) and (9) the mean values of e and i of an ensemble of solid bodies are found to change with time as

$$\frac{1}{e^2}\frac{de^2}{dt} = \frac{1}{\tau_c}\frac{1}{e^2 i(e^2+i^2)^{1/2}} - \frac{e+i+\eta}{\tau_g} , \tag{10}$$

$$\frac{1}{i^2}\frac{di^2}{dt} = \frac{1}{\tau_c}\frac{1}{i^3(e^2+i^2)^{1/2}} - \frac{e+i+\eta}{\tau_g} . \tag{11}$$

It is found from the above equations that in a relatively short period of time ($\simeq$ several times 10^4 yrs) both e and i attain equilibrium values given by

$$e \simeq i \simeq 1\times10^{-13}\ (an_s m/\rho)^{1/5}\ m^{4/15}\ \text{(c.g.s.)}. \tag{12}$$

Using the values of n_s and ρ in the model of Kusaka et al., we found that the equilibrium values of e and i at r=1 AU are 0.0004 (for $m=10^{18}$g), 0.003 (for 10^{21}g) and 0.014 (for 10^{24}g).

Putting these values of e and i into the equations for a, i.e., Eqs. (4), (6) and (8) we can estimate the time of radial migration which is composed of the two processes, i.e., (1) in-and out-wards diffusion ($a \rightarrow a/2$ or $2a$) due to the scattering effect and (2) inflow ($a \rightarrow a/2$) due to the gas drag effect. It will be seen that the interplay of scattering and gas drag shortens the migration time considerably compared to the case where only one of the two effects is present. The migration time for the case of, for example, $m=10^{21}$g is given in Table 3.

Table 3. Equilibrium values of e and i and time of radial migration.

	$e \simeq i$	Diffusion time	Inflow time
Earth region	0.003	1×10^{9}y	8×10^{6}y
Jupiter region	0.006	1×10^{10}y	2×10^{8}y .

Considering the change of m with time, we estimated the growth time of the planets (see Section 7). This growth time depends essentially on the above migration rate for solid bodies lying in regions distant from a protoplanet under consideration.

The above statistical treatment of the many-body problem was refined and extended (to the case where solid bodies has a mass spectrum of the form of the power law, m^{-k} dm) by Nakagawa (1978). He generalized the theory of Brownian motion to the case of non-isotropic velocity distribution and succeeded to find a self-consistent solution to the Fokker-Planck equation. Furthermore, Nakagawa calculated the accumulation of planetesimals and the change of the mass spectrum in very early stages where the mass increases from 10^{18} to 10^{22}g and the accumulation is due mainly to sticking of two bodies in direct collisions. He used the sticking cross section of the form,

$$\sigma = \pi\ (r+r')^2 \left\{ 1 + \frac{2G(m+m')}{|\vec{v}-\vec{v}'|^2 (r+r')} \right\} , \tag{13}$$

where the magnitude of the relative velocity, $\vec{v}-\vec{v}'$, is known from the distribution function obtained by him. He has neglected the time required for radial migration since the migration over a short distance is very rapid. The result shows that, if the accumulation starts from bodies with a same mass, 1×10^{18}g, the mass spectrum tends to have a form of $m^{-1.5}$ in a period of 3×10^4 yrs (see Fig. 3). The largest body in this spectrum is expected to grow to a protoplanet with mass of about 10^{25}g (as described in the next section) in about 10^5 yrs.

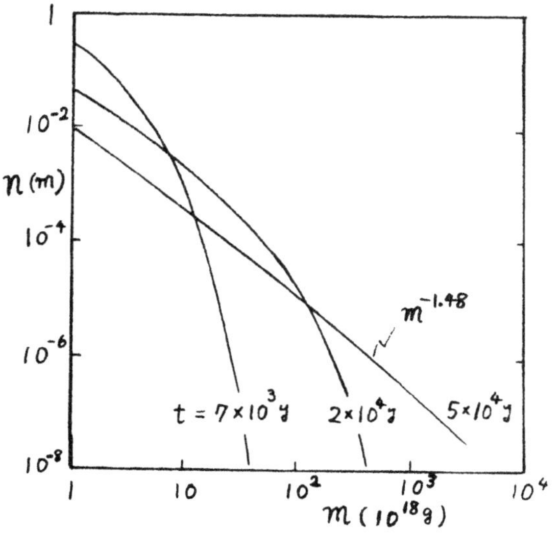

Fig. 3. Time variation of the mass spectrum of planetesimals in the Earth region. The ordinate denotes the number of planetesimals with mass between m and m+dm. At time t=0, all the planetesimals have the same mass, 1×10^{18} g.

6. CAPTURE OF PLANETESIMALS IN THE HILL SPHERE OF A PROTOPLANET

The radius h of the Hill sphere, i.e., the sphere of gravitational influence of a protoplanet (with mass M and distance a from the Sun) is given by $h=a(M/3M_{\odot})^{1/3}$. When M becomes greater than 10^{25} g, the escape velocity at the surface of the protoplanet becomes greater than the sound velocity of the nebular gas. Then, with M increasing, the nebular gas is attracted more and more towards the protoplanet and a surrounding atmosphere is formed. The gas density in the interior of the Hill sphere becomes higher than in the external regions. Consequently, the kinetic energy of a planetesimal which entered the Hill sphere is, more or less, dissipated by the gas drag effect and it tends to be trapped within the Hill sphere.

7. FORMATION OF THE PROTO-EARTH AND THE CORE OF JUPITER

The rate of further growth of a protoplanet is determined mainly by the rate of radial migration of planetesimals from distant regions and by the probability of their trap inside the Hill sphere. Hayashi et al. (1977) estimated that the Earth takes about 10^6 yrs to grow to the present mass, 1 M_E, and that a core of Jupiter (made of rocky and ice materials) of 10 M_E is formed in about 10^7 yrs.

8. PRIMORDIAL ATMOSPHERE OF THE EARTH

The structure of an atmosphere (in hydrostatic and thermal equilibrium) surrounding the proto-Earth was calculated by Hayashi, Nakazawa and Mizuno (1979). Boundary conditions are that the gas density ρ and temperature T take the nebular values at the surface of the Hill sphere. The opacity of the gas for thermal radiation is expressed in the form

$$\kappa(\rho, T) = \kappa_m + \kappa_g ,$$
$$\kappa_m = \kappa_{H_2} + \kappa_{H_2O} \simeq 100\,\rho \text{ (c.g.s.)}, \qquad (14)$$

where κ_m is the opacity due to molecular absorption and scattering and κ_g is that due to dust grains. The value of κ_g is considered as a parameter which lies in the range between 1 $cm^2 g^{-1}$ (the interstellar value) and 10^{-5} $cm^2 g^{-1}$ (for which κ_g can be neglected compared to κ_m). At the bottom of the atmosphere there is a release of gravitational energy of accreting planetesimals and the resulting energy outflow (i.e. the luminosity) in the atmosphere is givn by $L=GM\dot{M}/R$, where $\dot{M}$ is the mass accretion rate which is taken to be $1M_E/10^6$ yrs, M and R being the mass and radius of the proto-Earth composed of rocky and metallic materials.

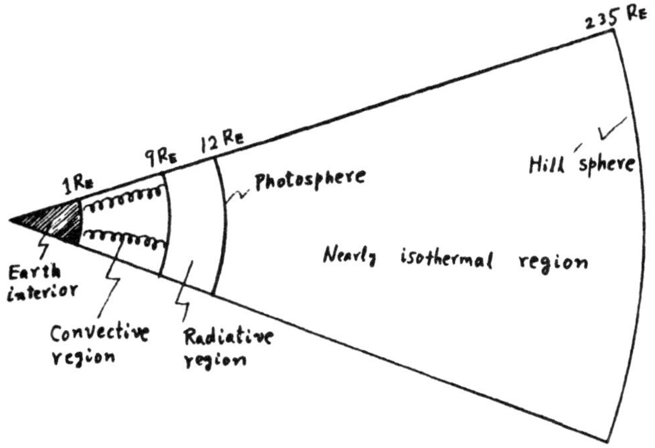

Fig. 4. Primordial Earth's atmosphere at a stage of $M/M_E=1$. The present Earth's radius is denoted by R_E.

The results of our calculations for a case of $\kappa_g=1\times10^{-4}$ cm^2g^{-1} are given in Table 4, where the subscript b denotes the bottom of the atmosphere. The atmosphere consists of radiative and convective regions (see Fig. 4).

Table 4. Structure of the primordial Earth's atmosphere.

M/M_E	T_b(K)	ρ_b(gcm^{-3})	p_b(atom.)	M_{gas}/M	L(erg/s)
0.1	690	1.4×10^{-4}	3.5	0.014	2.5×10^{24}
0.5	2590	2.2×10^{-3}	203	0.018	2.6×10^{25}
1.0	4100	6.1×10^{-3}	890	0.027	1.1×10^{26}

The existence of such a hot and dense atmosphere has two important effects, (1) melting of rocky and metallic materials in the Earth's interior which leads to the formation of the present core-mantle structure and (2) disolution of volatile gases into rocky materials as pointed out by Mizuno, Nakazawa and Hayashi (1980). It is well-known that the present atmosphere of the Earth is formed as a result of degassing from the interior and it is generally believed that this formation occured within a period of less than 5×10^8 yrs after the Earth grew to the present mass. Then, we have to consider that the primordial atmosphere was dissipated into outer space within the above period of time (see Section 10.1).

9. INSTABILITY OF ATMOSPHERES AND FORMATION OF THE GIANT PLANETS

The Earth stops its growth at a time when $M=1M_E$ because of the exhaustion of accreting planetesimals. How about the giant planets?

It is well-known that the mass of an interstellar isothermal cloud surrounded by a gas with constant pressure is limited (this is due to a kind of Jeans' instability). For the same reason, when the mass of a surrounding atmosphere becomes comparable to that of a protoplanet (being called a core, for simplicity), the self-gravity of the atmosphere becomes important and the atmosphere begins to collapse onto the planet because of gravitational instability.

The above problem was first studied by Perri and Cameron (1974) under the assumption of a wholly-adiabatic temperature gradient and they found that the critical core mass for the onset of instability is as large as 70 M_E for the case of Jupiter. Mizuno, Nakazawa and Hayashi (1978) studied the same problem but with a different model of the solar nebula and different treatments of boundary conditions and opacities. Very recently, Mizuno (1980) calculated the atmospheric structure in the same manner (but including self-gravity) as described in Section 8 and found that, for the growth rate $\dot{M}=10\times M_E/10^7$ yrs, the critical core mass lies in the range between $2M_E$ (for $\kappa_g=10^{-5}$ cm^{2}g^{-1}) and $10M_E$ (for $\kappa_g=1$ cm^{2}g^{-1}) for all the protoplanets of the solar system. Then, we have to consider that after the collapse of the atmosphere a considerable amount of the surrounding nebular gas is further accreted onto the planets to form the present giant planets.

10. ESCAPE OF THE NEBULAR GAS AND THE EARTH PRIMORDIAL ATMOSPHERE

It is required that the solar nebula is dissipated into outer space after the formation of the present Jupiter but befoer the formation of the present Earth's atmosphere. Then, the time of escape of the solar nebula (as shown in Table 1) should be in the range between 1×10^7 and 5×10^8 yrs.

10.1 Escape of the Earth's atmosphere.

First, we consider the escape of the Earth's atmosphere which occurs after the dissipation of the solar nebula. Sekiya, Nakazawa and Hayashi (1980) studied the effects of the solar radiation (EUV as well as visible) which is incident on the rotating Earth. They calculated the spherically-Symmetric steady outflow of a gas which starts with very small velocity from the photosphere of the atmosphere (which has a structure as described in Section 8), taking into accounts the effects of (1) cooling of a gas due to adiabatic expansion (2) heating due to absorption of the solar EUV and visible radiation and (3) cooling due to emission of radiation from dust grains and molecules. The outer boundary condition for the gas flow is that the flow should pass through a sonic point, i.e., a well-known condition for a gas outflowing into free space.

The results of calculations show that, if the flux of the solar EUV radiation during the stage considered (i.e. the T Tauri stage of the proto-Sun) is greater by a factor of 10^3 than the present value, the main constituents (H and He) of the atmosphere escape in about 1×10^8 yrs. Furthermore, almost all of heavy atoms such as Kr and Xe can also escape owing to the drag of outflowing H_2 molecules and He atoms. Then, as to the abundance of the rare gas elements, there is no contradiction between the existence of the primordial atmosphere and the composition of the present Earth's atmosphere.

10.2 Escape of the solar nebula.

The effect of strong solar wind on the escape of the nebula was studied by Horedt (1978) and Elmegreen (1978). The above results of Sekiya et al. indicate that more than 10 percent of the absorbed EUV energy is converted into the kinetic energy of the outflow. By means of such energy considerations, let us here estimate the escape time of the solar nebula as a whole. For the model of Kusaka et al., gravitational binding energy of the solar nebula is given by

$$E = -\int \frac{GM_\odot}{2r}\,\rho_s\, 2\pi r dr = -\,8.3\times10^{43}\ \text{erg}. \tag{15}$$

Then, the escape time of the nebula due to the irradiation of the solar wind is given by

$$t_{es} = \frac{-E}{\eta L_W \Omega/4\pi} = 3\times10^6\ \frac{0.1}{\eta}\ \frac{10^{-2}L_\odot}{L_W}\ \frac{0.25}{\Omega/4\pi}\ \text{yrs}, \tag{16}$$

where η is the efficienty of conversion of wind energy, L_W is the solar wind luminosity (or the particle luminosity) and Ω is the heliocentric solid angle subtended by a whole wind-aborbing surface layer of the nebula.

The value of $\Omega/4\pi$ is in the range 0.2–0.3 in the model of Kusaka et al. and probably we have $\eta \lesssim 0.1$. The particle luminosity of T Tauri stars is not well-known at present but probably it does not exceed 1/100 of the total luminosity (i.e., $\sim 1L_\odot$). Then, with $\eta=0.1$ we have $t_{es} > 3\times10^6$ yrs, corresponding to $L_W < 10^{-2}L_\odot$. This time-scale is consistent with our result (see Section 7) that the Earth and Jupiter are both formed during stages where the nebular gas is still existing.

11. FORMATION OF THE SATELLITES

As metioned in Section 6, the growth of the protoplanets is due to the accretion of planetesimals which, after entering the Hill sphere, gradually lose their kinetic energies (in interaction with gas molecules which fill the interior of the Hill sphere) until they reach the surface of the protoplanets. If the gas density in the Hill sphere is too high, planetesimals cannot survive as orbiting bodies but soon fall onto the planets. On the other hand, if the gas density and, then, the gas drag effect are too small, planetesimals cannot be trapped into bound orbits around the planets. Therefore, it is probable that the satellites are the remainder of planetesimals (already grown to their present masses) which are trapped in the Hill sphere at stages where the nebular gas is escaping and the gas density is decreasing to an appropriate level. In order to clarify this situation, the orbital motion of a planetesimal in the presence of a gas in the Hill sphere (in the Restricted Three-Body Problem) is now being computed by Nishida and by Komuro of our group.

12. CONCLUSION

It is very obvious that any theory of planetary formation should be self-consistent in itself and also should be tested by direct comparison with observations. In this respect, to promote the study of the origin of the solar system, collaborations of researches in very different fields of science, such as astrophysics, physics, chemistry, geology, geophysics and mineralogy, are indispensable.

REFERENCES

Adachi, I., Hayashi, C. and Nakazawa, K., 1976, Prog. Theor. Phys. 56, 1756.
Cameron, A.G.W., 1973, Icarus, 18, 407.
DeCampli, W.M. and Cameron, A.G.W., 1979, Icarus, 38, 367.
Elmegreen, B.G., 1978, Moon and Planets, 19, 261.
Goldreich, P. and Ward, W.R., 1973, Ap. J., 183, 1051.
Hayashi, C., 1972, Report of the Fifth Symposium on the Moon and Planets at the Institute of Space and Aeronautical Science, University of Tokyo.
Hayashi, C., Nakazawa, K. and Adachi, I., 1977, Publ. Astron. Soc. Japan 29, 163.
Hayashi, C., Nakazawa, K. and Mizuno, H., 1979, Earth and Planetary Science Letters 43, 22.
Horedt, G.P., 1978, Astron. Astrophys., 64, 173.
Kusaka, T., Nakano, T. and Hayashi, C., 1970, Prog. Theor. Phy., 44, 1580.
Mizuno, H., Nakazawa, K. and Hayashi, C., 1980, to be published in Earth Planet. Sci. Lett.
Mizuno, H., 1980, to be published in Prog. Theor. Phys.
Nakagawa, Y., 1978, Prog. Theor. Phys. 59, 1834.
Nakagawa, Y., Nakazawa, K. and Hayashi, C., 1980, in preparation.
Perri, F. and Cameron, A.G.W., 1974, Icarus, 22, 416.
Safronov, V.S., 1969, "Evolution of the Protoplanetary Cloud and Formation of the Earth and the Planets" (Moscow).
Sekiya, M., Nakazawa, K. and Hayashi, C., 1980, to be published in Earth Planet. Sci. Lett. and also in Prog. Theor. Phys.
Weidenschilling, S.J., 1977, Mon. Not. Roy. Astr. Soc., 180, 57.

DISCUSSION

Cameron: Did your calculations obtain differential velocities between grains that are low enough that the grains stick, and therefore accumulate rather than fragment, on collision?

Hayashi: Since gas friction is present, the relative velocities between grains are small enough.

Schatzman: If you apply McCrea's theory of formation of planetesimals (fall toward the equatorial plane, viscous drag, growth of dust particles by capture), you get planetesimals of about 1 meter in diameter which reach the equatorial plane in a very short time. What is the situation in the presence of turbulence?

Hayashi: We solved the growth and sedimentation equation by precise numerical computations, the details of which will be talked about by Nakagawa later in this session. That is, we obtained the mass spectrum as a function of t and z (height from the equatorial plane). The maximum size of a grain is about 1 cm. In our theory, a dust layer composed of these grains and gas molecules fragments into a number of planetesimals with masses of the order of 10^{18} g. We consider a situation where turbulence has already decayed.

Schatzman: Which rule did you use to determine the properties of the turbulence in the gaseous disc?

Hayashi: We investigated the density and pressure distributions as functions of r and z. The solar wind produces turbulent eddies only in the very outer layers of the disk where the gas pressure is very small.

Kippenhahn: Just a small objection to the chairman's remarks. Dr. Schatzman, you indicate that, in the case of rotation, sheer means a deviation from solid body rotation. In my review paper tomorrow I hope I can convince you that this is not the case.

Sugimoto: You treated the statistical mechanics of the planetesimals in a box and in a, e, and i coordinates. However, energy is fed into the system from the change in a, i.e. from the change in the gravitational force between a planetesimal and the sun. In this sense the system is thermodynamically open. Is a kind of Boltzmann distribution expected to be realized even in such an open system?

Hayashi: The Boltzmann distribution function in our case, namely, in the presence of a gravitational field, does depend on time, as was shown, for example, by Nakagawa (1978). We did not consider the planetesimals to be confined in a box.

Tayler: How much larger than the present radius of the sun is its radius when the planetesimals begin to form?

Hayashi: If we use, for example, the result of Ezer and Cameron (1965) for the pre-main-sequence evolution, the sun has a radius of about 10 $R_\odot$ when the planetesimals begin to form.

Durisen: Recent theoretical and observational work suggests that processes important for understanding the present dynamics and structure of Saturn's rings are similar to those which occurred in the planetesimal disk. These include gravitational scattering, interparticle collisions, collective gravitational effects, and perturbations by satellites. Observations of the rings can afford a direct test of theoretical models for such processes against a real system.

Hayashi: I agree with you completely.

Aizu: How about the angular momentum in your model?

Hayashi: I suppose that, at a stage when the collapse of the nebula is stopped by centrifugal forces, the nebula is oscillating both in the r- and z-directions. These oscillations, together with the magnetic force, if this force is strong enough, give rise to transfer of angular momentum outwards.

Bodenheimer: What are the main arguments in favor of the low-mass solar nebula that you use as opposed to the high-mass nebula of Cameron?

Hayashi: It will not be easy to form the planets of terrestrial type, all the satellites, and also cometary objects if we start only from the giant protoplanets.

MAGNETIC BRAKING, THE SOLAR NEBULA AND THE COMETARY CLOUD

L. Mestel
Astronomy Centre, University of Sussex

Biermann (1979) has revived the earlier suggestion of Cameron and of Donn that the proto-cometary cloud and the proto-solar nebula were contiguous but distinct cloudlets, forming e.g. as fragments within the same massive interstellar cloud. As in Goldreich and Ward (1973) and Biermann and Michel (1978), the cometesimals are thought to have formed in a layer of dust that settled in the equator of a cloudlet maintained at a moderate density by a combination of thermal pressure and centrifugal force due to a high angular momentum. The solar nebula had much less angular momentum and so could contract to about the radius of Pluto's orbit before achieving centrifugo-gravitational balance. In this paper it is noted that modest variations in the initial parameters of fragments forming within the same massive magnetic cloud can yield both high and low angular momentum cloudlets (Mestel and Paris 1979). A fragment of mass M greater than a critical mass M_c, defined in terms of its magnetic flux F by $GM_c^2 \sim F^2/\pi^2$, contracts in approximate mechanical equilibrium, at the rate determined by the magnetic transport of angular momentum, and with centrifugal force remaining comparable with gravity. Rapid flux-loss at molecular cloud densities leaves a weakly magnetic, rapidly rotating, low-density cloudlet which could be the locale of cometary formation. If $M<M_c$, the magnetic stresses both limit contraction and enforce corotation with the surroundings. As flux leaks out slowly, the cloudlet contracts in approximate magneto-gravitational equilibrium, with centrifugal forces becoming a steadily smaller fraction of gravity. At the molecular cloud phase, rapid flux-loss leaves now a slowly rotating cloudlet, which can therefore become the proto-solar nebula. Whether a cloudlet is super- or sub-critical in mass will depend on the details of the fragmentation process in the parent cloud, in particular on the amount of mass agglomeration down the field-lines.

REFERENCES

Biermann, L. 1979. Veröffentlichungen der Remeis-Sternwarte Bamberg, 12, 132.
Biermann, L. and Michel, K.W. 1978. The Moon and Planets, 18, 447.
Goldreich, P. and Ward, W.R. 1973. Astrophys. J., 183, 1051.
Mestel, L. and Paris, R.B. 1979. Mon. Not. R. Astr. Soc., 187, 337.

D. Sugimoto, D. Q. Lamb, and D. N. Schramm (eds.), Fundamental Problems in the Theory of Stellar Evolution, 129–130.

DISCUSSION

Mouschovias: It is certainly proper to study magnetic braking for a wide range of values for the free parameter(s) (e.g., M/M_{crit}). It is, however, fair to point out that actual physical conditions in the interstellar medium are such that, for densities less than a few x $10^3 cm^{-3}$, the free fall time is longer than the characteristic time for magnetic braking. Since, as I explained in my review, even observations of rapidly rotating, dense clouds show that the bulk of the angular momentum problem has been resolved by the time the density has reached about $10^4 cm^{-3}$ and since individual cloud masses are relatively small, it is $M/M_{crit} < 1$ which is relevant. Your statement about solutions for $M > M_{crit}$ could apply only to the later, denser ($>10^4 cm^{-3}$) phases of cloud contraction; in these stages, only a relatively small additional amount of angular momentum need be removed.

Mestel: A time of magnetic braking must refer to the mass that is being braked. I presume you mean that most of the actual clouds observed today are of sub-critical mass, and so are magnetically-supported long enough for effective corotation with the surroundings to be maintained. However, one can still speculate about a massive magnetic cloud, from which the proto-solar nebula and possibly the proto-cometary cloud condensed and with different parameters. There may also be an observational selection effect; clouds with $M > M_{crit}$ may be hard to find since they contract as they lose angular momentum and presumably fragment.

Kippenhahn: In order to avoid the impression that all work on rotation is worthless if there is the slightest magnetic field present, I would make the following remark. We observe effects in chemical abundances which can be explained by circulation caused by rotation. We hope to see whether this is the only possible explanation and use it as a test for the existence of circulation. But even if there are magnetic fields in all stars, it is important to know what kind of topology they have. One could think of complicated flow patterns which can change the topology of the fields separating the very interior from the outer regions. Only if you have magnetic field lines which connect the surface with the central regions is there transport of angular momentum from the central region outwards. I think it would be too simple to assume that whenever there is a magnetic field we have uniform rotation (or solid body rotation), and all the meridional circulation is suppressed.

Mestel: I did not imply that the magnetic field must suppress meridional circulation. On the contrary, I think that moderately strong fields keep the Eddington-Vogt-Sweet circulation going, by offsetting the advection of angular momentum. However, in singular regions it may no longer be consistent to ignore magnetic effects on hydrostatic, and so also on radiative, equilibrium. I agree that the detailed structure of the magnetic field is often crucial, and steady detachment of the field-lines of a contracting core from the expanding envelope will slow the outward transport of angular momentum. But, again, the long time-scale of normal stellar evolution works in favour of isorotation.

GROWTH AND SEDIMENTATION OF DUST GRAINS IN THE PRIMORDIAL SOLAR NEBULA

Yoshitsugu Nakagawa, Kiyoshi Nakazawa and Chushiro Hayashi
Department of Physics, Kyoto University, 606 Japan

Of the formation processes of the solar system, the process of growth and sedimentation of dust particles in the primordial solar nebula is investigated for a region near the Earth's orbit. The growth equation for dust particles, which are sinking as well as in thermal motion, is solved numerically in the wide mass range between 10^{-12} g and 10^{6} g.

The numerical simulation shows that the growth and sedimentation proceed faster than found in the earlier work owing to the co-operated interaction of the growth and the sedimentation; that is, in about 3×10^{3} yrs after the beginning of the growth and sedimentation a dust layer, composed of particles of centimeter sizes is formed at the equator of the solar nebula. Furthermore, the mass density of dust particles floating in the outer layers of the nebula is found to be of the order of 10^{-4} compared with that before the sedimentation. From these results, it can be estimated that in about 5×10^{3} yrs after the beginning of sedimentation the dust layer breaks up owing to the onset of gravitational instability.

DISCUSSION

Unno: What effect would the shape of the dust grain have, if it is chain-like as inferred from laboratory experiments?

Nakagawa: Here we assumed spherical dust grains. As you pointed out, dust grains can have a chain-like or plate-like shape. But this is the case when their sizes are much much smaller than a micrometer, according to laboratory experiments. Of course, we cannot expect the shapes of grains with sizes of micrometers to be strictly spherical. The sedimentation velocities of these non-spherical grains may be different from those we used. But I think the difference is rather small, and our conclusions cannot be altered in order of magnitude.

Tscharnuter: What are the effects of turbulence on the growth of the

D. Sugimoto, D. Q. Lamb, and D. N. Schramm (eds.), Fundamental Problems in the Theory of Stellar Evolution, 131–132.

particles and the time scale of sedimentation?

Nakagawa: We assumed that in the initial state any turbulent motion had already decayed. I think that turbulence, if it exists, will soon decay after the solar nebula reaches hydrostatic equilibrium, because there is no energy source exciting the turbulence.

FORMATION OF THE GIANT PLANETS

Hiroshi Mizuno
Department of Physics, Kyoto University
Kyoto 606, Japan

The structure of a gaseous envelope surrounding a icy/rocky core is studied in consideration of radiative transfer. It is found that when the core grows beyond a critical core mass, the envelope cannot be in equilibrium and collapses onto the core to form a proto-giant planet. The results are as follows (for details, see Mizuno 1980).
1) The critical core mass is smaller than that estimated by Perri and Cameron (1974) and Mizuno, Nakazawa and Hayashi (1978). 2) When the grain opacity in the envelope varies from 0 to 1 cm^2/g, the critical core mass changes from ~2 to ~12 Earth's masses. 3) The critical core mass is independent of the region in the solar nebula.
These are due to the existence of the radiative region in the envelope.

Result 3) is consistent with the recent theory of the structure of the giant planets that they have the common core masses (Slattery 1977; Hubbard and MacFarlane 1980). Also it is seen from result 2) that the terrestrial planets did not become the giant planets.

REFERENCES

Mizuno, H. 1980, Prog. Theor. Phys. 64, No.2, in press.
Perri, F. and Cameron, A.G.W. 1974, Icarus 22, 416.
Mizuno, H., Nakazawa, K. and Hayashi, C. 1978, Prog. Theor. Phys. 60, 699.
Slattery, W.L. 1977, Icarus 32, 58.
Hubbard, W.B. and MacFarlane, J.J. 1980, J. Geophys. Res. 85, 225.

DISCUSSION

Bodenheimer: What is the physical mechanism for the collapse of the envelope in your models?

Mizuno: Here, collapse means that there is no equilibrium configuration.

Bodenheimer: Why shouldn't the core mass continue to grow by accretion of planetesimals even after the critical core mass is reached?

D. Sugimoto, D. Q. Lamb, and D. N. Schramm (eds.), Fundamental Problems in the Theory of Stellar Evolution, 133–134.

Mizuno: If the escape of the solar nebula occurs soon after the collapse of the envelope, the core mass does not increase very much over the critical value.

Tscharnuter: Shouldn't the giant planets spin up considerably during the collapse of the envelope, since they separate from a differentially rotating nebula?

Mizuno: The angular velocity of the inner region of the envelope, where almost all of the envelope mass is contained, is very small compared with the Keplerian angular velocity at the boundary of the core.

DISSIPATION OF THE PRIMORDIAL TERRESTRIAL ATMOSPHERE DUE TO IRRADIATION OF SOLAR EUV

Minoru Sekiya, Kiyoshi Nakazawa and Chushiro Hayashi
Department of Physics, Kyoto University
Kyoto 606, Japan

When the Earth had grown to the present mass through accretion of the planetesimals in the solar nebula, the Earth was surrounded by a dense primordial atmosphere which was mainly composed of hydrogen and helium (Hayashi et al. 1979). Mass of the atmosphere was about 1×10^{26} g. We investigate the dissipation of this atmosphere due to the irradiation of solar EUV. The effect of solar wind is neglected. We assume that the flow of the escaping gas is spherically symmetric and steady. We impose the boundary condition that the flow velocity go through a sonic point. The results show that the primordial atmosphere is dissipated within a period of 5×10^8 yrs, which is the upper limit imposed from the theory of the origin of the present terrestrial atmosphere (Hamano and Ozima 1978), as far as the solar EUV flux is more than two hundred times as large as the present one. In this case, the rare gases contained in the promordial atmosphere are also dissipated owing to the drag effect (Sekiya et al. 1980).

REFERENCES

Hayashi, C., Nakazawa, K. and Mizuno, H. 1979, Earth Planet Sci. Lett., 43, 22.
Hamano, Y. and Ozima, M. 1978, Advance in Earth and Planetary, 3, ed. Alexander, E.C. and Ozima, M. (Center for Academic Publ. Japan), 155.
Sekiya, M., Nakazawa, K. and Hayashi, C. 1980, Earth Planet Sci. Lett., to be published.

D. Sugimoto, D. Q. Lamb, and D. N. Schramm (eds.), Fundamental Problems in the Theory of Stellar Evolution, 135.

EVOLUTION OF CLOSE BINARIES

A. V. Tutukov
Astronomical Council, USSR Academy of Sciences

1. INTRODUCTION AND SUMMARY

Most stars of our Galaxy's disk are double. The existing estimation of duplicity α_d range from $\sim$ 50-70 % (Abt, Levy, 1976, Abt, 1979) to $\sim$ 100 % (Kraitcheva et al., 1978). About half of all stars are close binaries (CB) and hence their components fill their Roche lobes during evolution. This accounts for the constantly increasing interest to their evolution. Several review papers were published in the last years: Paczynski (1971), Tutukov et al. (1975), van den Heuvel (1976), Massevitch et al. (1976), Thomas (1977), Paczynski (1979), Webbink (1979b), Yungelson and Massevitch (1980). The subject matter of close binary evolution is very wide now, therefore, I will limit myself to the review of modern scenarios for evolution of massive ($M \gtrsim 10\ M_\odot$) close binaries (MCBS) and close binaries of moderate ($M \lesssim 10\ M_\odot$) mass.

CB formation is a part of the fragmentation process determined mainly by rotation. Our catalog of spectroscopic binaries has now data on $\sim$ 1050 binaries and makes it possible, taking into account numerous selection effects, to find "innate" distributions of close binaries over mass of primaries M, ratio of mass q and large semiaxis (Tutukov, Yungelson, 1979). These distributions can be represented as $dN \approx 0.5(M/M_\odot)^{-2.3}\ d(M/M_\odot)$ per yr, $q \approx 1$, $d\alpha_d \approx 0.17\ d\ \ln a$ for stars with $1 \lesssim M/M_\odot \lesssim 40$. The remarkable absence of unevolved binaries with $M \gtrsim 1.5\ M_\odot$ and $a \lesssim 10\ R_\odot$ and the low number of binaries with $a \lesssim 10\ R_\odot$ in general give a good possibility to advance the theory of binary formation and early stages of evolution of binaries with $M \lesssim 10\ M_\odot$ (Tutukov, Yungelson, 1979).

Most evolutionary computations for close binaries were

D. Sugimoto, D. Q. Lamb, and D. N. Schramm (eds.), Fundamental Problems in the Theory of Stellar Evolution, 137–154.

performed under "conservative" assumptions M_t = const and $J = (Ga/M_t)^{1/2} M_1 M_2$ = const, i.e. the binary mass M_t and the orbital angular momentum J are constant during evolution. The real evolution is nonconservative of course. Examples of radiopulsar PSR 1913+16 in the close binary, cataclysmic variables and other stars show that close binaries can lose most part of its mass and the orbital momentum. The mass and momentum loss is also important for nonexplosive stages of evolution. The formalism and some analytical estimations on influence of mass and momentum loss on evolution were proposed by Tutukov and Yungelson (1971). Evolution of massive stars under mass loss was investigated also by Vanbeveren et al. (1979). It seems now that in most cases the "nonconservative" evolution does not change the qualitative picture - scenarios which were developed on "conservative" evolution computations mainly.

2. SCENARIO AS A MEANS OF THE THEORY OF STELLAR EVOLUTION

Observations of stars give information only about outer, optically thin part of stars. But all main processes of energy generation and transfer occur deeply inside stars. That leads to great difficulties in the purely theoretical approach to the stellar evolution consisting in the search of main determining physical causes based on observational data on their sometimes rather distant consequences. This property of stellar evolution theory makes difficult the fast and effective selection of models and frequently decreases prognostic value of these models. All that and complexity of the physical processes involved in stellar evolution practically exclude the axiomatic approach to the theory which is the most attractive from the logical point of view.

The scenario approach is one of effective means to overcome at least partly this obstacle. A scenario is a logically self-consistent picture of the main stages in development of a star and reason-consequence relation between them based on the whole accessible observational and theoretical information on the process under considera-tion. Modern scenarios are based from theoretical point of view on numerous evolutionary computations made in the last decades and from observational point of view on observa-tions in optical, UV, IR, and X-ray ranges of electromagnetic spectra. The attractive property of the scenario approach consists in the possibility of an operative and flexible reaction to new observational and theoretical information with the aim to have a fullest picture consistent with all fundamental observational and theoretical data.

3. EVOLUTION OF MASSIVE CLOSE BINARIES ($M_1 \gtrsim M_2 \gtrsim 10\ M_\odot$)

We will call MCBS double stars both components of which have masses above M_{SN}, where M_{SN} is the minimal mass of nonaccreting component of a binary exploding as a supernova. This value was estimated as $M_{SN} \approx 10^{+3}_{-2}\ M_\odot$ (Massevitch, Tutukov, 1980). The first variant of the scenario was proposed by van den Heuvel, Heise (1972) (to 1.5 stage) and Tutukov, Yungelson (1973) (through 1.6-1.9 to 1.11 stage) independently. This scenario unites into a single evolutionary sequence many objects related to evolution of MCBS of B and C types (see Fig. 1). Lifetimes and numbers of galactic MCBS in appropriate evolutionary stages are also approximately pointed there. This variant of scenario describes the behavior of B and C-types of MCBS which form about 90 % of all MCBS (Kraitcheva et al., 1978). Most part of the lifetime of such MCBS is spent in 1.1 stage. The usual space velocities of such stars are ~ 10 km/s and typical distances from the Galactic plane are ~ 80 pc. It is possible that most massive components with $M \gtrsim 30\ M_\odot$ can lose the noticeable part of their mass by stellar wind before the primary fills its Roche lobe.

Evolution of B-systems after filling the Roche lobe strongly depends on the input criteria of convective stability in the zone of variable molecular weight (Tutukov et al. 1975). If the gradient of molecular weight ∇_μ is taken into account (Ledoux criterion, L-models), then exchange occurs on Kelvin-Helmholtz time scale: $T_{KH} \approx 3 \cdot 10^7 (M/M_\odot)^2 (R_\odot/R)(L_\odot/L)$ years. The usual hydrogen abundance on the surface of remnant is ~ 0.2. If ∇_μ is not taken into account (Schwarzschild criterion, S-models), then the exchange time scale depends on initial mass (Tutukov et al., 1975). For stars with $M \lesssim 20\ M_\odot$ the exchange time scale is of the order of T_{KH} but for more massive stars this time scale is close to the core helium burning time which is about ten % of the main sequence time scale.

The product of exchange for L-models is a helium star with $M_{He} \approx 0.1 (M\ /M_\odot)^{1.4} M_\odot$ and $T_e \approx 10^5$ K. The effective temperature of the remnant of S-models with initial mass $M/M_\odot \gtrsim 13\ M_\odot$ is rather low $T_e \lesssim 3 \cdot 10^4$ K (Chiosi, Summa, 1970, Tutukov et al., 1975). But Kraitcheva (1974) found that mass loss with the observed rate $\sim 3 \cdot 10^{-6} M_\odot$/yrs leads to increasing the temperature of the remnant up to $\sim 10^5$K. Thus in both cases we can get high temperature and luminosity remnants resembling effective temperatures of WR stars. Mass loss explains existence of two types of WR stars - WN and WC stars - as a result of successive uncovering deep layers of stars affected by hydrogen and

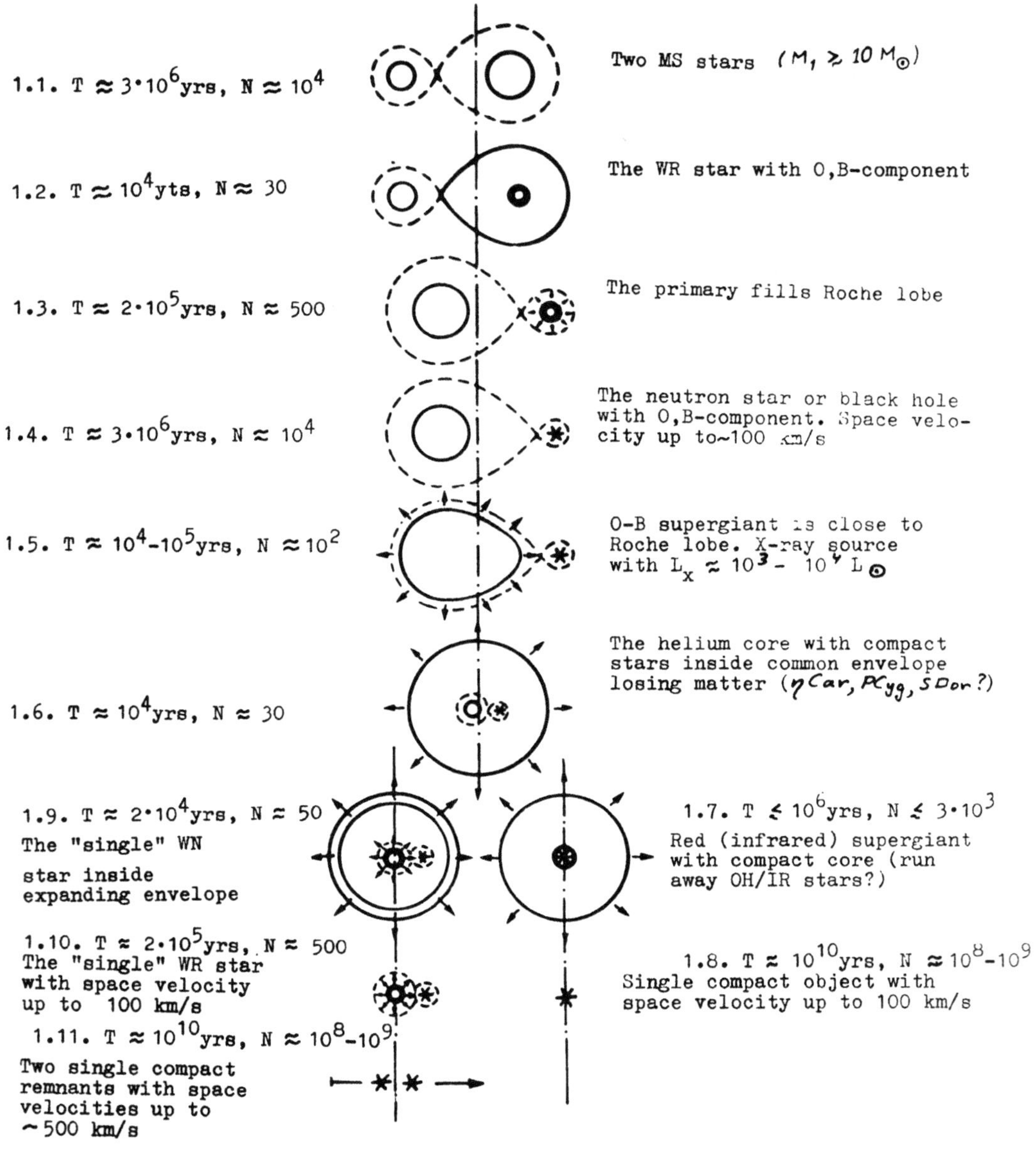

Fig. 1. Evolutionary scenario for massive close binaries .

helium burning (Tutukov, Yungelson, 1973). The comparison of relative numbers of observed WN and WC stars of the same absolute magnitude leads to the conclusion that the ratio of numbers of WR in WN stage and WC stage increases with luminosity (the initial mass) of the star. For WR stars from Moffat and Isserschtedt (1979) catalog $N_{WN}/N_{WC} \approx 3 \pm 0.7$ for $M_v \lesssim -6^m$ and only $\sim 0.8 \pm 0.1$ for $M_v > -6$. But to transform this ratio into the ratio of life times one needs reliable scales of bolometric corrections for WN and WC stars which are absent as yet. The relatively small number of WR stars and strong selection effects hinder determination of the low mass border of pre-WR stars. The present-day estimation of this value is $\sim 15\ M_\odot \pm 5\ M_\odot$. It is possible that WR stars can form in binaries only (Vanbeveren, Conti, 1979).

Cooling of the nucleus of a massive star by UFI neutrino shortens carbon, neon, silicon burning times to several thousand years only and after that exprimary explodes as a supernova leaving a neutron star with the mass $\sim 1.4\ M_\odot$. The system loses part of the matter but is conserved as a double obtaining the space velocity up to ~ 100 km/s, which is a typical velocity of "run away" OB-stars. Such a velocity is quite enough to reach during the optical component lifetime z-distance $\sim 10^3$ pc. The duplicity of such a star with a neutron star or black hole as a component is hardly discoverable due to high mass ratios. HD59543 (Gott, 1972), HD108 (Hutchings, 1975) are examples of such systems. Wide triple systems will be disrupted after the loss of the run away binary. The number of run away stars is comparable with the number of usual O,B-stars (Stone, 1979). It is remarkable that almost all stars with $M \gtrsim 50\ M_\odot$ are fast (Stone, 1979) and placed mainly near MS in the HR diagram (Humphreys, Davidson, 1979). It is possible that the presence of a close relativistic component prevents them from long blue and red supergiant stages (Tutukov, Yungelson, 1980b).

The stellar wind matter is swallowed up partly by the relativistic component which leads to the X-ray emission (Davidson, Ostriker, 1973, Tutukov, Yungelson, 1973b). Of such nature are X-ray sources in massive binaries like Cen X-3. The X-ray luminosity remains rather weak until the optical component approaches the Roche lobe. But the Roche lobe filling itself leads to switching out of the X-ray radiation due to high opacity. This determines the time scale of the X-ray stage (Ziolkowski, 1977). Numerical computations of MCBS evolution with the observed mass loss rate allow us to estimate the number of massive X-ray binaries with X-ray luminosity above $10^3\ L_\odot$. This number was found to be $\sim 10^2$ if the optical component is a MS star and ~ 10 if it is a supergiant (Massevitch et al., 1979). The X-ray stage is the longest one ($\sim 3 \cdot 10^5$ yrs) for binaries with the optical component mass $\sim 20\ M_\odot$. These estimations agree with observations. Accretion of the disk matter accelerates rotation of a neutron star in the time scale $T_\omega \approx 10^{4.5}/p/(L_x/10^{37})$ (yrs), which agrees with the observed acceleration of X-ray pulsars (Mason, 1979).

The filling of the Roche lobe by the optical component leads to the common envelope formation as the accretion rate for compact component does not exceed $\sim 10^{-8} M_\odot$/yr. The point L_2 has no dynamical significance, of course, in the absence of solid body rotation. The unique object SS433 may be a system just before the common envelope formation (van den Heuvel et al., 1980). The common envelope stage of MCBS was proposed by Paczynski (1974) and studied numerically by Taam et al. (1978), Tutukov and Yungelson (1979). The estimated number of common envelope stars in Galaxy is ~ 50. They may have high mass loss rate and luminosity. It is possible that η Car, P Cyg, S Dor and other stars of this type are common envelope binaries (Tutukov, Yungelson, 1979).

If the orbital period of MCBS just before the Roche lobe filling does not exceed $\sim 20^d$, then a supergiant with a neutron core (Thorne-Zytkow object) forms. If the orbital period of MCBS exceeds $\sim 20^d$, then after the thermal time of the optical star envelope ($\sim 10^4$ yrs) the close binary consisting of WR star and the neutron star forms inside the lost envelope. Remnants of that envelope can be observed $\sim 2 \cdot 10^4$ yrs as a bright nebula with the radius of ~ 1 pc around a "single" WN star (Massevitch et al., 1976). Several percent of all WR stars should have observed nebulae nearby. Nine of such "single" WN stars were found in our Galaxy and their properties agree well with theoretical predictions (Massevitch et al., 1976; Lozinskaya and Tutukov, 1980). One of them HD50896 (EZCMa) was proposed to be a single line binary with the mass of unseen component $\sim 1.4\ M_\odot$ (Moffat, Seggewiss, 1979) and z-coordinate ~ 280 pc (Moffat, Isserstedt, 1979). It is worth to add that accretion of the common envelope matter is negligible in comparison with mass loss. Otherwise heavy black hole formation would be inevitable which would exclude the system disruption during the second supernova explosion and formation of many pulsars in close binaries.

The absence of ring nebulae around WR+O,B systems indicates the absence of intense mass loss from the systems in 1.2 stage. So, the dissipation of the expanding nebula leaves a "single" WR star (1.10) with the unseen relativistic component as it was proposed by us (Tutukov, Yungelson,

1973a). But these stars conserve high space velocities and z-coordinates. Moffat and Isserstedt (1979) found that the $\langle z \rangle \approx 80$ pc (25 stars) for WR+O,B stars and $\langle z \rangle \approx 130$ pc (31 stars) for "single" WR stars. These results agree well with our predictions.

Exhaustion of nuclear fuel in the core of WR leads to a supernova explosion. If the presupernova mass is α M which is close to the helium remnant mass, and the mass of the relativistic remnant is β M , and initial masses of components are equal to M , then the condition of disruption of the conservative binary is $\alpha(2-\alpha)-\beta > 2\beta(2-\alpha)$. From the results by Weaver and Woosley (1980) $\beta \approx 1.4 M_\odot/M$. Now the condition of disruption for $M_{SN} = 10\ M_\odot$ is $\alpha \gtrsim 0.37$. Our computations give $\alpha \approx 0.1\ (M/M_\odot)^{0.4}$ which is not enough for disruption of such systems. To disrupt all MCBS the mass of helium remnant αM or M_{SN} should be increased ($M_{SN} \gtrsim 13\ M_\odot$). Some systems can survive as binaries like the pulsar in the close binary. As the semiaxes of MCBS decrease in the common envelope stage, young neutron stars (future pulsars) can get space velocities up to 500 km/s after disruption (Tutukov, Yungelson, 1979a).

Let us discuss disruption of wide massive binaries the components of which do not fill Roche lobes ($P \gtrsim 3$ yrs). Condition of disruption of such systems in apoastres of their orbits is: $(1-e)\ M_1 - (1+e)\ M_2 \gtrsim 2\ M_{1R}$, where M_1 and M_2 are masses of components before explosion and M_{1R} is the mass of remnant. The eccentricity of orbit e may be innate or due to the first supernova explosion. Now the condition of disruption of all systems is $e < (\alpha/\beta - 3)\ /\ (\alpha/\beta + 1)$, where αM is presupernova mass. The numerical analysis shows that several percent of wide systems can conserve as binaries with orbital period exceeding 5-10 yrs also after the second explosion. The pulsar in long period binary discovered by Manchester et al. (1980) could not form in such a way if the eccentricity of this system is really close to zero.

The lifetime of the Thorne-Zytkow object is limited to $\sim 10^6$ yrs probably mainly due to mass loss. Otherwise the number and total luminosity of (infra)red supergiants would be too high. Fast mass loss ($\dot{M} \gtrsim 10^{-5}\ M_\odot$/yrs) transforms such stars into infrared sources with high space velocities. Some of OH/IR stars have space velocities up to ~ 70 km/s according to Habing (1977) but their nature is not clear as yet.

In all cases single neutron stars (black holes) of high (as a rule) space velocities will form at the end of MCBS evolution. We supposed that radio pulsars consist of two kinematically different populations in our Galaxy (Tutukov, Yungelson, 1973). One of them consists of "slow" pulsars ($\langle z \rangle = 80$ pc) which are products of wide binaries and the other ($\langle z \rangle \approx 150$ pc) consists of "fast" pulsars which are products of MCBS. Predicted high space velocities agree well with the observed velocities (Hanson, 1979, Helfand, Tademaru, 1977).

The significant number of neutron stars may form in conservative evolution case in CB with $5.5 \lesssim M/M_\odot \lesssim 10$. The primary forms a degenerate dwarf with mass M_α. The secondary having accreted most part of the primary envelope becomes more massive than the minimal mass of supernova M_{SN} and at the end explodes as SN. Simple considerations show that the system will be destroyed at that moment if $M_\alpha/(\alpha - 2\beta) < M_{SN}$. This condition does not depend either on the initial mass ratio or mass loss which determines only the total number of such systems. If $\beta \approx 1.4\, M_\odot/M$ (Weaver, Woosley, 1980), the disruption condition for $M_\alpha = 1.2\, M_\odot$ transforms into $\alpha \gtrsim 0.4$ (if $M_{SN} = 10\, M_\odot$) or $M_{SN} \gtrsim 14\, M_\odot$ (if $\alpha \approx 0.1\,(M/M_\odot)^{0.4}$). So, if $\beta \approx 1.4\, M_\odot/M$, then to disrupt most of such binaries one needs to assume that $M_{SN} \gtrsim 14\, M_\odot$ or that real α exceeds the theoretical estimation of the value. The violation of the disruption condition leads to the neutron star formation in the close eccentric binary like PSR 1913+16.

4. EVOLUTIONARY SCENARIO FOR CLOSE BINARIES OF MODERATE MASS ($1 \lesssim M\ /M_\odot \lesssim 10$)

Borders of the mass range are determined by condition of degenerate dwarf formation in the end of evolution of components. This scenario remains rather poorly developed and Fig. 2 is only a preliminary sketch.

The analyses of distribution of spectroscopic binaries of different mass over semiaxes a have shown the absence of unevolved binaries with $M \gtrsim 1.5\, M_\odot$ and $a \lesssim 10\, R_\odot$ (Svechnikov, 1969, Kraitcheva et al., 1978). Kraitcheva et al. (1978) assume that it is the consequence of conditions of the close binary formation as a result of which only binaries with $a \gtrsim 10\, R_\odot$ can form irrespectively of mass. Stars with the mass smaller than $\sim 1.5\, M_\odot$ have convective envelopes and probably the hot stellar wind. Such wind with magnetic field leads to effective orbital momentum loss and to gradual drawing together of components (Mestel, 1967). If it is so, the relative number of unevolved binaries with $M \lesssim 1.5\, M_\odot$ and $a \lesssim 10\, R_\odot$ must be lower than the number of wider systems. That is the case. Popova et al. (1980) found that the number of double-line spectroscopic binaries

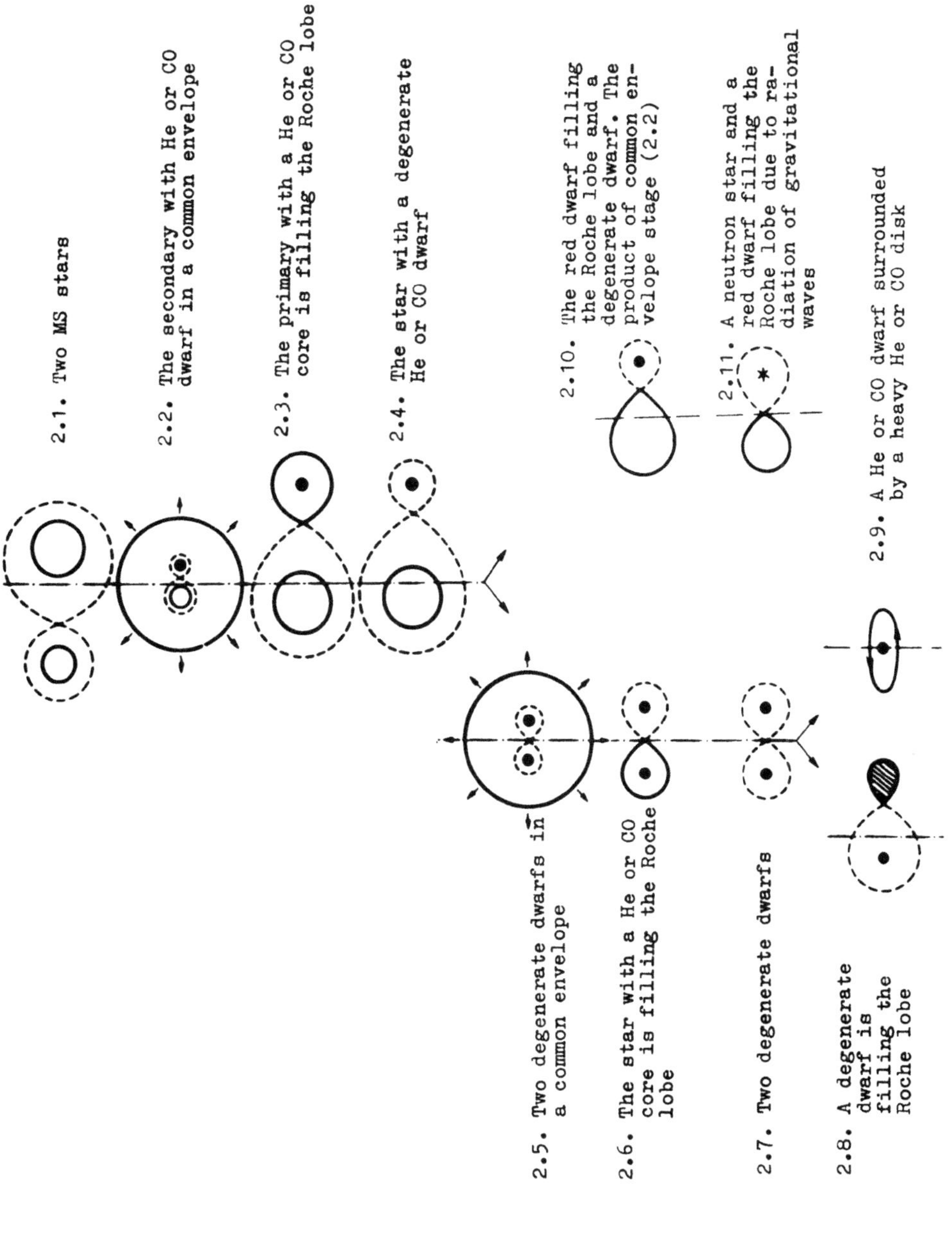

Fig. 2. Evolutionary scenarion for moderate mass close binaries ($1 \lesssim M_1/M_\odot \lesssim 10$).

with $0.6 \lesssim M/M_{\odot} \lesssim 1.5$ and $a \lesssim 15\ R_{\odot} M_{\odot}/M$ on unit of lg$a$ is about ten times lower than the number of wider binaries which supports the possibility of such evolution of dwarf binaries having a component(s) with convective envelopes. If both components at the moment of contact stay unevolved, their merging could be a result of such evolution with "blue struggler" formation (Paczynski, 1979). Slightly evolved components form, possibly, WUMa systems. Properties and evolution of these stars have been reviewed extensively elsewhere (Binnendijk, 1977, Webbink, 1977, Shu et al., 1979).

The exhaustion of hydrogen in the nucleus of the primary leads as usually to expansion of the envelope and to the Roche lobe filling. Mass exchange in the thermal time scale of the primary can lead to the common envelope formation (2.2). If the system survives that stage, then after the partial loss of the primary envelope it becomes a helium or carbon-oxygen dwarf. If a low mass star filling the Roche lobe has the deep convective envelope, then very close cataclysmic variable like binary may form as a result owing to the extensive orbital momentum and mass loss (Webbink, 1979a; Meyer-Hoffmeister, 1979). As in the case of MCBS in the common envelope stage (Tutukov, Yungelson, 1979a) during the short time the drag luminosity exceeding the Eddington luminosity is possible as well. This leads to the extensive mass loss. The total lifetime of B and C systems in the semidetached stage (2.3.) is of the order of the thermal timescale of the expanding envelope. The system consisting of He or CO dwarf and a MS star forms as a result (2.4). The lifetime of low mass ($M \lesssim 3\ M_{\odot}$) B systems in (2.3) stage is determined by hydrogen burning of hydrogen rich envelope surrounding degenerate helium core (Paczynski, 1967). Probably, classical Algol type systems with the highly overluminous primary are the (2.3) systems. Analysis of observations of algols shows that prealgols are B-type systems mainly (Mezzetti et al., 1980). It is the natural result of the absence of A-type systems with the initial primary mass $1.5 \lesssim M/M_{\odot} \lesssim 10$ (Tutukov, Yungelson, 1980a). The exhaustion of the envelope's hydrogen or the central helium flash terminates the semidetached stage.

Mass and angular momentum loss can sometimes create rather close binaries but up to the Roche lobe filling the presence of the compact degenerate component is almost undiscoverable. It is possible that most part of single-line spectral binaries are such systems. The optical components of these systems may be "blue strugglers". One more possibility to distinguish such systems is the rotational velocity. Van den Heuvel (1968) found the bimodal distribution of stars over rotational velocity. It is possible that stars with low rotational velocities are hidden evolved binaries. If the initial mass of the primary was above $\sim 3\ M_{\odot}$ it produces a CO-dwarf. Helium shell flashes in its envelope can enrich it by heavy s-process product elements. Part of the envelope matter may be accreted during flashes by the secondary MS-component. That can help to understand the cause of some abundance anomalies of peculiar stars of spectral type A (Tutukov, Yungelson, 1980b).

If the system (2.4) is wide, the optical component may be a red (super)giant not filling its Roche lobe. Such cold stars lose the matter by stellar wind and a part of that matter is intercepted by the compact CO-dwarf. In this case the CO-dwarf reactivates its shell sources which keeps its luminosity on a rather high ($10^2 - 10^3\ L_{\odot}$) level for a long time. Hot radiation of dwarf ionizes part of the extended envelope of the red (super)giant. Instability of nuclear burning in partly degenerate shells supplements the symbiotic star model (Tutukov, Yungelson, 1976). The increase of the CO-dwarf mass can lead to supernova explosion and to the neutron star formation. XRS with lifetime of $\sim 10^6$ yrs may be a result of such evolution. And the radio pulsar in a wide system with low eccentricity forms. It is possible that radio pulsar in wide binaries like PSR0820+02 with $p \approx 1700^d$ and $e \approx 0$ (Manchester et al., 1980) was formed this way. But there is another possibility for such systems' formation: the capture in two body collisions between a red giant and a neutron star in dense globular clusters.

The expansion of optical component leads to the second mass transfer stage (2.4). All B-type systems with $M/T_{KH} \gtrsim 10^{-6}\ M_{\odot}$/yrs plunge into a common envelope (2.5). The evolution in that stage was studied by Meyer, Meyer-Hoffmeister (1979). The lifetime in that stage is of the order of the thermal time for the envelope. The fast friction leads to the high luminosity and to mass loss. The friction of the double core in the common envelope may lead, as in the MCBS case, either to a coalescence of degenerate cores or to the loss of common envelope. But in moderate mass binaries case the gravitational energy of two compact degenerate cores is so high that the envelope can be dispersed. The double core of the young planetary nebula like UU Sge according to Paczynski (1974) form as a result. Such cores were discovered by Miller et al. (1976). About 10^4 years later the envelope disperses and two degenerate cooling dwarfs remain. Such systems are also found now: V 471 Tau (Nelson and Young, 1970), PG 1413+0.1 (Green et al., 1978). The total number of similar systems (2.7) in our Galaxy is large ($\sim 10^{10}$) but the probability to find low luminous compact dwarfs is very low, so only the youngest and closest of them may be discovered.

If the initial mass of the optical component was lower than $\sim 3\ M_{\odot}$, then a rather long semidetached stage (2.6) is possible. Another possibility to have the high mass exchange rate in dwarf binaries is exchange in thermal time scale and, possibly, the strong magnetic stellar wind.

Three main types of instabilities involved in accretion are usually discussed: dynamical instability of the convective low mass component losing matter (Bath, 1975), disk accumulation (Osaki, 1974) and thermonuclear runaway (Lebedinsky, Gurevich, 1947). The disk accumulation instability leads to dwarf nova explosions if an accreting star is a degenerate and to X-ray recurrent nova like Aqu X-1 if it is a neutron star. If we suppose that $\sim 10\ \%$ of time Aqu X-1 radiates on the Eddington limit (Charles, 1980), then the average accretion rate is $\sim 10^{-9}\ M_{\odot}$/yrs, which is close to the accretion rate in dwarf novae and X-ray bursters. It is possible that some classical novae with the low ($\sim 10^{-9}\ M_{\odot}$/yrs) rate of the mass exchange may display the dwarf nova-like activity like nova GK Per (Webbink, 1977b). The analytical theory of nuclear burning in thin degenerate envelopes of compact stars was developed (Sugimoto et al., 1980, Ergma, Tutukov, 1980a,b) which gives a possibility to estimate some properties of bursting sources without extensive computations.

The numerical theory of novae is now well developed (Gallagher, Starrfield, 1978). The first computations were performed for models with the overabundance of CNO elements (Starrfield et al., 1972) which were supported by observations of novae (Pottasch, 1959). But pure thermonuclear explosion cannot lead to the loss of the envelope during the flash. To lose the hydrogen rich envelope an additional mass loss mechanism is needed, like the hot stellar wind or the common envelope stripping (Starrfield, 1980). The CNO overabundance can be produced by occurring from time to time helium shell flashes (Ergma, Tutukov, 1980b).

If the accreting star is an old cold neutron star, then recurrent thermonuclear X-ray bursts are possible (Marashi, Cavaliere, 1976). The simple analytical model explains energy, the period of repetition of X-ray bursters (Ergma, Tutukov, 1980), and numerical computations by Joss (1978) give the light curve of bursts. But the problem of chemical kinetics and possible deviation from spherical symmetry of the burning shell remains a problem under active investigation.

The growth of CO dwarf can lead to supernova explosion (Whelan, Iben, 1973, Ergma, Tutukov, 1976) with, possibly, a neutron star formation in pair with low mass component (2.11). But this is unlikely for recurrent novae because in the course of the nuclear flash only several percent of hydrogen can burn and be accumulated by CO-core. The common envelope stage (2.2), radiation of gravitational waves and the magnetic stellar wind may be important for orbital angular momentum loss. A close binary (2.10) forms as a result. Similar systems (2.10, 2.11) may be also formed after unelastic collisions (Fabian et al., 1975) or due to exchange collisions of a low-mass close binary with a degenerate dwarf (a neutron, a black hole) in dense nuclear of globular clusters according to Hills (1979). The accretion of the most part of the low mass component makes the collapse of a neutron star with black hole formation inevitable in some cases. It is possible that part of stationary bulge X-ray sources are in the post-burster stage and consist of a low mass black hole and low mass red dwarf components.

The model of the bulge source is an old neutron star with the red dwarf component filling its Roche lobe. The radiation of gravitational waves provides for mass exchange with the rate of $\sim 10^{-10} - 10^{-9}$ $M_{\odot}$/yrs which is well enough for keeping the X-ray luminosity $\sim 10^{36} - 10^{37}$ ergs/s (Tutukov, Yungelson, 1979b). The total number of such binaries formed in the Galaxy lifetime is $\sim 10^3$ and according to their space distribution most of them were formed in globular clusters and in galo (van den Heuvel, 1980). Evolution of these binaries whose secondaries fill their Roche lobes may be influenced by two specific peculiarities induced by the strong X-ray irradiation of optical components. One-side heating can lead to the circulation and to the mixing of the secondary and to the strong evolutionary meaningful stellar wind. Such evolution seems not studied numerically so far.

The evolution of systems consisting of two degenerate dwarfs or a degenerate dwarf and a neutron star (2.7) is possible in the cosmological timescale only due to gravitational radiation (Tutukov, Yungelson, 1979b) if the orbital period is $p(\text{hours}) \lesssim 13\ (M_1 M_2/M_{\odot}^2)^{3/8}\ ((M_1+M_2)/M_{\odot})^{-1/8}$. The component of lower mass has the larger radius and fills its Roche lobe first. If $0.6 \lesssim q \lesssim 1$, the expansion of the filling Roche lobe component is not accompanied by the proper expansion of the Roche lobe itself. So this dwarf may be destroyed in the limit of orbital time scale and the heavy disk surrounding the ex-primary forms as a result. Further evolution of such systems strongly depends on the effectivity of angular momentum transfer in the disk. If that process is so effective that $\dot{M} \gtrsim 10^{-6}$ $M_{\odot}$/yrs, the extended helium or carbon-oxygen envelope may be formed. A single star forms as a result.

The gravitational wave radiation driven by evolution of 2.10, 2.11 systems was investigated by Faulkner (1974), Tutukov, Yungelson (1979b), White, Eggleton (1980). Gravitational wave radiation decreases the orbital period until the star losing mass is in thermal equilibrium or until its mass is more than $\sim 0.1\ M_\odot$. Further mass loss leads to expansion of secondary and as a consequence to increasing of the orbital period. Thus for binaries 2.10 and 2.11 evolving due to gravitational wave radiation there must exist the minimal orbital period $\sim 1^h$ (Paczynski, 1979b, Massevitch et al., 1980). It is remarkable that the minimal orbital period of cataclysmic variables really exists and is $\sim 82^m$ for WZ Sge. The existance of the minimal period may be considered as a good argument for gravitational wave radiation. But we need to point out now that magnetic stellar wind of the appropriate intensity influences dwarf binary evolution similarly. The well known absence of cataclysmic variables with $2^h \lesssim p \lesssim 3^h$ gives a good but still unused (as it seems) chance to advance the theory of dwarf binary evolution.

So the short review of modern theory of close binary evolution shows that the theory is now actively developed by astrophysicists of many countries. The process of advancing and selecting new ideas in the field is constantly supported by new observational data on the binaries in different evolutionary stages. Many types of stars are involved in investigation in connection with close binary evolution.

As a conclusion let me point out the main still unresolved problems related to close binary evolution. 1. Formation of close binaries and planetary system. Explanation of distribution of close binaries over mass of components, semiaxes and ratio of mass components. 2. Mass and orbital momentum loss during evolution of close binaries. 3. Evolutionary scenario for moderate mass close binaries ($M \lesssim 10\ M_\odot$). 4. Evolution in common envelope stages. 5. The stability of disk accretion, especially for dwarf close binaries.

The author is grateful to Prof. A. G. Massevitch and Drs. L. R. Yungelson and E. V. Ergma for stimulating discussions.

REFERENCES

Abt, H.A., Levy, S.G.: 1976, Astrophys.J.Suppl. 30, p.273.
Abt, H.A.: 1979, "The frequency of binaries on the main sequence", preprint.
Bath, G.T.: 1975, Month.N.R.A.S. 171, p.311.
Binnendijk, L.: 1977, Vistas in Astron. 21, p.359.
Charles, P.: 1980, Sky and Telescope 59, p. 188.
Chiosi, C., Suma, C.: 1970, Astrophys.Sp.Sci. 8, p.478.
Davidson, K., Ostriker, J.P.: 1973, Astrophys.J. 179, p.585.
Fabian, A., Pringle, J., Rees, M.: 1975, Month.N.R.A.S. 172, p.15.
Faulkner, J.: 1974, Proc. IAU Symp. 73, Eds. P. Eggleton, S. Mitton, J. Wheelan, p.193.
Gallagher, J.S., Starrfield, S.: 1978, Ann.Rev.Astron. Astrophys. 16, 171.
Gott III, J.R.: 1972, Astrophys.J. 173, p.227.
Green, R.F., Richstone, D.O., Schmidt, M.: 1978, Astrophys. J. 224, p.892.
Joss, P.C.: 1978, Astrophys.J.L. 225, L123.
Joss, P.C., Rappoport, S.: 1979, Astron. Astrophys. 71, p.217.
Habing, H.J.: 1977, Proc. IAU Coll. 42, Eds. R. Kippenhahn, J. Rahe, W. Stromeier, p.401.
Hanson, R.B.: 1979, Month.N.R.A.S. 186, p.357.
Helfand, D.J., Tademaru, E.: 1977, Astrophys. J. 216, p.842.
Hills, J.G.: 1975, Astron.J. 80, p.809.
Humphreys, R.M., Davidson, K.: 1979, Astrophys.J. 232, p.409.
Hutchings, J.B.: 1975, Astrophys.J. 200, p.122.
Ergma, E.V., Tutukov, A.V.: 1980a, Proc. IAU Symp. 88.
Ergma, E.V., Tutukov, A.V.: 1980b, Proc. IAU Coll. 53.
Kraitcheva, Z.T.: 1974, Nauch.Inform. 31, p.58.
Kraitcheva, Z.T., Popova, E.I., Tutukov, A.V., Yungelson, L.R.: 1978, Astron. Zh. 55, p.1176.
Lebedinsky, A.I., Gurevich, L.E.: 1947, Dokl.Akad.Nauk 55, No.9 (USSR, in Russian).
Lozinskaya, T.A., Tutukov, A.V.: 1980, Nauch.Inform.USSR Acad.Sci., No.49 (in press).
Meyer, F., Meyer-Hoffmeister, E.: 1979, Astron.Astrophys. 78, 167.
Manchester, R.N., Newton, L.M., Cooke, D.J., Lyne, A.G.: 1980, Astrophys.J.L. 236, L25.
Marashi, L., Cavaliere, A.: 1976, Highlights Astron. 4,p.127.
Mason, K.O.: 1977, Month.N.R.A.S. 178, p.81.
Massevitch, A.G., Tutukov, A.V., Yungelson, L.R.: 1976, Astrophys.Sp.Sci. 40, p.115.
Massevitch, A.G., Popova, E.I., Tutukov, A.V., Yungelson, L.R.: 1979, Astrophys.Sp.Sci. 62, p.451.
Massevitch, A.G., Tutukov, A.V.: 1980, "The star evolution", Moscow (in Russian).
Massevitch, A.G., Tutukov, A.V., Yungelson, L.R.: 1980, Proc. IAU Symp. 93.

Mestel, L.: 1967, Proc. Liège Coll., 1966, p.351.
Mezzetti, M., Ginricin, G., Mardirossian, F.: 1980, Astron. Astrophys. 83, p.217.
Miller, J.S., Krzeminski, W., Priedhorsky, W.: 1976, IAU Circ. No. 2974.
Moffat, A.F.J., Isserstedt, J.: 1979, "The Nature of Single-Line Population I Wolf-Rayet Stars: Evidence for High Space Velocity", preprint.
Nelson, B., Young, A.: 1970, Publ.Astron.Soc.Pac. 82, p.699.
Osaki, Y.: 1974, Publ.Astron.Soc.Japan 26, p.429.
Paczynski, B.: 1967, Acta Astron. 17, p.355.
Paczynski, B.: 1971, Ann.Rev.Astron.Astrophys. 9, p.183.
Paczynski, B.: 1974, private communication.
Paczynski, B.: 1979a, private communication.
Paczynski, B.: 1979b, review paper at IAU Assembly in Montreal, "Stellar evolution and close binaries".
Popova, E.I., Tutukov, A.V., Yungelson, L.R.: 1980(in prep).
Pottasch, S.: 1959, Ann.d'Astrophys., 22, p.412.
Shu, F.H., Lubow, S.H., Anderson, L.: 1979, Astrophys.J. 229, p. 223.
Starrfield, S., Truran, J.W., Sparks, W.M., Kutter, G.S.: 1972, Astrophys.J. 176, p.223.
Starrfield, S.: 1980, Proc. IAU Coll. 53, p.274.
Stone, R.C.: 1979, Astrophys.J. 232, p.520.
Sugimoto, D., Fujimoto, M.Y., Nariai , K., Nomoto, K.: 1980, Proc. IAU Coll. 53.
Svechnikov, M.A.: 1969, Catalog of orbital elements of close binaries, Sverdlovsk.
Taam, R.E., Bodenheimer, P., Ostriker, J.P.: 1978, Astrophys.J. 222, p.269.
Thomas, H.-C.: 1977, Ann.Rev.Astron.Astrophys. 15, p.127.
Thorne, K.S., Zytkow, A.: 1977, Astrophys.J. 212, p.832.
Tutukov. A.V., Yungelson, L.R.: 1971, Nauch.Inform. 20,p.86.
Tutukov, A.V., Yungelson, L.R.: 1973a, Nauch.Inform.27,p.58.
Tutukov, A.V., Yungelson, L.R.: 1973b, Nauch.Inform.27,p.70.
Tutukov, A.V., Yungelson, L.R., Kraitcheva, Z.T.: 1975, Mem.Soc.Astron.Ital. 45, p.879.
Tutukov, A.V., Yungelson, L.R.: 1976, Astrofizika 12, p.521.
Tutukov, A.V., Yungelson, L.R.: 1979a, Proc. IAU Symp. 83.
Tutukov, A.V., Yungelson, L.R.: 1979b, Acta Astron. 29, p.665.
Tutukov, A.V., Yungelson, L.R.: 1980a, Proc. IAU Symp. 88.
Tutukov, A.V., Yungelson, L.R.: 1980b, Nauch.Inform. 37 (in press).
Vanbeveren, D., Conti, P.S.: 1979, preprint.
Van den Heuvel, E.P.J.: 1968, Bull.Astron.Soc.Nether. 19, p.309.
Van den Heuvel, E.P.J., Heise, J.: 1972, Nat.Phys.Sci. 239, p.67.
Van den Heuvel, E.P.J.: 1976, Proc. IAU Symp. 73, p.35.
Van den Heuvel, E.P.J.: 1980, Proc. of Advanced Study Inst. "Extragalactic X-ray Astronomy".

Van den Heuvel, E.P.J., Ostriker, J.P., Patterson, J.A.: 1980, Astron.Astrophys. 81, p.7.
Yungelson, L.R.: 1973, Nauch.Inform. 27, p.93.
Yungelson, L.R., Massevitch, A.G.: 1980, "Mass exchange and evolution of close binary stars", Sov.Sci.Rev., Ed. R.A. Syunyaev.
Ziolkowski, J.: 1977, Ann.New-York Acad.Sci. 302, p.47.
Webbink, R.F.: 1977a, Astrophys.J. 215, p.851.
Webbink, R.F.: 1977b, Proc. IAU Coll. 42, Eds. R.Kippenhahn, J. Rahe, W. Stromeier, p.363.
Webbink, R.F.: 1979a, "The evolutionary significance of recurrent novae", preprint IAP-79-3, Illinois.
Webbink, R.F.: 1979b, "The formation of white dwarfs in close binary systems", preprint IAP-79-29, Illinois.
Wheaver, T.A., Woosley, S.E.: 1980, Bull.AAS 12, p.202.
Whelan, J., Iben, I.Jr.: 1973, Astrophys.J. 186, p.1007.
White, C.A., Eggleton, P.P.: 1980, Month.N.R.A.S. 190, p.801.

DISCUSSION

Chen: Consideration of the evolution of close binaries usually begins with two main-sequence stars. How could two stars with quite different masses be on the main sequence at about the same time?

Tutukov: The nuclear timescale and the Kelvin-Helmholtz timescale are so different, and the nuclear timescale depends on the mass of the star so strongly, that I hope there are no problems in having two main sequence components.

Sugimoto: Do you claim that a common envelope binary is more likely than mass loss from the system? If so, what is the reason, and how long are the timescales for formation of the common envelope and for mass loss?

Tutukov: I think now that the common envelope stage is unavoidable, at least for wide B and C systems. The reason is the following. The expansion of an evolved star occurs on the thermal timescale of its envelope, and one usually needs several additional thermal timescales to achieve solid body rotation. Only for such a rigidly rotating system is it possible to lose excess mass from the system. Therefore, the first stage of common envelope evolution will persist at least several thermal timescales. But when a compact body plunges into the deep interior, there are processes that likely occur faster— on timescales between the thermal and the orbital timescales. (For details, see Tutukov and Yungelson in Proc. IAU Symp. No. 83.)

Vilhu: Can you identify any good candidates where "spiralling in" is now occurring, or will soon occur?

Tutukov: Possible candidates where "spiralling in" will soon occur are massive close-binary X-ray sources. Possible candidates for the common envelope stage are P Cyg, η Car. Possible candidates for the post spiralling-in stage are cataclysmic variables and single-line WR stars of the WN types that lie inside remnants with expanding nebulosities.

THE FORMATION OF COMPACT OBJECTS IN BINARY SYSTEMS

E.P.J. van den Heuvel
Astronomical Institute, University of Amsterdam, the Netherlands, and
Astrophysical Institute, Vrije Universiteit, Brussels, Belgium

ABSTRACT

The various ways in which compact objects (neutron stars and black holes) can be formed in interacting binary systems are qualitatively outlined on the basis of the three major modes of binary interaction identified by Webbink (1980). Massive interacting binary systems ($M_1 \gtrsim 10\text{-}12\ M_\odot$) are, after the first phase of mass transfer expected to leave as remnants:
(i) compact stars in massive binary systems (mass $\gtrsim 10\ M_\odot$) with a wide range of orbital periods, as remnants of quasi-conservative mass transfer; these systems later evolve into massive X-ray binaries.
(ii) short-period compact star binaries ($P \sim 1\text{-}2$ days) in which the companion may be more massive or less massive than the compact object; these systems have high runaway velocities ($\gtrsim 100$ km/sec) and start out with highly eccentric orbits, which are rapidly circularized by tidal forces; they may later evolve into low-mass X-ray binaries;
(iii) single runaway compact objects with space velocities of $\sim 10^2$ to 4.10^2 km/sec; these are expected to be the most numerous compact remnants.

Compact star binaries may also form from Cataclysmic binaries or wide binaries in which an O-Ne-Mg white dwarf is driven over the Chandrasekhar limit by accretion.

1. INTRODUCTION

We examine the various ways in which compact stars can be formed in binary systems. For simplicity we assume that the supernova collapse of a stellar core always produces a neutron star - keeping in mind, however, that the cores of very massive stars may also collapse into black holes.

We summarize, in section 2, the observational evidence on compact stars in binaries and possible selection effects affecting this evidence. In sections 3 and 4 we outline the various ways in which

D. Sugimoto, D. Q. Lamb, and D. N. Schramm (eds.), Fundamental Problems in the Theory of Stellar Evolution, 155–175.

massive binaries can evolve through a first stage of mass exchange and leave compact remnants. In section 5 we consider the possible formation of compact objects in evolving binaries with one white dwarf component.

2. OBSERVATIONAL EVIDENCE ON COMPACT STARS IN BINARIES

A. X-ray binaries and binary pulsars

The binary X-ray sources that contain compact objects can - roughly - be divided into two groups, the massive ones ($M_S \gtrsim 15\ M_\odot$) and the low-mass ones ($M_S \lesssim 2\ M_\odot$) each of which can be subdivided further into several subclasses (M_S indicates the mass of the non-degenerate companion star). Notably, among the massive X-ray binaries there are two broad categories, the strong and permanent sources, in which the companion star is nearly filling its Roche-lobe (evidenced by the double-wave optical lightcurve) and is a giant or supergiant star; and the weak or transient ones, in which the companion is in most cases a rapidly rotating B-emission star (see table I a,b). In the latter case the binary periods are longer than ~ 20 days and the star is deep inside its Roche-lobe. As to the low-mass systems, there are only a few for which there is direct evidence of binary motion: these are listed in table II. Among these there are already two types, viz: the pulsating ones, with hard X-ray spectra (example: Her X-1) and the non-pulsating ones with softer spectra, such as Sco X-1 and Cyg X-2. The large X-ray luminosities of the latter systems indicate that these also contain neutron stars, presumably surrounded by an accretion disk as evidenced by their optical spectra (cf. Cowley 1980; Ziolkowski and Paczynski 1980).

The galactic bulge X-ray sources and the steady sources associated with bursters have X-ray and optical spectra similar to those of Sco X-1; the several tens of identified optical counterparts are always intrinsically faint stars and show the spectrum of a bright accretion disk, somewhat similar to the spectra of cataclysmic binaries (cf. the reviews by Cowley 1980 and Lewin and Clark 1980). In two cases of such sources, Aql X-1 and Cen X-4 the spectrum of a faint K-dwarf has been detected (Cowley 1980; van Paradijs 1980).

In view of this evidence, and because of low optical luminosities of the companions it seems plausible that the bursters and bulge X-ray sources are low-mass close binaries in which the companion to the compact star has a mass of 1 $M_\odot$ or less (cf. Joss and Rappaport 1979; Lewin and Clark 1980). In order to have accretion, the low-mass star should fill its Roche-lobe, implying (if star is unevolved) binary periods of less than about ten hours. The high X-ray luminosities of these sources (10^{36} - 10^{38} ergs/sec) imply that also here the compact stars must be neutron stars or black holes. Van Paradijs (1978) has provided convincing evidence that the bursters are neutron stars. Similar arguments can be put forward for the globular cluster X-ray sources in our galaxy (a fraction of which are also burst sources) as well as those in M31 (cf. Lewin and Clark 1980; van den Heuvel 1980). The strongly radio emitting peculiar X-ray binaries Cyg X-3,

Circinus X-1 and SS433 might form a separate category (table III). They are characterized by occasional strong radio outbursts with synchroton spectra and by large IR luminosities, probably indicating very dense stellar winds like those observed in Wolf-Rayet stars; Circinus X-1 and SS433 are surrounded by radio shells similar in appearance to supernova remnants. Finally, there are the three binary radio pulsars listed in table IV.

B. Reasons for the existence of the two groups of X-ray binaries; compact companions to stars with intermediate masses.

The existence of the two large groups does not mean that stars with masses between 2 $M_\odot$ and 15 $M_\odot$ cannot have compact companions but rather that only stars with $M \gtrsim 15\ M_\odot$ and $\lesssim 2\ M_\odot$ are able to provide accretion rates suitable for producing a relatively long-lived strong X-ray source (van den Heuvel 1975). The reasons are the following. Stars with $M \gtrsim$ 15-20 $M_\odot$ develop into blue supergiants and Of stars during later phases of H burning; also during helium burning they are blue supergiants; such stars have winds that are sufficiently strong ($\dot{M} \gtrsim 10^{-6}\ M_\odot$/yr) for feeding an X-ray source (Davidson and Ostriker 1973). In addition, in these stars beginning Roche-lobe overflow does not lead to a rapid growth of the mass transfer rate, especially not if strong stellar wind mass loss occurs (Ziolkowski 1977; Savonije 1978[1], 1979; McCray 1979). Consequently, such stars may remain close to their Roche-lobes for fairly long times (up to 10^5 yrs), powering their companions with Roche-lobe overflow accretion rates of $\sim 10^{-10}$ to $10^{-8}\ M_\odot$/yr, sufficient to produce a strong X-ray source. On the other hand, for $M \lesssim 15\ M_\odot$ stellar wind mass loss rates are below $10^{-8}\ M_\odot$/yr, insufficient for powering an X-ray source or for stabilizing beginning Roche-lobe overflow. The only mode of mass transfer available in this mass range is fully developed Roche-lobe overflow which for $M_S \gtrsim 2\ M_\odot$ leads to mass transfer on a thermal timescale at rates $\gtrsim 10^{-7}\ M_\odot$/yr which will quench the X-ray source (Shakura and Sunyaev 1973). Only for $M_S \lesssim 2\ M_\odot$ and especially if $M_S \lesssim M_C$ (M_C = mass of the compact star) where transfer can take place on a nuclear timescale, a long-lasting stage as a strong X-ray source ($\sim 10^{37}$ ergs/sec), powered by Roche-lobe overflow, is possible.

We conclude from the above that the existence of the two observed groups of X-ray binaries is just what one would expect if compact objects do exist as companions to stars *of any kind of mass*.

It seems therefore most reasonable to assume that also in the mass range 2 - 15 $M_\odot$ stars with compact companions do exist.

TABLE I. MASSIVE X-RAY BINARIES

a. Persistent strong sources

Source	Type	P_{orb}(d)	P_{pulse}	$m \sin^3 i$	e	Ref.
SMC X-1	BOI	$3^{d}.9$	$0^{s}.71$	0.8 + 12.5	0.00	(1)
LMC X-4	O8III-V	1.4	--	2.5 + 22.5	0.00	(2)
0900-40	B0.5Ib	9.0	283^{s}	1.4 + 21.3	0.09	(1)
Cen X-3	O6.5II-III	2.1	$4^{s}.84$	1.4 + 17.2	0.00	(1)
1223-62	B1.5Iab	35.0	698^{s}	1.4 + 31	0.44	(3)
1538-52	B0-1	3.7	529^{s}	2 + 20	?	(4)
1700-37	O6.5f	3.4	--	1.3 + 27.1	0.00	(1)
1907+09	O-BI	?	--	?	?	(5)
Cyg X-1	O9.7Iab	5.6	--	1.5 + 2.4 (i = 30°)	0.00	(1)

b. Weaker or transient pulsating sources

Source	Type	P_{orb}(d)	P_{pulse}	$m \sin^3 i$	e	Ref.
0115+634	BO	24.3	$3^{s}.6$	--	0.34	(6)
0352+309 (X Per)	O9.5III-Ve	581(?)	853^{s}	--	?	(6)
0535+262	BOe	> 17	104^{s}	--	?	(6)
1118-615	BOe	--	405^{s}	--	?	(6)
1145-619	B1Vne	--	297^{s} or 292^{s}	--	?	(6)
1258-613 (GX304-1)	BO-5V	--	272^{s}	--	?	(6)
1728-247 (GX1+4)	M6III + + hot star	--	$138^{s} \rightarrow 116^{s}$	--	?	(6)

(1) Conti (1978)
(2) Hutchings et al. (1978)
(3) Kelley et al. (1979)
(4) Hutchings (1980)
(5) Schwartz et al. (1979)
(6) Bradt et al. (1978)

3. EVOLUTION TOWARDS CORE COLLAPSE IN PRIMARIES OF CLOSE BINARIES

3.1. Evolution of close binaries

In binaries with periods up to several tens of years the envelope of the primary star will, at some stage of the evolution, overflow a critical surface (Roche lobe or tidal lobe) and be lost to the companion star or from the system. Such binaries, in which the stars interact during some stage of their evolution, we will call "close". The way in which the two stars interact depends on the evolutionary state of the core of the primary star at the onset of the mass transfer, on the structure of the envelope of this star at that moment, and also on the mass ratio of the components. The classification, by Kippenhahn and Weigert (1967), in terms of the evolutionary state of the core at the onset of the mass

TABLE II. LOW-MASS X-RAY BINARIES

Source	Sp. Type	P_{orb}	P_{pulse}	M_{opt}	M_x	L_x/L_{opt}	z	ref.
Her X-1	A-F	$1^d.70$	$1^s.2$	2.2	1.3	35	3 kpc	(7)
Sco X-1	accr. disk	$0^d.787$	--	<1	--	$>10^2$	400 pc	(7)
1627-673	accr. disk	41^m	$7^s.7$	<0.5	∿1.4	$>10^2$	--	(9)
2129+47	G-dwarf(?)	$5^h.2$	--	<1(?)	--	>10	--	(7)
Cyg X-2	F-giant	$9^d.843$	--	0.5-1.1	1.3-1.8	10-40	1.5 kpc	(7)
0620-003(transient)	K5-7V	8d(?)	--	--		80(outburst)	--	(7)

TABLE III. PECULIAR SOURCES WITH STRONG RADIO AND IR EMISSION

Source	Sp. Type	P_{orb}	P_{pulse}	M_{opt}	M_x	L_x/L_{opt}	z	ref.
Cir X-1	O-Be	$16^d.6$	--	--	--	10^{-2}	in gal.plane	(6)
SS433	{Wolf-Rayet like	$13^d.1$	--	--	--	10^{-3}	130 pc	(8)
Cyg X-3	--	$4^h.8$	--	--	--	?	in gal.plane	(6)

TABLE IV. BINARY RADIO PULSARS

Name	P_{pulse}	P_{orb}	e	Type of Companion	Ref.
PSR 0656+64	$0^s.196$	24^h41^m	0.00	Probably normal	Fowler (1980)
PSR 0820+02	$0^s.865$	3.2 yr	0.00	$M \sim 0.85\ M_\odot$ (?)	Manchester et al. (1980)
PSR 1913+16	$0^s.059$	7^h45^m	0.62	Compact star	Taylor et al. (1976)

(7) Cowley (1980); (8) Crampton et al. (1980).

(9) Middleditch et al. (1980)

transfer is particularly useful if one wishes to study the possible final evolutionary state of the primary star, i.e. the kind of remnant that will be left. We will first concentrate, in the next section, on this problem. On the other hand, if one wishes to know whether or not the system will be disrupted by the supernova of the primary star, one should know how much mass is captured by the other star, and how the orbital period is affected by the mass transfer. These factors will be discussed in section 4.

3.2. The final evolution of the primary star

For the definition of Kippenhahn and Weigert's classifications A, B and C we refer to earlier reviews (Plavec 1968; Paczynski 1971, Thomas 1977, Van den Heuvel 1978; Webbink 1980). Since case A is relatively rare ($\lesssim$ 15% of all systems) we will only concentrate here on the cases B and C. In these cases after the onset of the mass transfer the primary star loses practically its entire hydrogen-rich envelope (either to the secondary or - partly - from the system, cf. section 4) and only the core, consisting of helium and (in case C) heavier elements, remains. The further evolution of the primary star can, therefore, relatively simply be described in terms of the evolution of the helium core. Calculations of the evolution of helium stars by Paczynski (1971), Arnett (1978), Savonije (1978[2]) and Delgado and Thomas (1980) show the following results (cf. especially Sugimoto and Nomoto 1980):

a. In helium stars with $M \leq 2\ M_\odot$ the C + O core formed by helium burning degenerates and during helium shell burning the outer layers expand to giant size. In binaries this produces a second phase of mass transfer (case BB of binary mass transfer cf. Delgado & Thomas 1980) such that a degenerate C + O star with $M < M_{ch}$ is left, which cools off to become a C + O white dwarf (cf. De Loore and De Greve 1976);
b. Helium stars with $M \approx$ 2-3 $M_\odot$ ignite carbon off center under non-degenerate (or at least: not highly degenerate) conditions and undergo a series of carbon shell flashes (Miyaji et al. 1980; Nomoto 1980; Sugimoto and Nomoto 1980) which leave behind a growing degenerate O - Ne - Mg core. When the boundary of this core approaches the helium-burning shell, carbon burning dies out and just like in stars with a degenerate C + O core, helium shell burning causes the outer (helium) layers to expand to giant size. The mass of the degenerate O - Ne - Mg core is then in the range 1.2 - 1.4$M_\odot$. In a binary system, the extended outer layers will be lost to the companion in a second stage of mass transfer (case BB or BC, etc.) and one expects an 1.2 - 1.4 $M_\odot$ O - Ne - Mg white dwarf to be left.
c. $M \gtrsim 3\ M_\odot$. In these helium stars the C + O core produced by helium burning has a mass larger than the Chandrasekhar limit and Ne, O and Si are ignited under non-degenerate conditions; here the core is expected to evolve directly to an Fe-photodesintegration collapse, giving rise to a supernova and the formation of a neutron star. This happens regardless of whether or not still a part (or even most) of the envelope matter is transferred to the secondary star in a case BB (or BC, etc.) mass transfer. Helium stars more massive than 4 $M_\odot$ do not reach radii larger than a few $R_\odot$ before the final core col-

lapse, and therefore in a binary are not expected to lose further mass to their companions. On the other hand, in the mass range 3 - 4 $M_\odot$ still considerable expansion of the outer layers occurs; the 3 $M_\odot$ helium star considered by Arnett reaches R = 21 $R_\odot$ at Si-ignition. Hence, in a binary with a period of less than about 10 days it may still lose part (or most) of its envelope by case BB mass transfer before the final core collapse. Similarly, Delgado and Thomas (1980) found that a 4 $M_\odot$ He-star in a $1^d.49$ period binary with a 14 $M_\odot$ companion still loses some 1.3 $M_\odot$ to its companion before the final core collapse. The mass range 3 - 4 $M_\odot$ seems particularly important since, although direct evolution to a supernova collapse seems unavoidable, the collapsing stars may have lost a considerable part of their envelopes, and in some cases (notably for M $\sim$ 3 $M_\odot$) it is conceivable that they are almost bare cores of about a Chandrasekhar mass.

d. In helium stars more massive than about 60 $M_\odot$ the oxygen core evolves to a pair-creation collapse again leading to the formation of a compact object with a mass of a few solar masses (cf. Arnett 1978).

The various types of core evolution as a function of initial primary main-sequence mass M_{ms} are depicted in figure 1. The almost vertical dashed line indicates the approximate lower mass limit for evolution towards direct core collapse after one or two phases of mass transfer (B or BB, C or CC, etc.). For short binary periods ($\sim$ 10 days) the main-sequence mass required for producing a helium core $M_{core} \sim 3\ M_\odot$ is about 12 ($\pm$ 1) $M_\odot$ (the precise value depends on the initial abundances Y and Z and on the binary period and mass ratio; for P $\sim 4^d$ even initial masses as high as 14-15 $M_\odot$ may be required). For wide binaries (P $\gtrsim 100^d$) where the core still has time to grow considerably by hydrogen shell-burning before the onset of mass transfer, the required initial mass is $M_{ms} \simeq 10\ (\pm\ 1)\ M_\odot$. The thick fully drawn line, similarly, indicates the lower mass limit, as a function of binary period, for leaving behind a 2 $M_\odot$ helium core. For short binary periods this limit is about 10 ($\pm$ 1) $M_\odot$, for long periods about 8 ($\pm$ 1) $M_\odot$. (The above quoted uncertainties are tentative and were obtained by comparing results computed by various authors (cf. Webbink 1980). Primary stars in the hatched region between the two curves leave O - Ne - Mg white dwarfs with masses of $\sim$ 1.2 - 1.4 $M_\odot$, after a second phase of mass transfer (BB or BC, etc.).

3.3. Evolution of O - Ne - Mg white dwarfs in binaries towards an accretion-induced core collapse

Myaji et al. (1980), Sugimoto and Nomoto (1980) and Nomoto (1980) have pointed out that when the secondary star evolves away from the main sequence and begins to transfer mass to the O - Ne - Mg white dwarf, this dwarf may be driven over the Chandrasekhar limit and undergo an electron capture supernova collapse. They carried out calculations for a 1.2 $M_\odot$ O - Ne - Mg white dwarf with an acretion rate of helium of 4.10^{-6} $M_\odot$/yr (about the rate at which such a core grows by helium shell burning inside a red giant) and found that electron captures on ^{20}Ne and ^{24}Mg cause the core density to rise, followed by a weak O-deflagration, that cannot prevent further electron captures which induce a total

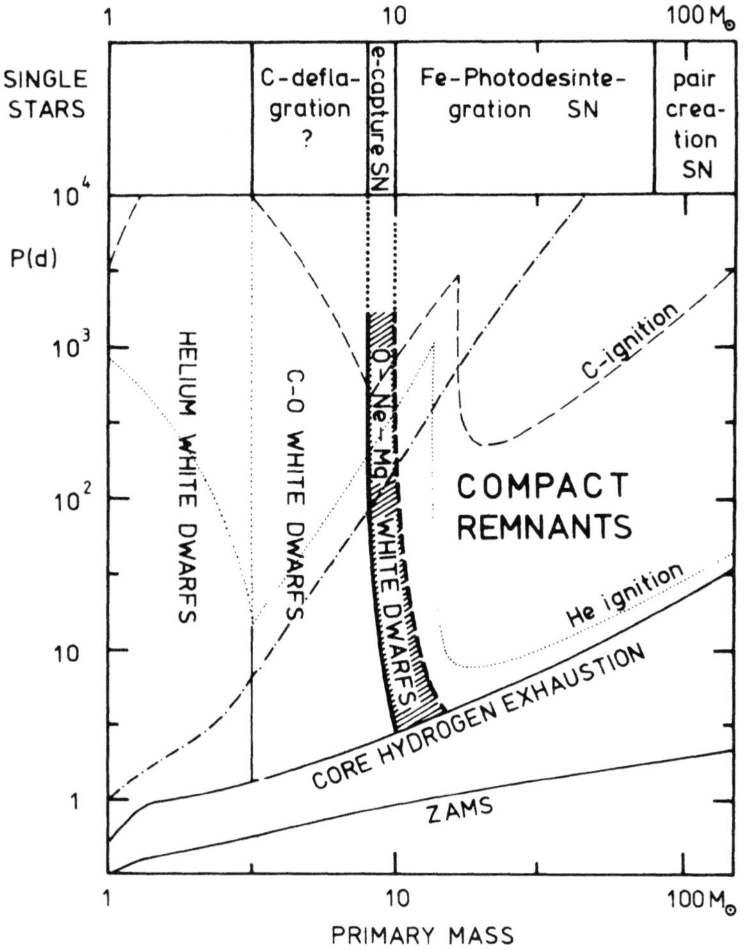

Figure 1. Classification of expected final evolutionary states of primary stars of close binaries as a function of primary mass and orbital period (partly after Webbink 1980). The orbital periods correspond to binaries in which the primary star just fills its Roche lobe (for mass ratio unity). At the top of the figure the expected final evolutionary states of single stars are indicate Dash-dot line indicates convective boundary (Hayashi-line).

collapse of the core. The reason why the collapse cannot be prevented is the reduction of the value of the Chandrasekhar mass due to the electron captures, which causes it to become almost equal to or even smaller than the actual core mass. Consequently, the O - Ne - Mg white dwarfs that result from the first stage of mass transfer in binaries with primary masses between 8 and 12 $M_{\odot}$ (the hatched area in figure 2) may (sometimes much) later, during the mass transfer from secondary to primary undergo an electron-capture SN. Since this may, if the secondary is a low-mass star, occur billions of years later, such SNe may also occur in old stellar systems. It seems attractive, therefore, to indentify this type of collapse with type I SNe (Sugimoto and Nomoto 1980; Nomoto 1980). In section 5 we will consider this type of SN model in more detail; it closely resembles the type I SN models proposed by Whelan and Iben (1973), Warner (1974) and Gursky (1976). Notice that, since the O - Ne - Mg white dwarfs are produced in quite a wide main-sequence mass range, this type of core collapse in binaries may be a fairly common type of event.

3.4 Precision of the boundaries between the various types of final evolution

Even for a fixed initial chemical composition the boundaries between the various types of final evolution sketched in figure 1 are, in fact, bands with a certain intrinsic width. This is due to the fact that the boundary masses also depend on the initial mass ratio - a factor which was assumed fixed in figure 1. Furthermore, the lower mass limit for evolving to a direct photodesintegration core collapse for helium stars in binaries may depend somewhat on the binary separation. Since helium stars in the mass range 2.6 to 3.0 $M_\odot$ are not expected to expand to full giant size (cf. Webbink 1980), in wide binaries perhaps even helium stars with masses as low as 2.6 $M_\odot$ may evolve to direct core collapse. Although all these uncertainies may shift the boundaries of the hatched region somewhat to higher or lower mass values, the total width of this region will remain as large as 2 to 2.5 $M_\odot$, which implies that O - Ne - Mg white dwarfs are expected to be produced in considerable quantities. At the top of figure 1 we have also tentatively indicated the approximate boundaries for various types of supernovae in single stars. The cores of single stars in the mass range ∿ 8 -10 $M_\odot$ (or possibly 7 - 12 $M_\odot$) evolve directly to electron-capture supernovae, whereas the primaries of binaries in this mass range terminate as O - Ne - Mg white dwarfs.

4. THE FATE OF THE ENVELOPE

4.1 Conservative evolution: formation of massive X-binaries

We consider primary stars which directly evolve to core collapse ($M_{core} \gtrsim 3\ M_\odot$). In binaries that evolve with conservation of total mass and orbital angular momentum ("conservative" mass transfer) the core will at the time of its collapse be the less massive component of the system. For circular orbits the conservative assumptions imply that the orbital radius a changes according to the equation (cf. Paczynski 1971):

$$a/a_o = \left[M_1{}^o(M - M_1{}^o)/M_1(M - M_1)\right]^2 \qquad (1)$$

where index zero indicates the initial situation, M is the total mass of the system and M_1 is the primary mass. Since explosion of the less massive component is unlikely to disrupt the system, even if the effects of impact, ablation and asymmetries in the SN mass ejection are taken into account (Sutantyo 1974, 1975; Wheeler et al. 1975; Fryxell and Arnett 1978; De Cuyper 1980) one expects that the compact stars in conservatively evolving systems will practically always remain bound after the explosion. Apparently this was the case in the massive X-ray binaries (Van den Heuvel and Heise, 1972; Tutukov and Yungelson 1973). Their most likely evolutionary history, through an intermediate stage as a Wolf-Rayet (WR) binary has been extensively summarized elsewhere (Van den Heuvel 1976, 1978; see especially: Tutukov, this volume).

4.2. Limitations to conservative evolution

De Grève et al. (1978) have shown that in order to explain the presently observed system parameters of the WR binaries, conservative mass transfer is not fully adequate as it would predict too large orbital periods as well as mass ratios M_{WR}/M_{OB} which are a factor 1.5 to 2 smaller than observed. In order to explain the combination of mass ratio and short binary period of systems like CQ Cep (P = $1^d.6$) and CX Cep (P = $2^d.1$), one has to invoke considerable angular momentum loss ($\geq$ 50%) from the systems presumably during the first phase of mass transfer (Flannery and Ulrich 1977; Kippenhahn and Meyer-Hofmeister 1977). This situation is similar to that for the Algol systems which also must have lost some 50 percent of their angular momentum during the preceding mass transfer (Ziolkowski 1976). The substantial mass loss by stellar wind during Of and WR stages is an important additional factor (cf. Van Beveren, this volume).

The outcome of the evolution with moderate losses of mass and angular momentum (i.e. less than two third of the total) is qualitatively still similar to that of conservative evolution - i.e.: the more evolved component is less massive than the secondary and the binary separation does not differ by more than a factor 2 to 3 from that in the conservative case. We will indicate this type of evolution as "quasi-conservative" (cf. Webbink 1980). Apparently, the existence of WR binaries and massive X-ray binaries can be understood in terms of such evolution.

4.3. Modes of envelope interaction

Quasi-conservative evolution is expected only in case that (a) the initial mass ratio of the system is not too low, (b) the separation is not too small and (c) the envelope of the mass-losing star is in radiative equilibrium. Although for the conditions (a) and (b) no precise limits can be set, it seems that for q $\gtrsim$ 0.5 and orbital periods longer than a few days the quasi-conservative approximation is probably adequate since in that case equation (1) does not induce a drastic reduction of the binary period during the exchange and there will remain enough room in the system to accommodate the rapidly swelling secondary. As in this case the thermal timescales of the two stars are not too much different the rate at which mass is transferred will be roughly similar to the rate at which the secondary can accommodate it.

Condition (c) follows from the fact that mass transfer from a radiative envelope (*Mode II* of binary interaction defined by Webbink 1980) is self-stabilizing, as it induces the envelope to shrink. Consequently, systems that simultaneously fulfil the conditions (a), (b) and (c) are expected to evolve quasi-conservatively, and to transfer mass at a rate of the order

$$\dot{M}_1 = M_1/\tau_{KH} \simeq -3.10^{-8}\ M_1^{\ 3}\ (M_\odot/yr) \tag{2}$$

where M_1 is expressed in solar units (cf. Paczynski 1971).
If conditions (a) and/or (b) are not fulfilled, the secondary will after the onset of the mass transfer rapidly swell to its Roche lobe such that a contact system forms surrounded by a common envelope (*Mode I* of Webbink 1979).
Convective envelopes and degenerate stars have the tendency to expand as a consequence of mass loss. Therefore, mass transfer from a convective envelope (Webbink's *Mode III*) has the tendency to grow catastrophically. The expanding envelope will in this case engulf the companion and again a common-envelope system will be formed.

4.4. Outcome of common-envelope evolution for $M_1 \lesssim 10-12\ M_\odot$

Although no precise calculations exist, nor seem possible at present, observations and speculations suggest that cataclysmic (CV) binaries with P $\sim$ 0.25 - 0.5 days are the outcome of common-envelope evolution of moderately wide to wide binaries consisting of a (sub)giant with a degenerate core ($M_{giant} \lesssim 10-12\ M_\odot$) together with a main-sequence dwarf (Paczynski 1976; Ostriker 1976; Ritter 1976). A variety of arguments (e.g. cf. Van den Heuvel 1976; Meyer and Meyer - Hofmeister 1979; Webbink 1980) strongly suggest that soon after the formation of the common envelope, rapid mass loss with high specific angular momentum will start. Suggestive trial calculation by Taam et al. (1978), Taam (1979), Meyer and Meyer - Hofmeister (1979) show that in all examined cases the secondary spirals down into the envelope of the primary on a timescale of order 10^3 - 10^4 yrs. The secondary has no time to accrete mass from the common envelope, this envelope is lost, and final binary periods of the order of a fraction of a day are expected. The precise reasons for the termination of the spiral-in are, however, not known. Notably, it is not understood why young post-spiral-in systems are always detached (Paczynski 1980). Examples are V 471 Tau in the Hyades (0.8 $M_\odot$ white dwarf + 0.8 $M_\odot$ K-dwarf, P = 12^h) and several double cores of planetary nebulae (e.g. UU Sge, P $\sim$ 12^h, Bond et al. 1978).

4.5. Common-envelope evolution in massive systems

We consider systems in which the primary has $M_{core} \gtrsim 3\ M_\odot$. If $q < 0.5$ or if the envelope of the primary is convective, or both, one expects that common envelope evolution will occur, and the two stellar cores will spiral-in on a short timescale (10^3 - 10^4 yrs). The resulting system after spiral-in is - in analogy to the CV binaries - expected to consist of the evolved core of the primary together with the unaltered secondary, in an orbital period of less than one day. Since the precise reasons for the termination of spiral-in are not known, we will tentatively assume that, like in the observed lower-mass post-spiral-in systems, the Roche lobe around the non-degenerate component has a radius of between 1 and 2 times the stellar radius. For core masses and companion masses of $\sim$ 3 to 5 $M_\odot$ the resulting post-spiral-in periods then are typically of order 0.5 - 1.0 days. The core will finally explode as a SN. We carried out trial calculations for the effects of the SN explosion on these close systems. We assumed that just the helium core of the primary remained after spiral-in and that for $M_{core} \gtrsim 4\ M_\odot$ this star underwent no further mass loss before exploding. In the mass range 2.6 - 4 $M_\odot$ the later expansion of the envelopes of helium stars (cf. § 3.2.c) may give rise to further spiral-in and mass loss. In the absence of precise calculations for this case we have tentatively assumed that at the time of the explosion core masses as low as 2.0 and 2.5 $M_\odot$ are possible in some cases and again calculated the SN effects. We assumed the remaining compact remnant to have a mass of 1.4 $M_\odot$, and carried out calculations for SN ejection velocities of 5.10^3 km/sec and 10^4 km/sec. We assumed instantaneous sherically symmetric mass ejection and estimated the impact effects following Wheeler et al. (1975) (but corrected for the overestimate - by a factor of order 5 - of the effects of backward blow-off of matter; cf. Fryxell and Arnett 1978). Table V lists some representative results. Since the systems are close and the companions have masses of the same order of those of the exploding cores, disruption is likely. Systems that remain bound always have high orbital eccentricities and high runaway velocities, of order 10^2 - 2.10^2 km/sec. In view of their short separation at periastron, their orbits will rapidly circularize by tidal forces. Figure 2 schematically illustrates this situation. Table V also lists - for bound systems - the resulting binary period P_t after tidal circularization and synchronization, calculated following Sutantyo (1975).

Intermezzo: Low-mass Population I X-ray binaries and their relation to runaway radio pulsars

Bound systems after tidal circularization always have orbital periods $\sim$ 1-2^d and runaway velocities $\sim 10^2$ - 2.10^2 km/s. As pointed out by Sutantyo (1975; cf. Van den Heuvel 1976) Her X-1 must be the result of such an evolution since its 2 $M_\odot$ companion is clearly a Population I object (age < 10^9 yrs); hence, the system must have originated in the galactic plane, and must have been shot out if this plane with a velocity $\geq$ 125 km/sec to reach its present z-distance of 3 kpc. The same applies to Cygnus X-2, where the luminosity of the F-giant indicates an original mass $\gtrsim 2\ M_\odot$, while its z-distance indicates a runaway velocity $\geq$ 100 km/sec. [The present mass of $\sim 1\ M_\odot$ of the F-giant suggests that later considerable mass transfer took place; in such a case an original binary period of 1 - 2^d may easily be transformed into a longer period like $\sim 9^d$]. From the fact that low-mass Population I X-ray binaries are very rare, whereas spiral-in evolution must be very common, occurring in at least one third of all massive stars (cf. § 4.7) we conclude that post-spiral-in systems with $M_{core} \gtrsim 3\ M_\odot$ are practically always disrupted by the supernova explosion. Hence, the vast majority of the remnants of such systems will be runaway (young) neutron stars together with runaway normal stars (mostly with $M \lesssim$ 4-5 $M_\odot$), with velocities of

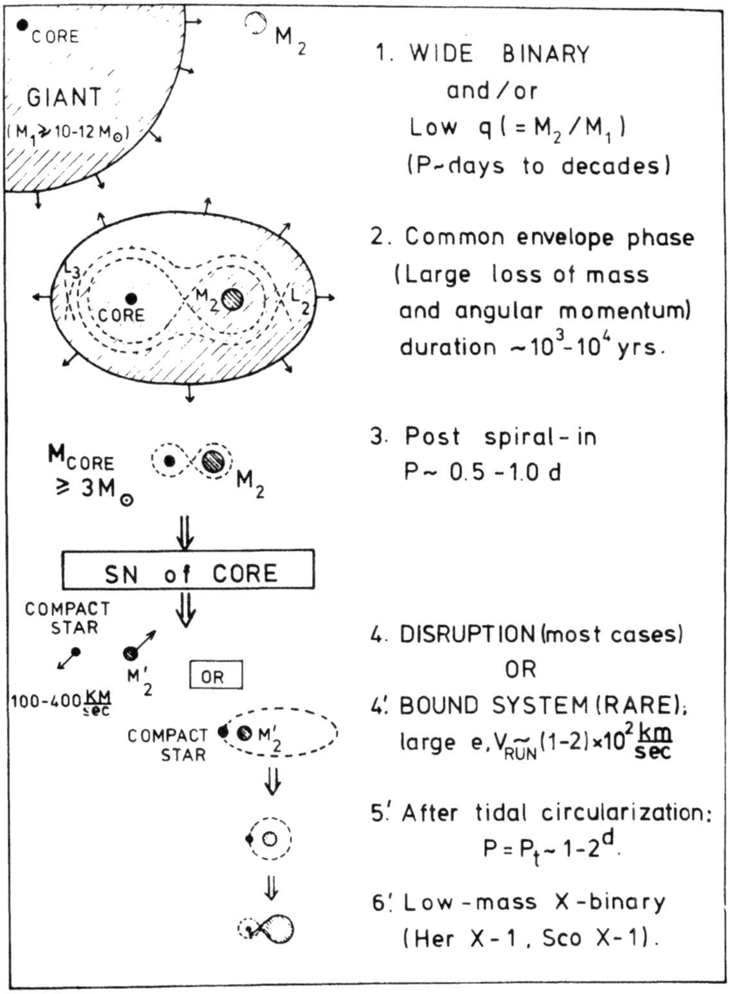

Figure 2. Outline of expected highly non-conservative evolution of massive close binaries with a long orbital period and/or a low initial mass ratio ($q \lesssim 0.5$). Explanation in the text.

order 10^2 - 4.10^2 km/sec. The condition that the bulk of the post-spiral in systems should be disrupted sets a lower limit to the SN ejection velocity. For $M_{core} = 2.5\ M_\odot$, $P_o = 12^h$, the condition that virtually no companions with $M_2 \lesssim 2\ M_\odot$ remain bound requires $V_{SN} \gtrsim 10^4$ km/sec. For $P_o = 24^h$ this limit becomes $\gtrsim 1.4 \times 10^4$ km/sec. Since P_o cannot be much smaller than $\sim 12^h$, we conclude that most probably $V_{SN} \gtrsim 10^4$ km/sec (notice that this ejection velocity of the mantle of a collapsing helium star is not necessarily related to an observable SN; P_o is the post-spiral-in period, before the SN).

4.6. The final evolution of quasi-conservative systems

Also the massive X-ray binaries will go through a spiral-in phase before terminating their lives (Van den Heuvel and De Loore 1973; Taam et al. 1978; Delgado 1979; cf. Tutukov, this volume), which will result in the formation of a very close system consisting of the compact star and the core of the massive star. Explosion of the stellar core will in most cases lead to disruption of the system, producing two runaway compact stars, one young and one old. The old one will, presumably, have been spun up back to a short rotation period by accretion during the X-ray phase (Smarr and Blandford 1976), so both neutron stars may be observable as radio pulsars. In the rare case that the system is not disrupted in the second SN, a close binary pulsar with a very eccentric orbit will result, resembling PSR 1913 + 16 (Flannery and Van den Heuvel 1975; cf. Tutukov, this volume). In the (probably rare) case that the companion has a helium core mass $\lesssim 3\ M_\odot$, spiral-in may result in a close binary consisting of a white dwarf and an old neutron star. Such a system will have a circular orbit and may, in principle, have any orbital period upwards from about one minute.

4.7. The expected incidence of runaway stars among radio pulsars

Statistical investigations by Abt and Levy (1978) show that over 2/3 of the main-sequence stars in the spectral range B2 - B9 have stellar companions with orbital periods $\lesssim$ 10 yrs. The overall distribution of mass ratios $q = M_2/M_1$ of these systems is roughly represented by $q^{-\frac{1}{4}}$ (down to the completeness limit of $q \simeq 0.2$). At closer scrutiny, the spectroscopic binary mass ratio distribution is double-peaked (Trimble 1974; Tutukov and Yungelson 1980), with one peak near $q = 1$ and the other at low q-values. Systems with short periods ($\lesssim 100^d$) tend to have q-values close to unity (i.e. $\gtrsim 0.5$; cf. Lucy, this volume; Massey and Conti 1980) and produce the peak near $q = 1$. Hence, systems with $q < 0.5$ will preferentially have long periods. These systems will undergo spiral-in evolution. Since, according to Abt and Levy's overall q distribution, systems with $q < 0.5$ represent over 50% of all spectroscopic binaries, one expects that over 1/3 of all stars with $M \gtrsim 10\text{-}12\ M_\odot$ undergo common-envelope evolution and become very short-period binaries before they explode as a SN. Consequently, over 1/3 of all stars with $M \gtrsim 10\text{-}12\ M_\odot$ will produce runaway pulsars with $V > 100$ km/sec. The other roughly one third (the shorter-period systems with $q \gtrsim 0.5$) will evolve quasi-conservatively and, at the end of their lives, release two runaway compact stars with runaway velocities in the range $10^2 - 10^3$ km/sec (the orbital velocity at periastron of PSR 1913 + 16 is > 500 km/sec). Since quasi-conservatively evolving systems produce *two* runaway pulsars, one expects that some three quarters of all stars with $M \gtrsim 10\text{-}12\ M_\odot$ produce runaway pulsars. Correcting for the fact that also single stars ($\sim$ 1/3 of all) in the mass range 8-12 $M_\odot$ are expected to leave pulsars, one roughly expects some 2/3 of all radio pulsars to be runaways with space velocities > 100 km/sec. This may explain the observed fact that the majority of the radio pulsars have runaway velocities of order 100 - 200 km/sec (Lyne 1980).

5. ACCRETION INDUCED COLLAPSE OF O-Ne-Mg WHITE DWARFS

5.1. Restrictions

In Miyaji et al.'s (1980) calculations a helium accretion rate of 4.10^{-6} $M_\odot$/yr induced core collapse. Such a high rate can only be accommodated in a *wide* binary since for (H) accretion rates $\geq 1.5 \times 10^{-7}$ $M_\odot$/yr a giant envelope forms. In a close system this envelope will be lost immediately along L_2 and L_3. Only if the companion has a mass $M_2 <$ 1.7 $M_\odot$ will the accretion rate, according to eq. (2), be $< 1.5 \times 10^{-7}$ $M_\odot$/yr, and may the binary be close. However, the helium envelope (produced by H-shell burning) expands much during He-shell burning and is probably lost before the core has grown sufficiently to collapse. Hence, in close systems the mass of the O - Ne - Mg white dwarf must already be very close to the Chandrasekhar limit in order to enable collapse to a neutron star. Consequently, only a minor fraction of all O - Ne - Mg white dwarfs in *close* systems is expected to become a SN. No quantitative estimate of this fraction can presently be given.

5.2. Possible relation to the bulge X-ray sources

The amount of mass ejected in the collapse of a white dwarf is not expected exceed a few tenths of a solar mass, and hence is unlikely to cause disruption of the system (cf. Van den Heuvel 1977, and table VI). After the explosion the system will be detached and it may take billions of years before the companion again fills its Roche lobe (either due to gravitational radiation losses or due to its interior evolution), and becomes an X-ray source. If the companion has a mass of 1.0 - 1.2 $M_\odot$ it needs some (5-10) x 10^9 yrs to leave the main-sequence, such that these systems will be very old when they become X-ray sources. A system like Sco X-1 may have formed in this way (its orbital period of $0^d.78$ excludes gravitational radiation from driving the mass transfer). The failure to explain the very high X-ray luminosities (10^{37} - 10^{38} ergs/sec) of the bulge sources in the galaxy and in M31 by gravitational-radiation-induced mass transfer (Li et al. 1979; Ostriker and Zytkov 1979) suggests that also here interior evolution of the companion may play a role.

5.3. Speculations on the origin of binary radio pulsars with circular orbits

In wide binaries (which provide the best seat for accretion-induced collapse) the companion will itself be a giant (presumably with a degenerate core) when the accretion starts. In that case the white dwarf will be engulfed by the giant's envelope and will spiral down into it. Since the white dwarf may collapse at any instant during spiral-in, the resulting neutron-star binary may in principle have any period between a minute and many years. Although the SN will blow off part of the giant's envelope, the remaining part will soon again expand to giant size. Tidal and other friction will rapidly circularize the orbit and spiral-in may or may not resume (depending on the extent of the envelope). When the giant finally ejects its envelope and cools down to a white dwarf a detached star binary with a circular orbit remains. Possibly this was the evolutionary history of PSR 0820 and PSR 0656

(Blandford 1980) although the latter one may also be the product of the scenario depicted in figure 2, in which case its companion would be a main-sequence star. An alternative scenario for PSR 0820 is one which starts from a wide X-ray binary (like G X 1+4) in which the companion to the neutron star has a mass $\lesssim 10-12\ M_\odot$ (cf. §4.6).

6. CONCLUSIONS

(i) It seems likely that the majority of the radio pulsars were formed in binaries that went through a common envelope stage; this explains their high space velocities. Because of its overwhelming importance also among the lower-mass binaries, it seems therefore that common-envelope evolution merits to be a main focus point for future research.

(ii) The evolution of stars in the mass range 8-12 $M_\odot$ (and of helium stars in the range 2.6 to 4 $M_\odot$) should be further explored. Especially the work of Chechetkin et al. (1980) requires further investigation - and some of our conjectures about this mass range may still change considerably.

The reader may judge for himself from the above sections that also on many other points our understanding of the evolution of close binaries is still very poor, and that much further investigation is required.

		$P_o = 0.5^d$ (after spiral-in)		$P_o = 1^d.0$
M_{core}	M_2	$V_{SN} = 10^4$ km/sec	5.10^3 km/sec	$V_{SN} = 10^4$ km/sec
	$1.2\ M_\odot$	disrupted	$M_2' = 1.1\ M_\odot$ $e = 0.8$ $P_f = 4^d.6$ } $P_t = 23^h.6$ $V_{run} = 165$ km/sec	Bound (disrupted if $V_{SN} \geq 1.4 \times 10^4$ km/sec)
$2.5\ M_\odot$	$2.0\ M_\odot$	$M_2' = 1.69\ M_\odot$ $e = 0.92$ $P_f = 17.3d$ } $P_t = 25^h.9$ $V_{run} = 213$ km/sec	$M_2' = 1.87\ M_\odot$ $e = 0.65$ $P_f = 2.18d$ } $P_t = 23^h.2$ $V_{run} = 161$ km/sec	$M_2' = 1.89\ M_\odot$ $e = 0.64$ $P_f = 4^d.16$ } $P_t = 45^h$ $V_{run} = 126$ km/sec
	$4.0\ M_\odot$	$M_2' = 3.74\ M_\odot$ $e = 0.60$ $P_f = 1.56d$ } $P_t = 20^h.0$ $V_{run} = 196$ km/sec	$M_2' = 3.87\ M_\odot$ $e = 0.44$ $P_f = 1^d.05$ } $P_t = 18^h.9$ $V_{run} = 153$ km/sec	Bound
$3.0\ M_\odot$	$2.5\ M_\odot$	$M_2' = 2.14\ M_\odot$ disrupted	$M_2' = 2.32\ M_\odot$ $e = 0.76$ $P_f = 3.88d$ } $P_t = 25^h.9$ $V_{run} = 196$ km/sec	Bound (disrupted if $V_{SN} \geq 1.4 \times 10^4$ km/sec)
$4.0\ M_\odot$	$2.0\ M_\odot$	$M_2' = 1.44\ M_\odot$ disrupted	$M_2' = 1.74\ M_\odot$ disrupted	$M_2' = 1.78\ M_\odot$ disrupted

Table V. Effect of SN explosions in some representative post-spiral-in systems; V_{SN} is the ejection velocity of the SN shell; the assumed mass of the compact remnant is 1.4 $M_\odot$. P_f, P_t, etc. are defined in figure 2.

Assumed remnant mass	$M_2 = 1.2\ M_\odot$, $P_o = 9^h.2$	$M_2 = M_\odot$, $P_o = 7^h.33$
$M_1' = 1.3\ M_\odot$	$M_2' = 1.18\ M_\odot$, $e = 0.24$ $P_t = 10^h.13$, $V_{run} = 64$ km/sec	$M_2' = 0.99\ M_\odot$, $e = 0.25$ $P_t = 8^h.02$, $V_{run} = 65$ km/sec
$M_1' = 0.9\ M_\odot$	$M_2' = 1.08\ M_\odot$, $e = 0.74$ $P_t = 16^h$, $V_{run} = 174$ km/sec	$M_2' = 0.87\ M_\odot$, $e = 0.79$ $P_t = 13.7h$, $V_{run} = 183$ km/sec

Table VI. Examples of SN effects in two CV binaries with a collapsing O-Ne-Mg white dwarf of mass 1.4 $M_\odot$ (for $V_{SN} = 10^4$ km/sec).

REFERENCES:

Abt, H.A. and Levy, S.G. 1978, Astrophys. J. Suppl. *36*, 241.
Arnett, W.D. 1978, in: "Physics and Astrophysics of Neutron Stars and Black Holes" (R. Giacconi and R. Ruffini, eds.), North Holl. Publ. Comp., Amsterdam, p. 356.
Blandford, R. 1980, I.A.U. Symp. Nr. 95, "Pulsars" (W. Sieber and R. Wiebelinski, eds.), Reidel, Dordrecht (in press)
Bond, H.E.,Liller, W. and Mannery, E.J. 1978, Astrophys. J. *223*, 252.
Bradt, H.V.,Doxsey, R.E. and Jernigan, J.G. 1978, "Positions and Identifications of Gal. X-ray Sources", MIT preprint CSR-P-78-54, also in: "Advances in Space Exploration, vol. 3 (1979)".
Chechetkin, V.M.,Gershtein, S.S.,Imshennik, V.S., Ivanova, L.N. and Khlopov, M. Yu. 1980, Astrophys. Space Sc. *67*, 61.
Conti, P.S. 1978, Astron Astrophys, *63*, 225.
Cowley, A.P. 1980, Proc. NATO Adv. Study Inst. "Compact Galactic X-ray Sources," (P. Sanford, ed.) Cambridge Univ. Press (in press).
Crampton, D.,Cowley, A.P. and Hutchings, J.B. 1980, Astrophys. J. 235, L. 131.
Davidson, K. and Ostriker, J.P. 1973, Astrophys. J. *179*, 585.
De Cuyper, J.P. 1980, in: "Fundamental Problems in Stellar Evolution" (D. Sugimoto et al., eds.), Reidel, Dordrecht this volume, p. 184.
De Greve, J.P., De Loore, C. and van Dessel Ed. 1978, Astrophys. Sp. Sc. *53*, 105.
Delgado, A. 1979, in: "Mass Loss and Evol. of O-type stars" (P.S. Conti and C. De Loore, eds.), Reidel, Dordrecht, 415.
Delgado, A. and Thomas, H.-C. 1980 , preprint Max Planck Institut für Astrophys. Garching b. München.
De Loore, C. and De Greve, J.P. 1976, in: "Structure and Evolution of Close Binary Systems" (P. Eggleton et al., eds.), Reidel, Dordrecht, p. 193.
Flannery, B.P. and Ulrich, R.K. 1977, Astrophys. J. 212, 533.
Fowler, L.A. 1980, (private communication).
Fryxell, B.A. and Arnett, W.D. 1978, Bull. Am. Astron. Soc., *10*, 448.
Gursky, H. 1976, in: " Structure and Evolution of Close Binary Systems" (P. Eggleton et al. eds.), Reidel, Dordrecht. p. 19.
van den Heuvel, E.P.J. 1975, Astrophys. J. *198*, L. 109.
van den Heuvel, E.P.J. 1976, in: "Structure and Evol. of Close Binary systems(P.Eggleton et al.,eds.),Reidel,Dordecht,p.35.
van den Heuvel, E.P.J. 1977, Annals N.Y. Acad. Sciences *302*, 14.
van den Heuvel, E.P.J. 1978, in: " Physics and Astrophys. of Neutron Stars and Black Holes, R. Giacconi and R. Ruffini (editors), North. Holl. Publ. Co. Amsterdam, p. 828.
van den Heuvel, E.P.J. 1980, in "X-Ray Astronomy" (R. Giaconni and G. Setti, eds.) Reidel, Dordrecht, p. 115.

van den Heuvel, E.P.J. and Heise, J. 1972, Nature Phys. Sc. *239*, 67.
van den Heuvel, E.P.J. and De Loore, C. 1973, Astron. Astrophys. *25*, 387.
Hutchings, J.B., Crampton, D., and Cowley, A.P. 1978, Astrophys. J. *225*, 548.
Hutchings, J.B. 1980, in: " Proc. NATO Adv. Study Inst. on Compact Gal. X-ray Sources" (P. Sanford, ed.) Cambridge Univ. Press (in press).
Joss, P.C. and Rappaport S. 1979, Astron. Astrophys. 71, 217.
Kelley, R., Rappaport, S. and Petre, R. 1979, Astrophys. J. (in press).
Kippenhahn, R. and Meyer-Hofmeister, E. 1977, Astron. Astrophys. *54*, 539.
Kippenhahn, R. and Weigert, A. 1967, Zeits. Ap. 65, 251.
Lewin, W.H.G. and Clark, G.W. 1980, in: "Proc. 9th Texas Symp.on Relativ. Astrophys. Ann. New. Y. Acad. Sc. *336*, 451.
Li, F., Rappaport, S., Joss, P.C., McClintock, J.E. and Wright, E. 1980, Astrophys. J. (Lett.), in press.
Manchester, R.N., Newton, L.M., Cooke, D.J. and Lyne, A.G. 1980, Astrophys. J. *236*, L 25.
Massey, P. and Conti, P.S. 1980, Astrophys. J. (in press).
Mc Cray, R. 1979, (priv. communication).
Meyer, F.and Meyer-Hofmeister, E. 1979, Astron. Astrophys. *78*, 167.
Middleditch, J., Mason, K.O., Nelson, J. and White, N. 1980, preprint
Miyaji, S., Nomoto, K., Yokoi, K. and Sugimoto, D. 1980, Publ. Astron. Soc. Japan, 32, 303.
Nomoto, K. 1980, Proc. Workshop on "Type I supernovae", Univ. of Austin, Texas (in press).
Ostriker, J.P. 1976, in: IAU Symp. 73 "Structure and Evólution of Close Binary Systems" (P.Eggleton et al., eds.) Reidel, Dordrecht.
Ostriker, J.P. and Zytkov, A. 1980, In preparation.
Paczynski, B. 1971 a, Ann. Review Astron. Astrophys. 9, 183.
Paczynski, B. 1971 b, Acta Astronomica 21, 1.
Paczynski, B. 1976, in: IAU Symp. 73 "Structure and Evolution of Close Binary Systems" (P.Eggleton et al., eds.) Reidel, Dordrecht, p 75.
Paczynski, B. 1980, Acta Astronomica (in press).
van Paradijs, J. 1978, Nature *274*, 650.
van Paradijs, J. 1980, I.A.U. Circ. No. 3487 (19 june).
Plavec, M. 1968, Advances Astron. Astrophys. *6*, 201.
Ritter, H. 1976, Mon. Notices Roy. Astr. Soc. *175*, 279.
Ritter, H. 1980, ESO Messenger (in press), ESO, Garching b. München.
Savonije, G.J. 1978[1], Astron. Astrophys. *62*, 317.
Savonije, G.J. 1978[2], Ph. D. Thesis, Univ. of Amsterdam.
Savonije, G.J. 1979 , Astron. Astrophys. *71*, 352.
Schwartz, D.A. et al., 1979, Preprint Center for Astrophys. (Harvard Univ.).
Shakura, N.I. and Sunyaev, R.A. 1973, Astron Astrophys *24*, 337.
Smarr, L. L. and Blandford, R. 1976, Astrophys. J. 207, 574.
Sugimoto, D., and Nomoto, K. 1980, Space Sc. Reviews, *25*, 155.
Sutantyo, W. 1974, Astron. Astrophys. *31*, 339.
Sutantyo, W. 1975, Astron. Astrophys. *35*, 251.
Taam, R.E. 1979, Astrophys. Letters *20*, 29.

Taam, R.E. Bodenheimer, P. and Ostriker, J.P. 1979, Astrophys. J. *222*, 269.
Taylor, J.H. 1980, (priv. communication).
Taylor, J. H., Hulse, R.A., Fowler, L.A., Gullahorn, G.E. and Rankin, J.M. 1976, Astrophys. J. 206, L 53.
Thomas, H.-C. 1977, Annual Rev. Astron. Astrophys. *15*, 127.
Trimble, V. 1974, Astron. J. *79*, 967.
Tutukov, A.V. and Yungelson, L.R. 1973, Nautsnie Inform, 27, 58.
Tutukov, A.V. and Yungelson, L.R. 1980, "Close Binary Stars" (editors: M. Plavec et al.) Reidel, Dordrecht, p. 15.
Warner, B. 1974, Mon. N. Roy. Astr. Soc. *167*, 61 p.
Webbink, R.F., 1980, in: I.A.U. Colloq. Nr. 53, "White Dwarfs and Variable Degenerate Stars", Reidel, Dordrecht
Wheeler, J.C., Mc Kee, C.F. and Lecar, M. 1975 Astrophys. J. *200*, 145.
Whelan, J. and Iben, I 1973, Astrophys. J. 186, 1007.
Ziolkowski, J. 1976, in: " Structure and Evol. of Close Binary Systems" (P. Eggleton et al., eds.) Reidel, Dordrecht, p. 321.
Ziolkowski, J. 1977, Annuals N.Y. Acad. Sci. *302*, 47.
Ziolkowski, J. and Paczynski, B. 1980, Acta Astronomica (in press).

DISCUSSION

Sugimoto: I think there are too many ways in that very close binaries containing a compact object can be formed. We do not know how much specific angular momentum is carried away with escaping mass at the time of "shrinkage" of the binary separation or of "spiralling in" of the compact object. Once this is known, it will impose stringent restrictions on the evolution of such binary systems. For example, if we apply the results of calculations of single particle losses from the L_2 point, the separation shrinks so much in most cases that the stars to coalesce. I believe this issue is one of the fundamental problems of binary evolution and I would like to hear your opinion about it.

Van den Heuvel: I am afraid that, in reality, there is no case in which a single cause for "spiralling in" dominates. For example, tidal instability is always expected to occur, either before or after the common envelope is formed (cf. Meyer and Meyer-Hofmeister 1979). As soon as the common envelope forms, there will be a great deal of friction which will accelerate the "spiralling" etc. It would, of course, be nice if one could simply say "mass is lost and carries off a factor $(1+\beta)$ times the specific angular momentum of the system" and if β could then be calculated in some simple way. We could then just introduce the function β into the evolution program, and the calculation of "spiralling in" would be simple. Unfortunately there are, as far as I can see, no such simple ways. One will have to make a combined hydrodynamics plus evolutionary code which also allows for tidal effects, and do a complete calculation. The function β will then follow from such a calculation, but will not be an input parameter. Unfortunately, nature is not always as mathematically simple as one might wish.

Wheeler: Type I supernovae explode at a rate of about 10^{-2} per year in the Galaxy. If they all produce a neutron star in a low-mass binary system, a low-mass binary X-ray source should be created. I have heard it argued that 10^{-2} per year is much greater than the estimated production rate for low-mass X-ray sources. Can you comment on this?

Van den Heuvel: As you saw in my last table, if $\simeq 0.5\ M_\odot$ is ejected in the supernova explosion at a velocity $v_{ej} \sim 10^4$ km/sec, the orbital period (after tidal circularization) is almost doubled, and the system is very detached. If the non-degenerate star has a mass $\leq 1\ M_\odot$, nuclear evolution will not increase its radius to the radius of the Roche lobe for 10^{10} yrs. Also, the timescale for reducing the orbital period enough that the star fills its Roche-lobe is of the same order. Consequently, it may well take more than the age of the galaxy before such a system turns into a low-mass X-ray binary. (All of this depends sensitively, of course, on v_{ej} and on the amount of ejected mass; but for the above parameters, the situation is as I have described it.) Therefore, there may be millions of such detached systems which have not yet become X-ray sources, and the low-mass (bulge) X-ray sources may be just the tip of the iceberg. Under these circumstances, it is not possible to say anything sensible about the formation rate of these objects.

Schatzman: Is it not right that the mass interval within which a white dwarf can become a supernova is very small? In this case, the probability of finding such a white dwarf companion is very small, and would explain the low SN rate.

Van den Heuvel: According to the recent work of Nomoto (preprint 1980), white dwarfs of the required composition may be formed in a fairly broad mass range, between 1.2 $M_\odot$ and 1.4 $M_\odot$ (see also Sugimoto and Nomoto Space Sci. Rev. 25, 155, 1980). Accretion may then bring them to the Chandrasekhar limit. This implies progenitors roughly in the mass range 8-10 $M_\odot$, again a fairly broad region. If one would wish to identify such collapsing white dwarfs with Type I supernovae, one would (because of their frequency) indeed wish them to originate from a fairly broad mass range. Hence, such an identification might well be consistent with Nomoto's results.

Lamb: As you know, there is some controversy about whether Cyg X-2 is a neutron star (Cowley, Crampton, and Hutchings 1979) or a degenerate dwarf (Branduardi et al. 1980). Could you elaborate on your mention of an evolutionary scenario by which a neutron star in a binary system with a companion mass $M_\odot \cong 0.5\ M_\odot$ and a period $P \cong 10^d$ could form, and comment on any implications for the nature of Cyg X-2?

Van den Heuvel: There are indeed ways to obtain neutron star X-ray binaries with periods of the order of 10 days and a non-degenerate companion of $\sim 0.5\ M_\odot$. For example, following further the evolution of a system such as Hercules X-1 in the way outlined in my talk, one would rather straightforwardly attain the system parameters of Cygnus X-2. So, even if a low-mass X-ray binary is formed with $P \sim 1$ to 2 days, evolution may considerably change its binary period later on, and there are many possibilities for the final configuration. Of course, this does not at all rule out the possibility that the Cygnus X-2 system contains a white dwarf.

Tayler: I have been surprised not to hear black holes mentioned this afternoon. What is the current view about the probability of black hole formation as a result of close binary evolution?

Van den Heuvel: From the theoretical point of view, I have no clear opinion. Arnett's work on helium stars shows that up to a mass of some 60 $M_{\odot}$ such stars would always leave neutron stars of roughly a Chandrasekhar mass. Since such helium stars must have been cores of stars more massive than 80 to 100 $M_{\odot}$, one would need a very massive star in order to terminate with a black hole, and such stars are extremely rare. On the other hand, from the observational point of view, 16 of the 17 known massive X-ray binaries are either pulsing or have mass functions low enough to fit a neutron star. The same holds for the low-mass X-ray binaries and for the bulge sources. For the bursters among the latter sources, independent evidence strongly suggests that we are dealing with neutron stars, and the same holds for the globular cluster X-ray sources (a large fraction of which are also bursters). Therefore, after observing X-ray sources for nearly a decade with more than half a dozen X-ray satellites, we still have only one reasonably strong black hole candidate, namely Cygnus X-1. In my opinion, this seems to tell us that, if stellar-mass black holes exist at all, their formation in close binaries must be extremely rare.

THE RAPIDLY ROTATING GIANTS OF THE FK COMAE-TYPE

B. W. Bopp, Ritter Observatory, University of Toledo, USA
S. M. Rucinski, Warsaw University Observatory, Poland

We present new spectroscopic and photometric observations and discuss a small but important group of rapidly rotating G - K giants (FK Com=HD 117555, UZ Lib=BD-8°3999, HD 199178 possibly others) which are photometrically variable, show slightly redshifted and variable Ca II H and K emission and variable Hα emission, show rotationally broadened spectral lines (50-100 km/s), and yet show no large radial velocity variations. Possible explanations of the properties of these stars are discussed and coalescence of a W UMa binary system seems to be the most probable evolutionary state.
The most extreme of the group, FK Com is discussed in detail. Its type, G2IIIa (Keenan, priv.comm.) implying M_V=-1 (the IR indices are consistent with a giant) suggests a reduction of log g (relative to Sun) -1.5 to -2.5; this can be only marginally reconciled with the dimensions of about 5 R_0 resulting from the photometric period 2.4 d. and V_{rot} sin i about 100 km/s. The photometric variability (0.05 in I, 0.07 in R) is most probably due to spots asymmetrically distributed in longitude. The brightness minima observed by Chugainov in 1966 and 1974 and the new minimum observed in 1979 (JD 2443949.025) can be phased with one period 2.3995±0.0002 d. The erratic behavior of rotationally broadened spectral lines precludes obtaining a radial velocity orbit but an upper limit of 25 km/s for any periodic variations implies that the secondary must be a very low mass object. The strong Hα emission is variable in relative intensity of its double peaks but has a constant full width in excess of 20 A. The full half width at base implies rotational velocities of the order of 570 km/s; the half separation of peaks implies 300 km/s. Existence of an excretion disc of the type suggested by Webbink is possible.

DISCUSSION

Vilhu: How did you notice these 3 stars? Just by looking for broad-lined giants which are not in any binary-star catalogues?

Ruciński: These stars are relatively bright, and the unusual nature of

D. Sugimoto, D. Q. Lamb, and D. N. Schramm (eds.), Fundamental Problems in the Theory of Stellar Evolution, 177–178.

FK Com (then identified only by HD-number) has been suspected since Merrill described its remarkable spectrum in 1948. Our intention is to bring attention to the fact that two similar stars exist and together they might form a group of objects quite important for the theory of stellar evolution (coalesced binaries? products of common-envelope evolution?)

Chen: How did you determine the giant nature of FK Comae?

Ruciński: Its giant nature is inferred from its spectral classification (line-intensity ratios) and infrared colors, which extend to a few microns. Obviously, this way we can determine only the pair of quantities (T_e, log g) and not the three parameters (M, L, R).

Kippenhahn: What are the ratios of centrifugal to gravitational force for these objects?

Ruciński: We do not know the masses and we have only rough ideas about the radii of these objects. So the ratio of the two forces is unknown. It is entirely possible that the spectral features of giants are indeed caused by the decrease of the surface gravity due to fast rotation. For example, FK Com might be less bright and distant than its luminosity class would indicate. The spatial velocity inferred from its proper motion is about 320 km/s per one kpc of distance to the star, whereas its spectroscopic parallax places it at 0.3-0.6 kpc, making its galactic velocity somewhat too large for a normal population I object.

ROTATION AND THE EVOLUTION OF THE MASS-ACCRETING COMPONENT IN CLOSE BINARY SYSTEMS

Wim Packet
Astrofysisch Instituut, Vrije Universiteit Brussel,
Pleinlaan 2, B-1050 Brussels, Belgium

When mass is transferred between the two components in a close binary system, the matter falling towards the secondary star can gain a considerable amount of angular momentum. Eventually a ring or disk around this star is formed. The star can increase its angular momentum by accreting part (or the whole) of this matter. We have examined the ensuing changes in rotational velocity of the mass-accreting star. A simplified calculation (assuming accretion from a ring, rigid rotation of the star, and taking the stellar radius as well as its radius of gyration constant) shows that the star reaches its break-up velocity after increasing its original mass by only a few percent.

More detailed evolutionary calculations, using a spherically symmetric evolution code, have been performed for a number of theoretical systems. The variation of the radius and the radius of gyration can be taken into account in this way. These computations confirm the simplified calculations: the stars are spun-up to their break-up velocity soon after the beginning of mass transfer and before they grow into contact.

This result has a number of important consequences: spherically symmetric accretion models can no longer describe the accreting star satisfactorily; contact configurations may be avoided in a number of systems; mass loss from the system may occur everywhere in the orbital plane; tidal effects will probably be very important during the mass exchange process.

A more detailed account of this research will be published in Astronomy and Astrophysics.

This investigation is supported by the National Foundation of Collective Fundamental Research of Belgium (F.K.F.O.) under No.2.9002.76.

D. Sugimoto, D. Q. Lamb, and D. N. Schramm (eds.), Fundamental Problems in the Theory of Stellar Evolution, 179–180.

DISCUSSION

Sugimoto: Isn't break-up velocity the limiting value of rotational velocity which can be attained by accretion?

Packet: In our computations, the stars attained rotational velocities much higher than their break-up velocity. This effect, which is clearly unphysical, comes from two of our assumptions: 1) rigid rotation of the star and 2) accretion continuing undiminished when the star approaches its critical velocity. Both assumptions are not fulfilled almost certainly in nature. Relaxing them and assuming that the mass accretion rate goes to zero in some way as ω approaches ω_{cr} will give an asymptotic approach to the break-up velocity. However, the purpose of our work was not to spin stars up to super-break-up velocities, but to show that rotation is a most important feature in the evolution of mass-accreting stars and should be taken into account.

Ruciński: In Algol, a well known mass-transfar binary, the episodes of mass transfer can be rather well located in time and the primary (mass-gaining) component seems to be able to resume synchronous rotation (to within 1 km/s) on a time scale of weeks. Could you comment on this?

Packet: Algol is in the slow stage of mass transfer and the amount of matter exchanged during a mass-transfer event is very small. Thus the amount of angular momentum that the accreting component gains will also be small, and can be transfered back to the orbital motion in a short time. In binaries like U Cep, and probably β Lyr, the phenomenon is much larger, and will be still larger during the first stage of mass transfer as I have shown. In such cases, the resynchronization time-scale may be comparable to or longer than the mass-loss timescale of the primary.

CONTACT BINARY EVOLUTION AND ANGULAR MOMENTUM LOSS

Osmi Vilhu and Timo Rahunen
Observatory and Astrophysics Laboratory
University of Helsinki
SF-00130 Helsinki 13, Finland

Some fundamental problems connected with the evolution of W UMa stars are considered. While no generally accepted theory for the evolution of these systems exists, different scenarios lead to single stars on a nuclear or thermal time scale, or even to dwarf novae. The cycling and contact discontinuity models for zero age systems have gained much attention during the last few years. The contact discontinuity hypothesis has been heavily criticized on physical grounds, and the cycling at small mass ratios will probably be too violent leading to overcontact. On the other hand, there is increasing evidence of strong magnetic activity in short period solar type binaries, including W UMa stars (spots, flares, strong chromospheres and coronae etc.). This points to enhanced dynamo action inside rapidly rotating components of solar type close binaries. Extrapolating from single stars one finds that this may efficiently brake the orbital rotation. With an angular momentum loss rate of about 10^{43} g cm^2 s^{-1} per year corresponding to the thermal time scale of the secondary the scenario, where the angular momentum loss controls the zero age contact evolution, seems at least possible. This scenario needs an (hypothetical) equilibrium process between the degree of contact and magnetic activity, damping the angular momentum loss if the contact becomes too thick, so that marginal contact will be preserved. If the angular momentum loss time scale is longer (comparable to the nuclear time scale of the primary), the system is likely to evolve towards more extreme mass ratios and with less violent cycling. (The complete paper will be published elsewhere.)

DISCUSSION

Nariai: Do you take into account spin angular momentum?

Vilhu: Yes. At large values of the mass ratio $q = M_2/M_1$, its effect is negligible; but at very small values ($q < 0.2$), it is essential for very close binaries.

D. Sugimoto, D. Q. Lamb, and D. N. Schramm (eds.), Fundamental Problems in the Theory of Stellar Evolution, 181–182.

Roxburgh: I should perhaps caution stellar astronomers who wish to derive the properties of stellar winds from solar type or late type systems by extraporating from the solar wind. We do not know whether the solar corona is heated by acoustic waves or by Joule heating. We do not know how the strength of a dynamo generated by differential rotation depends on the rotational velocity. We do not know how the fraction of the surface of the sun (or star) that has magnetic field lines that open to the interstellar medium depends on the field strength. Our models of the solar wind are based on physical assumptions that are often invalid in the models so derived. At large radii, the solar wind becomes weakly collisional and we do not have an adequate theory of weakly collisional plasmas. We can not explain the observed properties of the solar wind at the earth. Since we can not yet understand and explain the mass loss from the sun, where we have detailed measurements, I must express my reservations about extrapolating from simple models of the solar wind to models of stellar winds.

Vilhu: I agree with you, Ian. I used the Wilson-Skumanich sun-Hyades-Pleiades Ω^3-relation and its extrapolation to high rotational velocities just as an order of magnitude estimate to see what is potentially possible. And this gives for a typical W UMa star $dJ/dt > 10^{44}$ g cm^2 s^{-1} per year. If the true value is much below this, say $\sim 10^{42}$-10^{43} g cm^2 s^{-1}, then it is likely to be one of the dominant factors determining the evolution of these close binaries. There is increasing evidence of period-magnetic activity relations for solar-type close binaries that is even stronger than for single stars (e.g. Ca K-line reversals, Stawikowski and Glebowski 1980), and UV surface fluxes from chromospheres and transition regions (Dupree et al. 1979). For both dynamo theories and evolution of very close binaries, it is important to check and to extend these relations.

NON-CONSERVATIVE EVOLUTION OF MASSIVE O-TYPE CLOSE BINARIES WITH GALACTIC AND WITH MAGELLANIC CLOUD CHEMICAL ABUNDANCES

D. Vanbeveren
Astrophysical Institute, Vrije Universiteit Brussel,
Pleinlaan 2, B-1050 Brussels, Belgium.

The general evolutionary pattern of massive O type close binaries evolving according to a case B mode of mass transfer, including mass loss by stellar wind prior to Roche lobe overflow (RLOF) at rates appropriate for O type stars, only marginally depends on the choice of the initial chemical composition whether the galactic or the MC abundances are used (the difference never exceeds 10%). The theoretical results are compared to the observations, O type binaries describing the evolutionary phase prior to RLOF, WR type binaries describing the helium burning phase after RLOF. The large mass loss by stellar wind in WR stars considerably affects the evolution during the latter phase. The comparison yields the following conclusions:
a) from the ZAMS up to the WR stage, 50%-60% of the initial primary mass is leaving the system corresponding to at least 70%-80% of the total mass lost by the primary due to stellar wind and RLOF;
b) during the WR phase the star is losing approximately half of its mass;
c) the average mass ratio for binaries prior to the supernova explosion equals 3, i.e. the exploding star is 3 times less massive than its companion.

DISCUSSION

Tutukov: Selection effects lead to an overestimation of the relative number of massive close binaries with low mass ratios. The reason is that most low mass-ratio binaries are single-line stars. The minimal semi-amplitude K_{min} of discoverable single-line binaries is several times lower than that for double-line binaries, according to an analysis of the observations. Did you take this selection effect into account when determining the distribution of unevolved binaries as a function of mass ratio?

Vanbeveren: We did not include this selection effect. If the number of O type binaries with mass ratio close to unity is really increased when the selection effect is taken into account, then in order to match observations of WR stars the amount of mass lost from the system during Roche lobe overflow has to be even larger than the 80% proposed in my talk.

D. Sugimoto, D. Q. Lamb, and D. N. Schramm (eds.), Fundamental Problems in the Theory of Stellar Evolution, 183.

ASYMMETRIC SUPERNOVA EXPLOSIONS : THE MISSING LINK BETWEEN WOLF-RAYET BINARIES, RUN-AWAY OB STARS AND PULSARS

Jean-Pierre De Cuyper
Astrofysisch Instituut, Vrije Universiteit Brussel, Belgium.

The WR binary systems, consisting of a WR star and an O or B star companion, are supposed to be the progenitors of the massive X-ray binaries. The missing link is generally accepted to be the SN explosion of the WR star which leaves a pulsar remnant. As most pulsars originate from single stars, observations of their proper motions indicate that they receive at their birth a "kick" velocity of about 100 km s^{-1}.
We assume this velocity to be due to the asymmetry of the SN explosion. This asymmetry, together with the loss of the SN shell and its impact on the OB star, may cause to disrupt the remaining system.
For the ten best known WR binaries we evaluated the survival probability P after an instantaneous SN explosion, leaving a 1.5 $M_\odot$ collapsar with a random orientated kick velocity of 75 kms^{-1} (case a) and 150 kms^{-1} (case b) respectively. The influence of the impact is found to be marginal. The run-away velocity of the remaining system $\bar{v}_g$ and of the disrupted OB star $\bar{v}^{\infty}_{OB}$ are comparable and of the same order of magnitude, but smaller than the initial orbital velocity of the OB companion; which decreases for increasing values of the initial orbital period. They are found to be independent of the kick velocity.

Systems with an initial period of less than a few weeks stay together for case a and have $.85 > P > .7$ for case b. Indicating that most of the high velocity OB stars ($v_g > 60$ kms^{-1} have a collapsed companion. For systems having an initial period of the order of months : $.9 > P > .7$ (case a) and $.6 > P > .4$ (case b). Hence single run-away OB stars are less nummerous and most of them have a low runaway velocity ($\bar{v}^{\infty}_{OB} < 40 kms^{-1}$). These results are in agreement with the observed bimodal run-away velocity distribution of OB stars and the observed properties of massive X-ray binaries. In case of disruption a single pulsar is formed with a runaway velocity up to 100 kms^{-1} (case a) and 200 kms^{-1} (case b) resp; independent of the initial orbital period. Few of them have negligable runaway velocity. High velocity pulsars (such as PSR0450-18 : $\bar{6}50 kms^{-1}$) are found to originate from systems consisting of a "single" WR star and an old pulsar companion, which are disrupted by the second SN explosion. In the opposite case a binary pulsar (such as PSR 1913+16) is formed. Here the survival probability is low depending also on the previous mass loss and the asymmetry of the SN explosion.

D. Sugimoto, D. Q. Lamb, and D. N. Schramm (eds.), Fundamental Problems in the Theory of Stellar Evolution, 184.

EVOLUTION OF LOW MASS BINARIES UNDER THE INFLUENCE OF GRAVITATIONAL RADIATION

A.G. Massevitch, A.V. Tutukov, and L.R. Yungelson
Astronomical Council, USSR Academy of Sciences

Evolutionary changes of masses and periods under the influence of gravitational radiation (GR) are computed for binaries with main-sequence or degenerate hydrogen-helium (H), helium (He) and carbon (C) secondaries. Tracks in the P-M_2 and P-$\dot{M}_2$ planes are determined. The orbital period of systems P with a non-degenerate dwarf filling its Roche lobe decreases until the time-scale of GR ($\sim 10^9$-10^{10}yrs) becomes shorter than the thermal time-scale τ_{KH} of the mass losing component M_2. When M_2 becomes $\sim 0.1\ M_\odot$, $\dot{P}$ begins to increase. This could account for the existence of the minimal $P \simeq 1.3^h$ and for the accumulation of observed cataclysmic binaries (CB) below $P \leq 2^h$. The GR can be the driving force for evolution of cataclysmic binaries: TT Ari, OY Car, Z Cha, and WZ Sge as their $\dot{P}$, M_2 and $\dot{M}_2$ indicate. The rate of mass exchange driven by GR is $\sim 10^{-9}$- 10^{-10} $M_\odot$/yrs. It is enough to feed the X-ray bursters and related objects.

In a binary with a degenerate secondary and initial $M_2/M_1 \gtrsim 0.6$ the secondary may transform into a disk around the primary during several orbital periods, liberating the energy of $\sim 10^{50}$ ergs. It appears that close binaries with both components degenerate may be the most powerful source of background GR near the Earth: $F \simeq 10^{-5} - 10^{-6}$ erg/cm^2/sec.

The secondaries of CB have deep convective envelopes. Stellar winds are inherent in such stars, which may lead to a substantial angular momentum loss through magnetic braking. If the time-scale of the angular momentum loss is $10^{-9} - 10^{-10}$ yrs, the resulting evolution of CB will be similar to that dominated by GR.

DISCUSSION

Van den Heuvel: Can you explain the very high X-ray luminosities of the ~20 X-ray sources in the galactic bulge, which have $L_X > 10^{37}$ erg/sec, requiring 10^{-9} $M_\odot$/yr mass transfer? To my recollection, gravitational radiation-driven mass transfer from a red dwarf companion can never give

D. Sugimoto, D. Q. Lamb, and D. N. Schramm (eds.), Fundamental Problems in the Theory of Stellar Evolution, 185–186.

more than a few times 10^{-10} $M_\odot$/yr.

Tutukov: The mass exchange rate strongly depends on the assumed mass-radius relation. But up to now theory and observation can not provide us with reliable information about it. Our models show that a mass exchange rate in the range 10^{-9} - 10^{-10} $M_\odot yr^{-1}$ is possible. To me, it seems possible that a black hole can sometimes be one component of such a system. It is possible also that a magnetic stellar wind might sometimes provide the mass exchange.

Van den Heuvel: I would like to point out that Ostriker and Żytkov, and Rappaport and Joss are independently doing similar calculations and are obtaining similar results, namely when the companion mass becomes $\lesssim 0.1$ $M_\odot$ the system expands again.

Tutukov: That is natural because of increasing interest in low mass binaries in the last few years.

Sato: What is the final state of a system evolving due to the emission of gravitational radiation? Is it the collapse or the destruction of the binary system?

Tutukov: Systems initially having main-sequence components and evolving due to gravitational radiation finish their evolution as binaries with component masses ~ 0.03 $M_\odot$ + ~ 1 $M_\odot$ and a period $\sim 1^h.5$. Close binaries with degenerate components and having an initial mass ratio $q \gtrsim 0.6$ transform into single stars surrounded by a heavy disk in a few orbital timescales. The same systems, but having $q \lesssim 0.6$, remain binaries with component masses $\sim 3 \times 10^{-3}$ $M_\odot$ + $\sim M_\odot$ (see Figure 1 in my review talk).

BLUE STRAGGLERS AS LONG-LIVED STARS

J. Craig Wheeler
Department of Astronomy, University of Texas at Austin
Hideyuki Saio
Joint Institute for Laboratory Astrophysics
Michel Breger
Department of Astronomy, University of Texas at Austin

The existence of blue stragglers in old open clusters with apparent mass more than twice the mass of the turnoff argues against simple binary mass transfer as the mechanism of their origin. The excess of blue stragglers to the red of the termination of the core hydrogen burning main sequence suggests that blue stragglers are not evolving normally. Stellar evolution models invoking mixing in an extended core region can account for the distribution of blue stragglers in the H-R diagram. Such models live longer, brightening and evolving further to the red before core hydrogen exhaustion than do normal stars. The distribution of blue stragglers in NGC 7789 is consistent with a range of mixed core mass fraction ~30-90 per cent and a narrow range in mass ~1.7-2.1 $M_\odot$. Such evolution will result in a class of helium rich stars which have lived longer than normal and whose total mass exceeds the Chandrasekhar limit.

DISCUSSION

Vilhu: Your mass of ~2$M_\odot$ is about the same as that found for W UMa-stars of the W-type. There exists one scenario (Webbink) according to which these contact binaries may evolve by gradual mixing of the primary, while the system evolves towards a more extreme mass ratio.

Wheeler: Such coalescence may be relevant to blue stragglers, but I would expect it to produce a basically normal star. Once the mixing ceases, the star should evolve normally and it is then subject to the constraint that its luminosity should not be greater than that of a normally evolving star with twice the turnoff mass, and it should evolve very rapidly after central hydrogen exhaustion. As I have said, some blue stragglers in NGC7789 violate both of these restrictions. Bear in mind that the ~2$M_\odot$ I derived here followed from continuous mixing. A single mixing episode followed by inhomogeneous evolution is a different case, and one which apparently disagrees with observation.

Sugimoto: If the mixing is common and if the amount of the mixing depends

D. Sugimoto, D. Q. Lamb, and D. N. Schramm (eds.), Fundamental Problems in the Theory of Stellar Evolution, 187–189.

sensitively on various parameters, why aren't all stars blue stragglers?

Wheeler: All I can say is that, if the present hypothesis of evolution with a large mixing core of fixed mass is relevant, then results say that only a narrow mass range is involved. Any theory which accounts for this mixing should explain the narrow mass range or should give an alternative mass distribution which accounts self-consistently for the observations.

Mouschovias: As a non-expert on the subject, I understand that the evolution of a star is determined by a small number of physical parameters. Can any expert in this room point to anything funny happening to stars with mass $2M_{\odot}$? If so, can that possible physical effect cause the required mixing?

Wheeler: There is a whole zoo of possible mixing mechanisms. Depending on the circumstances, rotation, magnetic fields, and composition gradients can either help or hinder mixing. The difficulties in starting from first principles to study mixing encouraged me to take the alternate approach of searching for an empirical justification for mixing and then looking for hints of the physical process. If the narrow mass range we have derived is valid, it is a very important clue. For instance, Press has suggested that an outer convective zone could generate inwardly propagating sonic waves whose energy could perturb the central stellar regions. Press was interested in the solar neutrino problem but, for my purpose, I can imagine that as one goes up in mass a narrow range is selected by the competition between the growing inner CNO burning core and the decreasing outer convective envelope such that the inner region which mixes is amplified. Alternatively, Ian Roxburgh reminds me that the ^{3}He instability discussed by Gough et al. and others could play a role. Quantitatively, either of these very different physical processes would, a priori, be expected to operate more efficiently at a somewhat lower stellar mass than we have formally derived here.

Schatzman: What do we know about the rotational velocity of blue stragglers?

Wheeler: Little. Blue stragglers should be systematically restudied with modern instruments. Several open cluster blue stragglers are apparently slow rotators. Alternatively, Rogers has reported some relatively rapid rotators among field blue stragglers, which are selected by their blue colors and low metallicity.

Tayler: Is there a significant difference between your minimum blue-straggler mass and the turnoff mass? If not, continuing in your completely ad hoc manner, could we not suggest that your stars have mixed shortly after reaching the turnoff so that lower mass stars have not yet had time to become blue stragglers?

Wheeler: I will not try to defend the difference between the minimum blue straggler mass and the turnoff mass. I do think the average blue straggler mass we have derived is significantly larger than the turnoff. I am not certain whether the spread I showed is real. If stars of

$M \sim 1.5M_{\odot}$ underwent mixing just after turnoff, their subsequent evolution would depend on whether the extensive mixing continued or whether "standard" helium-rich central hydrogen burning ensued. If extensive mixing continued, the result would be similar to the present models but the timescale would be changed a little. A given value of X_c would be reached somewhat sooner. The present results should then follow and they are apparently inconsistent with $M \sim 1.5M_{\odot}$. If the mixing is transient and the star begins central hydrogen burning with a higher helium mass fraction in an ordinary small convective core, it will evolve fairly quickly but more or less normally off the main sequence. I would guess the result would not reproduce the observed distribution of the blue stragglers satisfactorily. If such stars mix again upon central hydrogen exhaustion, and continue to do so repeatedly, the outcome is difficult to intuit. Each time the star regains the main sequence, it will be brighter. Perhaps such repeated mixing at central hydrogen exhaustion would reproduce the observed blue stragglers. I can not say.

GENERALIZED THEORY OF SHELL FLASH AND ACCRETING WHITE DWARFS*

Daiichiro Sugimoto and Shigeki Miyaji
College of General Education, University of Tokyo
Tokyo 153, Japan

* This paper was presented by D. Sugimoto

ABSTRACT

Shell flashes take place both in deep interior of red giant stars and near surface of accreting white dwarfs. Theories of shell flashes have been thus far presented piece by piece in different papers. It is the purpose of the present review to construct and generalize them in order to reach better understanding. A non-linear yet almost analytical theory is presented which treats the development of the shell flash in finite amplitude. Recurrence of the shell flashes is also shown to be well understood as a non-linear oscillation in dissipative system which tends to be its limit cycle. As a result strength of the peak energy-generation and recurrence time of the shell flashes are related with mass of the accreting white dwarfs, accretion rate, etc.

1. INTRODUCTION

Schwarzschild and Härm (1965) discovered first that helium burning in a thin shell is unstable even when electrons are non-degenerate therein. They have advanced a linearized theory to obtain a stability criterion against the shell burning (Schwarzschild and Härm 1965). Soon thereafter, Weigert (1966) and Rose (1966) showed numerically that the instability results in recurrent thermal pulses.

About the same time Hayashi, Hōshi and Sugimoto (1965) constructed a model of less massive star in which a helium shell flash begins in an electron degenerate helium zone surrounding a carbon-oxygen core. By extending the model into a shell flash of *finite* amplitude, Sugimoto and Yamamoto (1966) showed the followings; (1) the shell burning is unstable even after the electron-degeneracy has been lifted, (2) effect of radiation pressure is essential in quenching the unstable nuclear shell-burning, and (3) a strong convection appears in the helium zone and it might reach the bottom of the hydrogen-rich envelope to trigger mixing between the helium zone and the envelope. However, they did not properly recognize that their shell flash was a different manifestation of the same instablity that Schwarzschild and Härm (1965) had met with.

D. Sugimoto, D. Q. Lamb, and D. N. Schramm (eds.), Fundamental Problems in the Theory of Stellar Evolution, 191–206.

Now it is known that Sugimoto and Yamamoto's (1966) equilibrium model overestimated the degeneracy in the helium zone and that the helium shell flash would have actually begun at somewhat lower density. Nevertheless, general understanding of the shell flash of *finite* amplitude has already been obtained in their study. Later on such theory was somewhat more advanced by Hayakawa and Sugimoto (1968) and was applied for nova-like explosion in which the hydrogen shell-burning was saturated by beta decays in CNO cycles.

After Schwarzschild and Härm (1965) many numerical computations are done. Indeed the shell flash, which is sometimes called flicker, thermal pulse, relaxation oscillation etc., has now a wide variety of applications. Helium shell-flash takes place deep in the interior of red giant stars, which is discussed in relation with the origin of carbon stars and the *s*-process nucleosynthesis (Sackmann 1980 and papers referred therein). Hydrogen shell-flash takes place near the surface of accreting white dwarfs, which is related with nova explosion. Such theory of nova explosion was greatly advanced by constructing detailed models (Nariai, Nomoto and Sugimoto 1980 and papers referred therein), by computing detailed nuclear reaction networks, and by comparing them with observed novae in detail (Gallagher and Starrfield 1978 and papers referred therein). Relatively strong helium shell-flash takes place in accreting helium or carbon-oxygen white dwarf which is related also with explosive phenomena (Taam 1980 and papers referred therein). The shell-flash near the surface of neutron stars is discussed in relation with X-ray bursters (Joss 1980 and papers referred therein).

Despite such wide applications and many numerical computations the physics involved in the shell flash of *finite* amplitude has not come to wide understanding. (One can cite many wrong statements concerning the shell flash if one wishes.) Of course, it is not only for the shell flash, but it is a widespread tendency in the theory of stellar *structure* and evolution. We are not saying about physical processes taking place locally under the condition at given temperature and density, which are rather well understood. On the contrary the stellar structure is non-linear and non-local in the sense that physical situations in other shells of the star affect those in a specific shell. Nevertheless we can advance almost analytical theories relatively easily in the case of the shell flash, which have been published piece by piece in different papers. This is the reason why such theories have not been understood even among the specialists. Therefore, it seems useful to summarize them. When they are posed with concepts of thermodynamics of dissipative open system, they will be clearly understood as will be shown in the following sections.

2. RELATIONS BETWEEN SHELL FLASHES IN RED GIANT STARS AND THOSE IN ACCRETING STARS

The shell flash was first noticed to occur in the bottom of the helium zone, which surrounds the carbon-oxygen core and which is immersed deep in the hydrogen-rich envelope. However, the pressure $\underline{P}$ at the bottom of the hydrogen-rich envelope (subscript H) is much lower than the pressure at the bottom of the helium zone (subscript He); for example, P_H/P_{He} is typically as small as 3×10^{-3}. Therefore, the existence of the hydrogen-rich envelope has practically nothing to do with the stability and further development of the helium shell-flash, except for possible mixing between the envelope and the helium zone.

Therefore the situation can be discussed in more generalized context, in which helium is being accreted onto a carbon-oxygen white dwarf or hydrogen is being accreted onto a helium or carbon-oxygen white dwarf. Models of accreting white dwarfs are specified by three parameters, i.e., the mass of the white dwarf M_1, the accretion rate dM/dt, and a parameter specifying thermal status of the core. The last parameter is typically represented by the intrinsic luminosity $L^{(0)}$ of the white dwarf just before onset of the accretion, or by the time elapsed between the formation of the white dwarf and the onset of accretion (Sugimoto et al. 1979; Fujimoto and Sugimoto 1979a).

In the case of red giant stars, these parameters are reduced into a single parameter, the core mass. The mass of white dwarf should be equal to the core mass. The accretion rate corresponds to the growth rate of the helium zone which is determined by the energy generation rate due to the hydrogen shell-burning L_H by

$$dM_1/dt = L_H/X_e E_H \ . \tag{1}$$

Here X_e is the concentration of hydrogen in the envelope, E_H is the energy release from unit mass of hydrogen, and L_H is determined by the core-mass to luminosity relation (Paczyński 1970) which is shown in Figure 1. The third parameter is also specified by the evolutionary history of the star which is also parametrized by the core mass. Therefore, the shell flash in red giant stars can be treated as a special case of the generalized shell flash in accreting stars. In Figure 1 parameters, which were applied in existing model computations, are plotted both for hydrogen and helium shell flashes.

Interesting cases of the hydrogen shell-flash are concerned only with accreting stars. They can also be discussed in the same framework as in the helium shell-flash, since the mechanism and essential features are common. Differences between them lie only in the local physics, i.e., nuclear reaction rate, equation of state and opacity.

3. THERMAL HISTORY OF ACCRETION AND ITS EFFECT ON THE DEVELOPMENT OF THE FLASH

When gas is accreting, the accretion rate is limited by the Eddington's critical accretion rate which is given by

$$(dM/dt)_{cr} = 4\pi c R_1/\kappa \ . \tag{2}$$

Here κ is the opacity and R_1 is the radius of the (corresponding) white dwarf. The critical accretion rate is also shown in Figure 1.

If gas is accreting at a rate close or higher than the critical accretion, the gas will be just piled up on the stellar surface. Then, the stellar radius will increase and the value of the critical accretion rate will also increase as seen in equation (2) until the latter exceeds the given rate of accretion. Though not exactly for the supercritical accretion, such sequences of events were computed for accreting main-sequence stars (Kippenhahn and Meyer-Hofmeister 1977; Neo et al. 1977). [In many papers such increase in stellar radius is described as an expansion, but in actual a Lagrange shell is contracting and newly added shells are just being piled up (Neo et al. 1977).]

For accreting white dwarf Nomoto, Nariai and Sugimoto (1979) computed a case of rapid accretion at the rate that is just equal to the critical accretion rate for its initial radius. This accretion rate is higher than the mean rate at which the hydrogen is processed into helium according to equation (1). Therefore the envelope is accumulating and the star evolves into a red giant.

For lower accretion rates the hydrogen/helium zone grows in time. The bottom of this zone is gradually compressed which tends to increase the temperature thereof. On the other hand heat diffuses out of this zone which tends to lower the temperature. As the accumulated mass ΔM becomes larger, the former effect weighs over the latter (see e.g. Nomoto and Sugimoto 1977), and at last the nuclear burning ignites when a certain mass ΔM_{ig} has been accumulated.

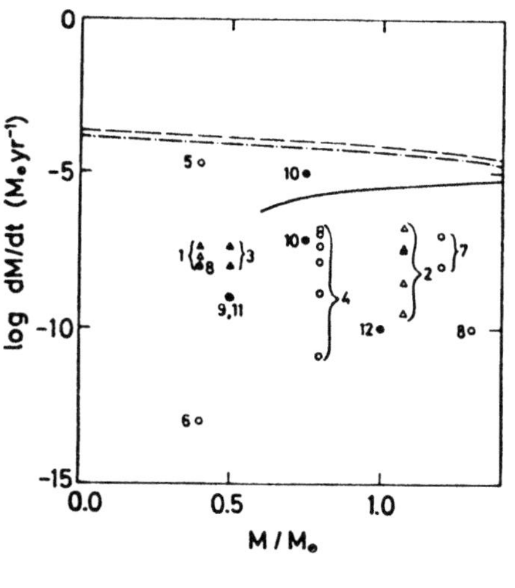

Fig. 1. Masses and accretion rates for some of existing computations. Triangles and circles are the helium- and hydrogen-shell flashes, respectively. Filled marks indicate that the computations were stopped at a stage preceding the peak of the flash. Also shown are the accretion rates corresponding to the core-mass to luminosity relation (solid curve), and the critical accretion rates for hydrogen-rich gas (dash-dotted) and for helium gas (dashed). References are as follows; 1. Nomoto and Sugimoto (1977), 2. Fujimoto and Sugimoto (1979b), 3. Taam (1980), 4. Paczyński and Żytkow (1978), 5. Nomoto et al. (1979), 6. Sugimoto et al. (1979), 7. Sion et al. (1979), 8. Nariai et al. (1980), 9. Giannone and Weigert (1967), 10. Rose (1968), 11. Redkoborodyi (1972), 12. Taam and Faulkner (1975).

The timescale of the compression is inversely proportional to the accretion rate, while the timescale of the heat diffusion depends mainly on ΔM and is relatively insensitive to the accretion rate. Therefore, ΔM_{ig} is smaller when the accretion rate is higher (Sugimoto et al. 1979). Its actual value can be calculated by taking account of thermal history of accretion. The temperature rise by compression is sometimes called *heating* erroneously, but the entropy of the mass element is decreasing in actual. Anyhow, such competition between the compression and the heat diffusion (and also neutrino loss) are concerned not only with the accreting white dwarf but also with various models; the growth of the stellar cores leading to helium flash in the center, to carbon-deflagration supernova, and to electron capture supernova, and also the accreting helium or carbon-oxygen white dwarf leading to strong flash in a shell or to helium detonation (see Sugimoto and Nomoto 1980 for more details).

In the case of the accreting white dwarf such thermal history of accretion was computed in the earliest work (Giannone and Weigert 1967) and in some of the later works (Taam and Faulkner 1975, Taam 1978). However, in such works the succeeding phase of the shell flash was computed only in the relatively early stages preceding the peak of the energy generation. In most of the later works, in which nova explosion was extensively discussed, the shell flash was computed through the peak (see, e.g., Gallagher and Starrfield 1978 and papers referred therein). However, the preceding thermal history of accretion was not computed in obtaining their initial model. Instead the mass ΔM_{ig} was arbitrarily assumed and the initial temperature distribution in the hydrogen-rich envelope was assumed to be in thermal equilibrium which corresponds to an infinitely slow accretion (see, e.g., Truran et al. 1977).

Recently the whole process of the shell flash has been computed starting from the onset of accretion through the peak of the flash upto the stage of envelope expansion including the transition phases from quasistatic accretion through dynamical stages (Nariai et al. 1980). Such computations are important in two respects. First, ΔM_{ig} can be determined only by means of such computation of accretion (see, e.g., Sugimoto et al. 1979). Second, the recurrence of the flash can be discussed only when the whole process of accretion through flash is computed (Paczyński and Żytkow 1978, Sion et al. 1979).

We have discussed that the value of ΔM_{ig} is determined by the three parameters of the problem, i.e., M_1, dM/dt and $L^{(0)}$, of which the latter two are particularly important. However, further development of the shell flash and its strength, in particular, are almost exactly determined only by the values of M_1 and ΔM_{ig} as discussed by Sugimoto et al. (1979). In other words, once ΔM_{ig} is properly chosen, the preceding thermal history of accretion is unimportant any more in determining the further development of the shell flash in finite amplitude. In this sense computations with even an arbitrarily assumed value of ΔM_{ig} will be a good approximation except for initial stages of the flash and for the recurrence of the flash, if and only if appropriate value of ΔM_{ig} is chosen.

4. GRAVOTHERMAL SPECIFIC HEAT AND GENERALIZED STABILITY CRITERIA

Schwarzschild and Härm (1965) were the first to develop the linearized stability theory for nuclear shell burning. It compares the change in the nuclear reaction rate with the corresponding change in the diffusion, both of which are induced by perturbation in specific entropy δs, for example, upon model in thermal equilibrium. Here the thermal equilibrium implies that the nuclear energy generation rate ε_n just balances the rate of the heat diffusion dL_r/dM_r. In order to compute the change in ε_n, i.e. $\delta\varepsilon_n$, the temperature change $\delta \ln T$ or the pressure change $\delta \ln P$ in the burning shell should be computed. This is accomplished by solving linearized equations of stellar structure in terms of Green's function, in principle.

Schwarzschild and Härm (1965) discussed also approximate nature of the Green's function and its relation with the notion of *thin* shell. However, such discussions were not fine enough in the sense that the notion of the thin shell was not formulated quantitatively. In the bottom of the thin helium zone, helium shell-burning is unstable. In the bottom of the hydrogen-rich envelope which was accreted on a white dwarf, the hydrogen shell-burning is unstable also. On the contrary, in the deep interior of the red giant stars, the hydrogen shell-burning is stable, though it is much thinner in mass than in the case of the accreted envelope. Here, the thinness of the shell was formulated as $\Delta M(H_p)/M_r << 1$, i.e., mass contained in unit scale hight of pressure $\Delta M(H_p)$ is much smaller than mass contained interior to its shell M_r. However, the important parameter for the instability is the thinness in radius, i.e., $H_p/r << 1$ as will be shown below in this section.

After Schwarzschild and Härm (1965), some works were done improving the approximations (Hōshi 1968; Unno 1970; Dennis 1971) but within essentially the same framework. On the other hand, Sackmann (1977) analyzed detailed numerical computations, and tried to extract some *empirical* relations for the shell flash of finite amplitude. Sugimoto and Fujimoto (1978) elaborated the non-linear theory for the shell flash of finite amplitude to well analytical and accurate one. Here we will discuss the shell flash mainly along their theory but in more generalized fashion by adding some new developments.

The hydrostatic equilibrium of stellar structure can be computed when the relation between the pressure $\underline{P}$ and the density ρ is given as in the case of polytropes for instance. Alternatively, when the distribution of the specific entropy $s(M_r)$ is given, we can compute the hydrostatic equilibrium of the star irrespectively of its thermal state. Therefore, the theory of stability criterion is easily extended to include the case *out of thermal equilibrium*, i.e., the case of finite amplitude.

The energy equations are given by

$$dL_r/dM_r = \varepsilon_n - \varepsilon_\nu + \varepsilon_g \ , \qquad (3)$$

$$\varepsilon_g \equiv -(T\, ds/dt + \sum_k \tilde{\mu}_k\, dN_k/dt). \tag{4}$$

Here, ε_ν is the neutrino loss rate, ε_g is the so-called gravitational energy release or, more exactly, the rate of heat energy flowing out of an element of unit mass, and $\tilde{\mu}_k$ and N_k are the chemical potential of a particle and the number of particles of the k-th kind which are contained in unit mass of matter. In what follows we will take account of neither ε_ν nor $\tilde{\mu}_k dN_k$ for brevity; they can be easily included when necessary. We call the case to be in thermal equilibrium when ε_g vanishes.

When a perturbation in specific entropy $\delta s(M_r)$ is applied, the temperature changes as much as

$$\delta \ln T = (1/c_g^*)\,\delta s\ , \tag{5}$$

$$\frac{1}{c_g^*} \equiv \frac{1}{c_p} + \left(\frac{\partial \ln T}{\partial \ln P}\right)_s \frac{d \ln P}{ds} \tag{6}$$

Here c_p is the usual thermodynamic specific heat, and $d \ln P/ds$ is the change of pressure $\delta \ln P(M_r)$ divided by the change in the specific entropy $\delta s(M_r)$ of the shell at M_r. The change $\delta \ln P(M_r)$ is called hydrostatic readjustment and is affected not only by δs in the same shell but also by those in other shells. Mathematically, such effect is described by means of Green's function as

$$\delta \ln P(M_r) = \int_0^M G(M_r, M_r')\,\delta s(M_r')\,dM_r'\ . \tag{7}$$

Schwarzschild and Härm's (1965) discussion was essentially the same as obtaining the Green's function. Later on it was explicitly formulated by Henyey and Ulrich (1972) in relation with the shell flash and by Hachisu and Sugimoto (1978) in relation with gravothermal catastrophe of selfgravitating system. Off-diagonal part of $G(M_r, M_r')$ represents the effect of the hydrostatic readjustment throughout the star. Therefore c_g^* defined in equation (6) will be called *gravothermal specific heat*.

In practical problems the change $T\delta s$ in the time interval δt comes from the nuclear energy generation and the heat diffusion, i.e.,

$$T\delta s \equiv (T\delta s)_0 + (T\delta s)_1\ , \tag{8}$$

$$(T\delta s)_0 = [\varepsilon_n^{(0)} - dL_r^{(0)}/dM_r]\,\delta t, \tag{9}$$

$$(T\delta s)_1 = [\delta\varepsilon_n - d\delta L_r/dM_r]\,\delta t\ , \tag{10}$$

where the zero-th order term $(T\delta s)_0$ does or does not vanish according to the thermal equilibrium or non-equilibrium, respectively. The first order term $(T\delta s)_1$ can be written down in terms of $\delta \ln P$, $\delta \ln r$, $\delta \ln T$, $d(\delta \ln T)/dr$ and derivatives of ε_n and κ.

Table 1. GENERALIZED STABILITY CRITERIA AND SEQUENCE OF EVENTS ALONG THE SHELL FLASH.

Phase/ Stage	stability	c_g^{*-1}	$(T\delta s)_0$	$(T\delta s)_1$	Region in Figure 2	Y		
1	unstable	+	0	+	I	~1	onset	main pulse
2	unstable	+	+		I	~1	growing	
3	neutral	0	+		I/II	~1	peak	
4	stable	-	+		II	~1	decaying	
5	stable	-	0	+	II		depletion of fuel	
6	neutral	-	0	0	II/III	~0	stagnation	
7	unstable	-	0	-	III	~0	onset	sub-pulse
8	unstable	-	-		III	~0	growing	
9	neutral	0	-		III/IV	~0	peak	
10	stable	+	-		IV	~0	decaying	
11	stable	+	0	-	IV		steady accretion	
12	neutral	+	0	0	IV/I	~1	stagnation	

When c_g^{*-1} and $T\delta s$ have the same sign, the shell burning is thermally unstable, i.e., the temperature of the burning shell is increasing as a result of perturbation [when $(T\delta s)_0 = 0$] or as a result of the *existing* deviation from thermal equilibrium [when $(T\delta s)_0 \neq 0$]. There are twelve cases for the signs of c_g^{*-1} and $T\delta s$. They are summarized in Table 1 along the sequence of events which will be discussed in detail in the next section. Usually only the cases with $(T\delta s)_0 = 0$ are referred to in the stability criterion of the linearized theory.

As seen in equations (8)-(10) the sign of $T\delta s$ is determined as a result of competition between the nuclear reaction and the heat diffusion. It was discussed by many authors and is easily understood. More important is the notion of c_g^{*-1}. We may ask what characteristics of the stellar structure determines the sign of c_g^{*-1}. It is determined by the sign and the magnitude of $d\ln P/ds$, i.e., by the effect of the hydrostatic readjustment. Thus it depends on the geometry in the hydrostatic equilibrium and on the equation of state.

For the plane-parallel configuration the pressure at a shell is determined only by the weight of the overlying layers. In other words it does not depend on the value of the entropy. Therefore, $d\ln P/ds$ should vanish, and c_g^{*-1} is always positive as seen in equation (6). On the contrary $d\ln P/ds$ is negative at the center of the spherical star. However, its magnitude depends on the equation of state. For ideal gas plus radiation pressure its absolute value is large enough to make c_g^{*-1}

negative. When electrons are degenerate, its absolute value is so small that c_g^{*-1} is positive [see, e.g., Sugimoto and Nomoto (1980) for more detail].

In the case of the burning shell which we are discussing, the configuration is intermediate between the plane-parallel and the spherical symmetry. Its degree will be parametrized by means of the ratio of the radius of the spherical shell to the pressure scale height, i.e.,

$$V \equiv r/H_p = -d\ln P/d\ln r = GM_r\rho/rP \ . \qquad (11)$$

The plane-parallel configuration is obtained in the limit of $V \to \infty$ and the spherical one corresponds to $V \to 4$ [see Sugimoto and Nomoto (1980) for reasoning].

Sugimoto and Fujimoto (1978) have shown that c_g^{*-1} is expressed analytically as a function of V_1 and the polytropic index N_1 of the burning shell (denoted by the subscript 1) when the mass contained in the overlying layers ΔM is small, i.e., when $\Delta M/M_r << 1$. It is to be noticed that the results depend only slightly upon the polytropic index in the overlying layers. In doing so they solved the stellar structure equation analytically for the envelope of $\Delta M/M_r << 1$. Its solution gave also a relation among ΔM_1, V_1, r_1 and P_1. The value of r_1, i.e., the radius of the core is well approximated by that of zero-temperature white-dwarf with mass equal to the core mass $M_r = M_1$. Therefore, if the density of the shell ρ_1 *or* the specific entropy of the shell s_1 is given, all physical quantities are known including the value of c_g^{*-1}. Along the evolutionary change s_1 is increasing when the nuclear energy generation is dominant over the heat diffusion. Then the layers overlying the burning shell expand to make H_p larger. Therefore the evolutionary sequence can be followed by decreasing the value of V_1.

5. LIMIT CYCLE OF NON-LINEAR OSCILLATION IN DISSIPATIVE SYSTEM

Sequence of events and recurrence of shell flashes are clearly understood in a plane such as Figure 2a, where the change of the temperature in the burning shell is schematically shown against its value of V_1. Numbers attached to the curve indicate specific phases or stages summarized in Table 1. According to the signs of c_g^{*-1} and $T\delta s$ the plane is divided into four regions, I - IV. Since the sign of $T\delta s$ depends on the concentration of the nuclear fuel $\underline{Y}$, the boundary between the regions shifts as a result of depletion of the nuclear fuel. Two cases of different values of $\underline{Y}$ are shown in Figure 2a. Directions of evolution are also shown by arrows.

Several remarks should be given. In Phase 5 the nuclear shell-burning is in stable equilibrium. Therefore, the burning shell should stay at the corresponding point unless there would be any change in physical parameters. In actual, helium is being depleted and the curve of $T\delta s = 0$ shifts toward the upper-right. From Stage 6 on, the diffusion is dominant over the nuclear energy generation, and the point moves *unstably* through

Phase 8. After Stage 9 the shell is cooled down, and the accretion brings the point back to the initial stage of the histeresis.

In Figure 2b a histeresis curve is taken from a model, which is repeating thermal pulses of steady-state amplitude in the deep interior of a red giant star (Fujimoto and Sugimoto 1979a). In such a model a surface convection zone in the hydrogen-rich envelope penetrates into the helium zone during Phases 7-9. It reduces the mass of the helium zone and shifts the histeresis curve towards larger value of V_1 so that it even crosses the curve in Phase 2. During Phase 11 the mass of the helium zone turns to increase again by the hydrogen shell-burning, which makes V_1 smaller.

For clarity of discussion it is better to turn back to the idealized case of Figure 2a which would be obtained if the recurrence of the

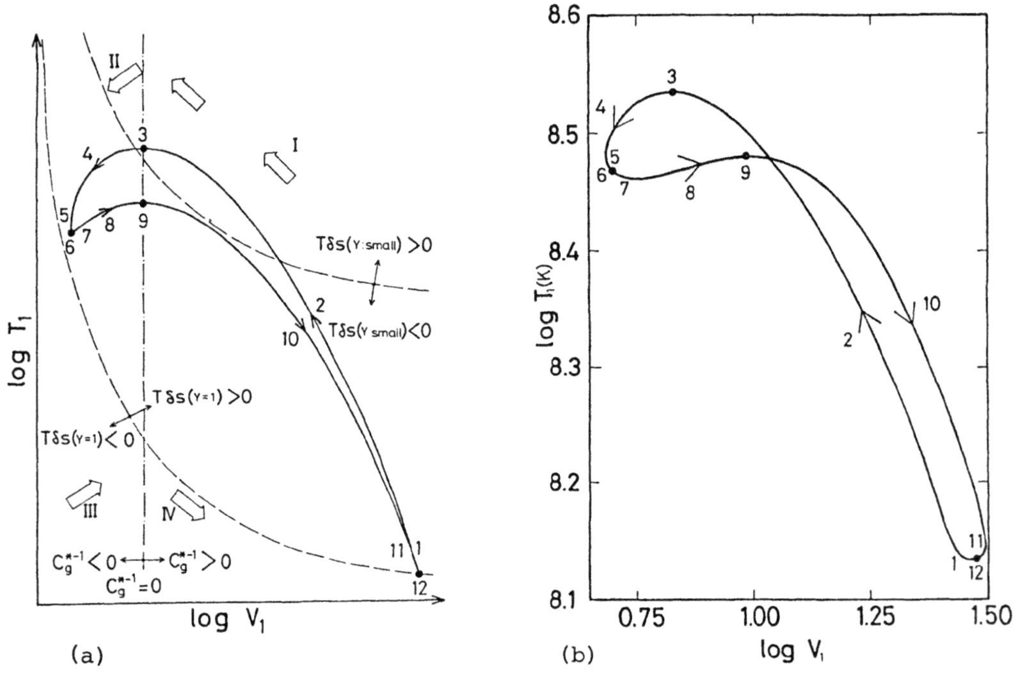

Fig. 2. Evolutionary change of the temperature in the burning shell is drawn by solid curve against V_1 thereof, which will be called histeresis curve. See the text for more details.
(a) Schematic diagram: Dash-dotted line corresponds to $c_g^{*-1} = 0$. Dashed curves correspond to $T\delta s = 0$ for $Y = 1.0$ and for small $\underline{Y}$ as indicated in the figure. These line and curve divide the plane into four regions I-IV. The boundary between the regions I/II and III/IV shifts up as nuclear fuel $\underline{Y}$ is depleted. Note that the histeresis curve encloses the point where both c_g^{*-1} and $T\delta s$ for small $\underline{Y}$ vanish simultaneously. Numbers attached along the histeresis curve are the phase/stage number in Table 1.
(b) Histeresis curve obtained for a model of helium shell-flash in red giant stars.

hydrogen shell-flash be computed for accreting white dwarf for instance. For relatively small $\underline{Y}$, the singular point, where both c_g^{*-1} and $T\delta s$ vanish, lies inside the histeresis curve. This is an unstable singular point. Therefore the recurrent shell flash meets with the necessary condition for a limit cycle of non-linear oscillation in dissipative system. Actually Fujimoto and Sugimoto (1979a) obtained two types of solutions. When the initial model is taken inside (outside) of the histeresis curve of the steady state, the peak of the shell flash becomes stronger (weaker), i.e., the histeresis curve is diverging (converging), as it recurs. Then it tends to the limit cycle that they called *steady-state solution.*

6. TWO BRANCHES IN THE SOLUTIONS OF STELLAR STRUCTURE

As seen in Table 1 and Figure 2a, there are two typical stages which are in thermal equilibrium. They are Stage/Phase 1 and 5 which are in unstable and stable equilibrium, respectively. This implies that there are two solutions for given values of the envelope mass ΔM_1 and the core mass M_1. We can consider a linear series of thermal equilibrium solutions with the total mass $M = M_1 + \Delta M_1$ fixed as illustrated in Figure 3. The parameter of the series is either M_1 or ΔM_1. The two equilibrium solutions cited above imply that there are two branches in the linear series, which we shall call white-dwarf branch corresponding to Stage/Phase 1 and red-giant branch corresponding to 5. (Such situation has been known long since in relation with evolution off the red-giant branch of the globular cluster into a hot white dwarf, which takes place when almost all of hydrogen in the envelope is exhausted.) In this context the shell flash is understood to be a phenomenon in which the star makes a transition from the white-dwarf branch to the red-giant branch by removing the condition of thermal equilibrium (see Figure 3). Then, the effects of cooling and of accretion bring about a complete transition back to the white-dwarf branch through the interpulse phase (Phases 6-12).

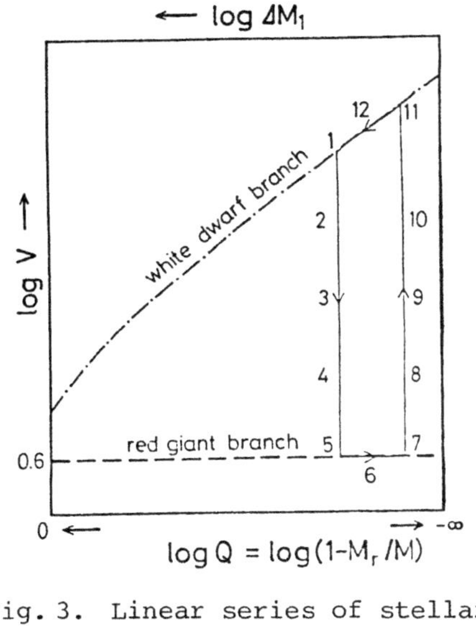

Fig. 3. Linear series of stellar structure in which an electron-degenerate core is surrounded by a relatively thin envelope. There are two branches of solutions as indicated in the figure. A cycle of thermal pulse is superposed by solid curve. See the text for more details.

Even for practical purpose such understanding is important as will be discussed below. In the white-dwarf branch the configuration in the envelope is almost plane-parallel with large values of $\underline{V}$. In other words the pressure at the bottom of the envelope is close to the weight per unit area of the overlying layers, i.e., $f \simeq 1$, in the expression

$$P = f(GM_1/r_1^2)(\Delta M/4\pi r_1^2). \quad (12)$$

Therefore c_g^* is positive and the shell burning is unstable.

On the contrary the red-giant branch is characterized by $f << 1$. It can be stated in another expression: In the bottom of the envelope, i.e., in the burning shell, mass contained in unit scale height of pressure $\Delta M(H_p)$ is much smaller than the mass contained in the whole envelope, since equation (12) is rewritten as

$$\Delta M(H_p) = -(dM_r/d\ln P)_1 = f\Delta M. \quad (13)$$

Therefore, the pressure at the bottom of the envelope is much lower than the weight of the overlying cone. The configuration is in spherical symmetry rather than plane-parallel and thus c_g^* is negative and the shell burning is stable.

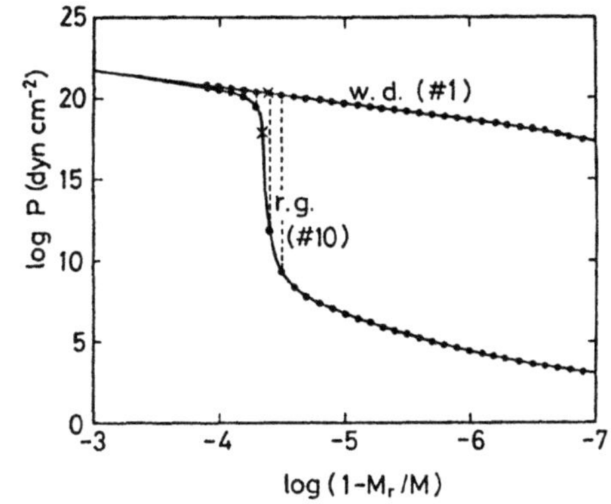

Fig. 4. Pressure distributions in solutions of white-dwarf branch (#1) and of red-giant branch (#10). Note that there is a sharp drop in pressure in the red-giant solution. See the text for more details.

How small is the value of f in Phases 5-7? It depends upon the height of the peak in the shell flash. In a typical helium shell-flash depicted in Figure 2b (Fujimoto and Sugimoto 1979a) the peak temperature reaches $T = 3.3 \times 10^8$ K (Stage 3) and f decreases down to 0.30 (Stage 5). In a nova model of accreting white dwarf of 1.3 $M_\odot$ with normal abundance of the CNO elements (Nariai et al. 1980), the peak temperature becomes as high as $T = 3.5 \times 10^8$ K even for the *hydrogen* shell-burning and f becomes as small as 0.0055. In order to compute such stellar structure we need mesh points 180 times finer than for a white-dwarf model.

It is instructive to illustrate solutions of the nova model discussed above. Their pressure distributions are shown in Figure 4 for solutions in the white-dwarf branch (#1) and in the red-giant branch (#10). As for the latter we see that the pressure drops very sharply in the region just outside the hydrogen-burning shell. This was computed by using mesh pints with the separation even as fine as $\Delta\ln(1-M_r/M) = 0.001$ near the hydrogen-burning shell. In another typical nova model, Sparks et al. (1978) used mesh points with $\Delta\ln(1-M_r/M) = 0.223$. (In other works mesh points are not described.) Their mesh points are superposed on Figure 4. As seen in this figure they are fine enough for models near the white-dwarf branch, while they would be too coarse for models in the red-giant branch if such a solution were the case.

Here we should give two warnings to those computing models near the red-giant branch. First, the fine mesh points are necessary, in particular, near the bottom of the envelope. Second, if one has used a relatively coarse mesh points and obtained a relatively slow decrease in pressure, it does not always mean that the accurate solution is actually of the type having such a slow decrease in pressure. Such nature of the red-giant envelope was relatively well known before the computer age, but it has been almost forgotten and now proper attention is not being paid for.

7. STEADY STATE SOLUTION AND RECURRENCE OF THE FLASH

In numerical computations of thermal pulses, the pulse height grows or decays pulse by pulse. However, the discussion in section 5 suggests that the pulse height corresponding to the limit cycle will have a limiting amplitude. The change in the pulse height results from a change in the thermal structure (distribution of temperature) near the outer edge of the core. Therefore the limiting amplitude will result from the model which are thermally well relaxed before the onset of the flash.

It is convenient to introduce a variable of mass fraction

$$q \equiv M_r/M(t) \ , \tag{14}$$

where the total mass $M(t)$ increases in time at the accretion rate. Then every quantity can be rewritten as a function of (q, t) instead of (M_r, t). In particular, equation (4) is divided into homologous term $\varepsilon_g^{(h)}$ and non-homologous term $\varepsilon_g^{(nh)}$, i.e.,

$$\varepsilon_g = \varepsilon_g^{(h)} + \varepsilon_g^{(nh)} \ , \tag{15}$$

$$\varepsilon_g^{(h)} \equiv \frac{d \ln M(t)}{dt} T\left(\frac{\partial s}{\partial \ln q}\right)_t , \quad \varepsilon_g^{(nh)} \equiv -T\left(\frac{\partial s}{\partial t}\right)_q . \tag{16}$$

Before the onset of the flash there is a long quiescent phase of accretion. In this phase the heat loss by diffusion dL_r/dM_r in equation (3) is balanced by $\varepsilon_g^{(h)}$, i.e., by the addition of entropy due to the inward propagation of a Lagrangian shell in q-coordinate. [Though $\varepsilon_g^{(nh)}$ is essential during the flash, the energy released by the flash diffuses out in a relatively early stages of the accretion phase.]

If we neglect $\varepsilon_g^{(nh)}$, the differential equations of stellar structure do not contain the time coordinate explicitly any more. Therefore they are reduced to ordinary differential equations. When the accretion rate and the envelope mass ΔM_1 are specified, they are solved as a boundary value problem. We obtain a higher temperature at the bottom of the envelope for a larger ΔM_1. For $\Delta M_1 = \Delta M_{ig}$ the temperature is just high enough to ignite unstable nuclear shell burning. As a result the value of ΔM_{ig} is obtained for each set of the accretion rate and the core mass.

Fujimoto and Sugimoto (1979a) computed such steady-state solution for the helium shell-flash in the deep interior of the red giant star and showed also that such steady-state solution yields the limit cycle of thermal pulses as discussed in section 5. Nariai and Nomoto (1979) computed such steady-state solution extensively for accreting helium white dwarfs. The recurrence period of the flash is obtained by

$$\tau_{rec} = \Delta M_{ig}/(dM/dt) \ . \tag{17}$$

As seen in Figure 1 of their paper ΔM_{ig} is smaller and thus τ_{rec} is still shorter for higher accretion rate. The smaller ΔM_{ig} yields the weaker flash (Sugimoto et al. 1979). Thus such model explains the tendency existing between ordinary novae (with very long τ_{rec}) and recurrent novae.

It is appropriate to note here about the role of mass flow in the (q, t)-coordinate for the limit cycle. If we describe equations (9) and (10) at constant $\underline{q}$ instead of constant M_r, then $\varepsilon_g^{(h)}$ and $\delta\varepsilon_g^{(h)}$ are to be included, respectively, in the right hand sides of these equations. Then $(T\,\delta s)_{q,0} = 0$ becomes the steady state with $\varepsilon_g^{(nh)} = 0$ instead of the thermal equilibrium state with $\varepsilon_g = 0$. Though it will be somewhat more complicated, such a formulation will lead to a more precise description of the limit cycle, especially for stages ii through 1. However, it is out of the scope of this brief review.

8. GENERAL FEATURE OF SHELL FLASH OF ACCRETING WHITE DWARFS

From the discussions in the preceding sections we can make the following statements concerning the shell flashes. We shall summarize here only those which have well analytical basis. In some papers do appear results of numerical computations and statements which are in contradiction with our statements. In such cases numerical computations and/or their interpretations should be examined more closely.

1) Mass contained in the envelope ΔM_{ig} is determined by the thermal history of accretion, i.e., by the mass of the white dwarf M_1, the accretion rate dM/dt, and the thermal structure of the white dwarf at the onset of the accretion.

2) ΔM_{ig} is larger for smaller M_1, for smaller dM/dt and for more cooled white dwarf in advance of the onset of the accretion.

3) After many cycles of recurrent shell flashes, it reaches a limit cycle of non-linear oscillation in dissipative system. In such a limit ΔM_{ig} is determined only by M_1 and dM/dt.

4) When the flash grows up to a finite amplitude, heat diffusion becomes negligible. Then further development of the flash is described only by the amount of nuclear energy thus far released. Therefore, the peak temperature, for instance, is determined only by the values of ΔM_{ig} and M_1. (Detailed expression of the nuclear reaction and its absolute value determine only the timescale to reach the peak temperature.) The peak temperature is higher for larger M_1 and for larger ΔM_{ig} because the pressure at the bottom of the envelope is higher.

In the statements above any dynamical effects have not been taken into account though expected in some cases of strongest flashes.

The authors would like to thank Professor K. Nariai, Dr. K. Nomoto and Dr. M.Y. Fujimoto for extensive discussions. This work was supported in part by Scientific Research Fund of the Ministry of Education, Science and Culture (564088).

REFERENCES

Dennis, T.R.: Astrophys. J., 167, 311.
Fujimoto, M.Y. and Sugimoto, D.: 1979a, Publ. Astron. Soc. Japan, 31, 1.
Fujimoto, M.Y. and Sugimoto, D.: 1979b, "White Dwarfs and Variable Degenerate Stars", eds. H.M. Van Horn and V. Weidemann, p. 285.
Gallagher, J.S. and Starrfield, S.: 1978, Ann. Rev. Astron. Astrophys., 16, 171.
Giannone, P. and Weigert, A.: 1967, Z. Astrophys., 67, 41.
Hachisu, I. and Sugimoto, D.: 1978, Prog. Theor. Phys., 60, 123.
Hayakawa, S. and Sugimoto D.: 1968, Astrophys. Space Sci., 1, 216.
Hayashi, C., Hōshi, R., and Sugimoto, D.: 1965, Prog. Theor. Phys., 34, 885.
Henyey, L. and Ulrich, R.K.: 1972, Astrophys. J., 173, 109.
Hōshi, R.: 1968, Prog. Theor. Phys., 39, 957.
Joss, P.C.: Paper presented in this symposium.
Kippenhahn, R. and Meyer-Hofmeister, E.: 1977, Astron. Astrophys., 54, 539.
Nariai, K. and Nomoto, K.: 1979, "White Dwarfs and Variable Degenerate Stars", eds. H.M. Van Horn and V. Weidemann, p. 525.
Nariai, K., Nomoto, K., and Sugimoto, D.: 1980, Publ. Astron. Soc. Japan, 32, 473.
Neo, S., Miyaji, S., Nomoto, K., and Sugimoto, D.: 1977, Publ. Astron. Soc. Japan, 29, 249.
Nomoto, K., Nariai, K., and Sugimoto, D.: 1979, "White Dwarfs and Variable Degenerate Stars", eds. H.M. Van Horn and V. Weidmann, p. 529.
Nomoto, K. and Sugimoto, D.: 1977, Publ. Astron. Soc. Japan, 29, 765.
Paczyński, B.: 1970, Acta Astron., 20, 47.
Paczyński, B. and Żytkow, A.N.: 1978, Astrophys. J., 222, 604.
Redkoborodyi, Y.N.: 1972, Astrofizika, 8, 261.
Rose, W.K.: 1966, Astrophys. J., 145, 496.
Rose, W.K.: 1968, Astrophys. J., 152, 245.
Sackmann, I.-J.: 1977, Astrophys. J., 212, 159.
Sackmann, I.-J.: 1980, Astrophys. J., 235, 554.
Schwarzschild, M. and Härm, R.: 1965, Astrophys. J., 142, 855.
Sion, E.M., Acierno, M.J., and Tomcsyk, S.: 1979, Astrophys. J., 230, 832.
Sparks, W.M., Starrfield, S., and Truran, J.W.: 1978, Astrophys. J., 220, 1063.
Sugimoto, D. and Fujimoto, M.Y.: 1978, Publ. Astron. Soc. Japan, 30, 467.
Sugimoto, D., Fujimoto, M.Y., Nariai, K., and Nomoto, K.: 1979, "White Dwarfs and Variable Degenerate Stars", eds. H.M. Van Horn and V. Weidemann, p. 280.
Sugimoto, D. and Nomoto, K.: 1980, Space Sci. Rev., 25, 155.
Sugimoto, D. and Yamamoto, Y.: 1966, Prog. Theor. Phys., 36, 17.
Taam, R.E.: 1978, Astrophys. Letters, 19, 47.
Taam, R.E.: 1980, Astrophys. J., 237, 142.
Taam, R.E. and Faulkner, J.: 1975, Astrophys. J., 198, 435.

Truran, J.W., Starrfield, S.G., Strittmatter, P.A., Wyatt, S.P., and Sparks, W.M.: 1977, Astrophys. J., 211, 539.
Unno, W.: 1970, Publ. Astron. Soc. Japan, 22, 239.
Weigert, A.: 1966, Z. Astrophys., 64, 395.

DISCUSSION

Kippenhahn: If I remember the history correctly, Härm & Schwarzschild first found the instability numerically and explained it as a thermal runaway. It was Weigert who first found that it is a recurrent phenomenon. In these thermal pulses, convective regions come and go and change the chemical composition by mixing. From the theory which you just described, one gets a very good insight into the mechanism which drives the pulses but the theory does take into account the convective mixing. For practical application, one would like to know how the thermal pulses influence the evolution. One can do that by computing through thousands of thermal pulses. Is there some hope that analytical work could help us to predict the cumulative effects of many pulses (including the accompanying mixing effects) without following the evolution using a computer pulse by pulse?

Sugimoto: Your description of the history is right. Concerning the effects of convection, there are two points. During the stages where the convection zone is confined within the helium layer, for example, there is no additional problem other than just specifying the polytropic index to be the adiabatic one. You are talking rather of the effect of mixing between the helium layer and the hydrogen-rich envelope. It may take place in two different modes, depending on the conditions. Firstly, convection in the helium zone penetrates up into the hydrogen-rich envelope just after the peak of the flash. Secondly, the surface convection zone penetrates down into the helium layer in the early part of the interpulse accretion phase. When each takes place, the mass of the helium layer changes and the histeresis curve changes somewhat, as illustrated in the text. When we compute such things numerically for the limit cycle, however, we obtain the same amount of mixing for each consecutive pulse. Therefore, we can easily compute the behavior which should result after thousands of pulses. We can estimate the influence of the pulse without following the evolution pulse by pulse using a computer.

Wheeler: Have you considered the effect on the shell-flash problem of the difference between the Ledoux and the Schwarzschild criteria for convective instability? I believe Barkat's group in Jerusalem has found that if the Ledoux criterion is applied, the hydrogen shell ignites in the convective envelope, not in a radiative zone, and there are no shell flashes at all.

Sugimoto: As I have discussed, hydrogen shell-burning is always stable for the red-giant structure because of the spherical geometry and the negative gravothermal specific heat. The stability of helium shell-burning should have almost nothing to do with the structure of the hydrogen-rich envelope and thus almost nothing to do with the stability criteria for convective instability. The solutions obtained by Barkat's group deserve further investigation. If their numerical results and, in particular, their U-V curves are shown, we can diagnose at once the reason for the behavior they have found.

THERMONUCLEAR PROCESSES ON ACCRETING NEUTRON STARS

Paul C. Joss
Department of Physics
and
Center for Theoretical Physics
Massachusetts Institute of Technology

I. INTRODUCTION

The observed properties of X-ray burst sources have recently been reviewed by Lewin and Clark (1980) and Lewin and Joss (1980). About thirty-five such sources are presently known, and they have a spatial distribution reminiscent of stellar Population II (see Figure 1). The salient features of these sources include burst rise times of $\lesssim$1s, decay time scales of $\sim$3-100 s, peak luminosities of $\sim 10^{39}$ ergs per burst, spectra that can generally be well fitted by blackbody emission from a surface with a constant effective radius of $\sim$10 km and a peak temperature of $\sim 3 \times 10^7$ K, and "tails" of softer X-ray emission that may persist for several minutes after the burst maximum. Profiles of bursts from some typical burst sources are shown in Figure 2. The intervals between bursts from a given source may be regular or erratic and are typically in the range of $\sim 10^4$-10^5 s; many sources undergo burst-inactive phases that can last for weeks or months. Most burst sources are also sources of persistent X-ray emission, and the ratio of average persistent luminosity to time-averaged burst luminosity is typically $\sim 10^2$ during burst-active phases. (The properties of the "Rapid Burster," MXB1730-335, are different from those of all other known burst sources and will be discussed separately in §VI below.) There are few correlations among the burst flux, burst intervals, and persistent X-ray flux from any given source, and the detailed burst shapes vary from one source to another and often vary with time in a given source.

A theoretical model for X-ray burst sources ideally should account for all these observational properties. Nearly all models that have been proposed so far assume that the bursting phenomenon involves accretion of matter onto a collapsed object (degenerate dwarf, neutron star, or black hole). In all cases, the collapsed object serves at least one of two functions: its deep gravitational potential well allows the release of large amounts of gravitational energy in the form of X-radiation by the accreting matter, and its small dimensions permit the X-ray emission to be released on the short time scale of an X-ray burst.

D. Sugimoto, D. Q. Lamb, and D. N. Schramm (eds.), Fundamental Problems in the Theory of Stellar Evolution, 207–227.

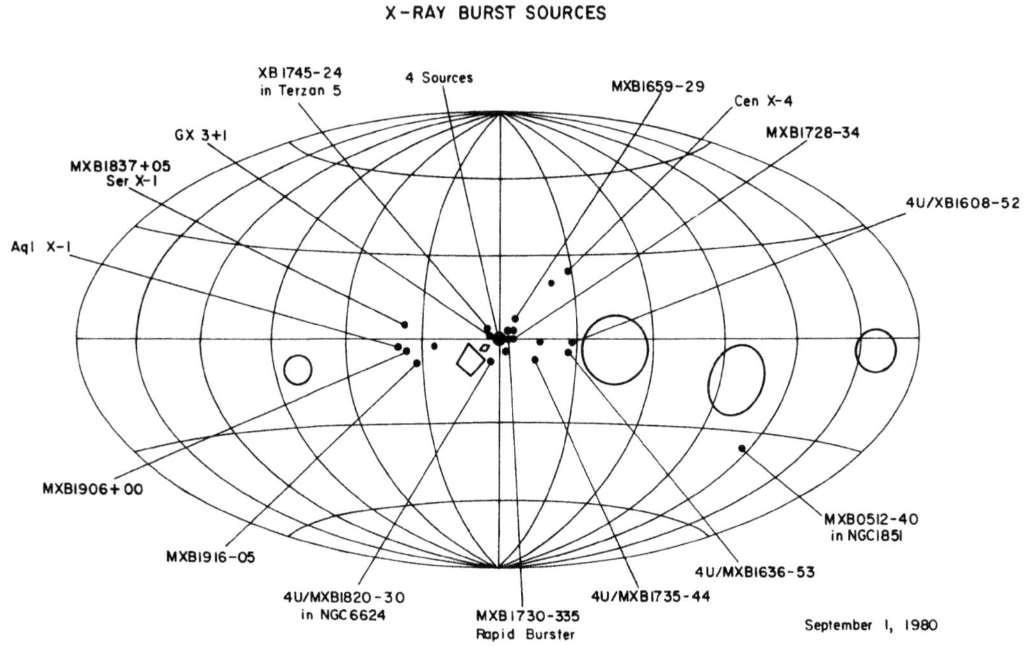

Figure 1. Map of 31 X-ray burst sources with accurately known positions as of September 1, 1980, in galactic coordinates (from Lewin and Joss 1980). The concentration of these sources in the direction of the galactic center, together with the association of several of them with globular clusters, strongly suggests an identification with an old stellar population (see Lewin and Joss 1980).

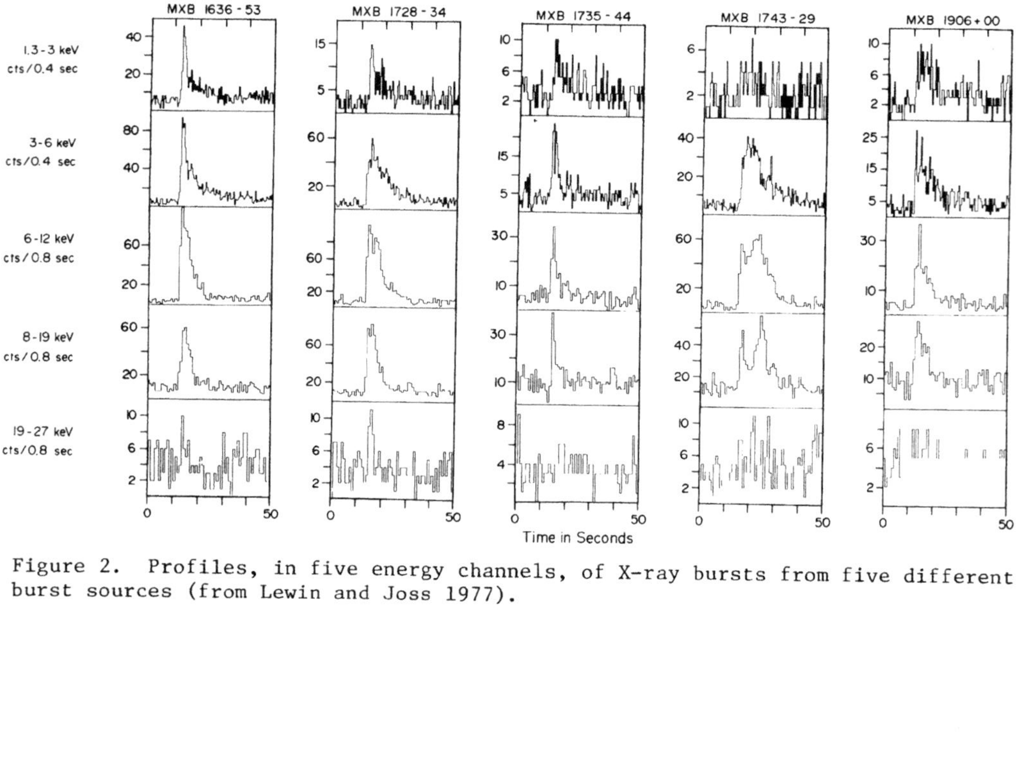

Figure 2. Profiles, in five energy channels, of X-ray bursts from five different burst sources (from Lewin and Joss 1977).

The proposed models can be broken down into two broad classes: (1) those invoking instabilities in the accretion flow onto a collapsed object, and (2) those that invoke thermonuclear flashes in the surface layers of an accreting neutron star. Models in class (1) were reviewed by Lamb and Lamb (1977). Although a few new models in this class have been proposed in the past several years, relatively little progress has been achieved in understanding the complex hydrodynamics and radiative transfer that all such models entail (see, however, Cowie, Ostriker, and Stark [1979] for a discussion of physical constraints on many of these models). During the past few years, the greatest amount of progress has instead been achieved in developing the thermonuclear flash model, which has proved to be amenable to detailed numerical computations. As will be documented below, these calculations have been remarkably successful in accounting for the general properties of the bursts from most X-ray burst sources. Moreover, the theoretical work to date strongly suggests that the characteristics of X-ray bursts should be capable of imparting substantial constraints upon the properties (masses, radii, internal temperatures, and so forth) of the underlying neutron stars.

II. HISTORICAL DEVELOPMENT

When X-ray pulsars were discovered in 1971 (Giacconi et al. 1971; Lewin, McClintock, and Ricker 1971; Schreier et al. 1972; Tananbaum et al. 1972), it was immediately recognized that these objects were probably neutron stars that were undergoing accretion from binary stellar companions. Soon thereafter, in 1973, Rosenbluth et al. (1973) pointed out that nuclear fusion in the surface layers would be an independent source of energy that might be radiated from the neutron-star photosphere. A few years later, in 1975, Hansen and Van Horn (1975) demonstrated that over a wide variety of conditions, the nuclear burning ought to be unstable and should lead to thermonuclear flashes.

Hansen and Van Horn (1975) noted that the energy released in such flashes might produce variable X-ray emission from the neutron star. However, they also discovered that the characteristic time scale for thermonuclear runaway was usually $\lesssim 1$ s. This was considerably shorter than most of the time scales of variability from X-ray sources that were then known (with the exception of the periodic pulses from X-ray pulsars). Van Horn and Hansen (1974) attempted to construct a hydrogen-flash model for the transient X-ray sources (which typically have rise times of a few days and decay time scales to weeks to months), but they were forced to resort to extremely low-mass ($\lesssim 0.15\ M_\odot$) neutron stars to get sufficiently thick hydrogen-rich envelopes and sufficiently long runaway time scales.

Following the discovery of X-ray bursts by Grindlay et al. (1976) in September 1975, a number of possible explanations for this phenomenon were soon advanced. Among the early proposals was the suggestion by Woosley and Taam (1976) and

Maraschi and Cavaliere (1977) that X-ray bursts result from thermonuclear flashes on accreting neutron stars. This suggestion spurred more detailed investigations by Joss (1977), Lamb and Lamb (1978), and Taam and Picklum (1978) into the physics of nuclear flashes on accreting neutron stars and their possible relation to X-ray bursts. Subsequently a number of authors, including Joss (1978), Taam and Picklum (1979), Joss and Li (1980), Fujimoto, Hanawa, and Miyaji (1980), and Taam (1980) have presented the results of detailed numerical computations of flashes of this type. Other discussions of various aspects of thermonuclear flashes on accreting neutron stars have been presented by Czerny and Jaroszyński (1979), Ergma and Tutukov (1980), Hoshi (1980), and Barranco, Buchler, and Livio (1980).

III. THE OVERALL PHYSICAL PICTURE

Consider a neutron star undergoing accretion from a binary stellar companion. The freshly accreted matter will be rich in hydrogen and/or helium. However, at depths $\gtrsim 10^4$ cm beneath the surface of the neutron star, the density is sufficiently high that nuclear statistical equilibrium will be swiftly achieved; the predominant nuclei will have maximal binding energies, with atomic weights of ~60. (Still deeper in the star, these nuclei dissolve into a fluid in which neutrons are the primary constituent.) Hence, the accreting matter must pass through a series of nuclear burning shells as it is gradually compressed by the accretion of still more material. If the core of the neutron star is sufficiently hot or the accretion rate is sufficiently high, the temperature in the surface layers will be high enough that the burning will proceed via thermonuclear reactions, rather than electron capture or pycnonuclear reactions (which are driven by high densities rather than high temperatures). A sketch of the resultant structure of the neutron-star surface layers is given in Figure 3.

It was first realized by Hansen and Van Horn (1975) that these burning shells will tend to be unstable to thermal runaway. The instability, known as the "thin-shell instability", was first discovered in a different context by Schwarzschild and Härm (1965). The existence and strength of the instability are a direct result of the strong temperature dependence of the thermonuclear reaction rates. In the case of neutron-star envelopes, the instability is further enhanced by the partial degeneracy of the burning material. A cogent and thorough technical discussion of this type of instability has been given by Giannone and Wiegert (1967) (see also Barranco, Buchler, and Livio 1980).

The p-p chains are insufficiently temperature-sensitive to produce a thermal runaway in the hydrogen-burning shell of a neutron star. The instability of this shell is thus largely quenched by the saturation of the CNO cycle at very high reaction rates (Joss 1977; Lamb and Lamb 1978). The saturation results from the appreciable lifetimes (~10^{2-3} s) of the beta-unstable nuclei N^{13}, O^{14}, O^{15} and F^{17} that participate in the cycle. For low neutron-star core temperatures ($\lesssim 1 \times 10^8$ K) the shell can

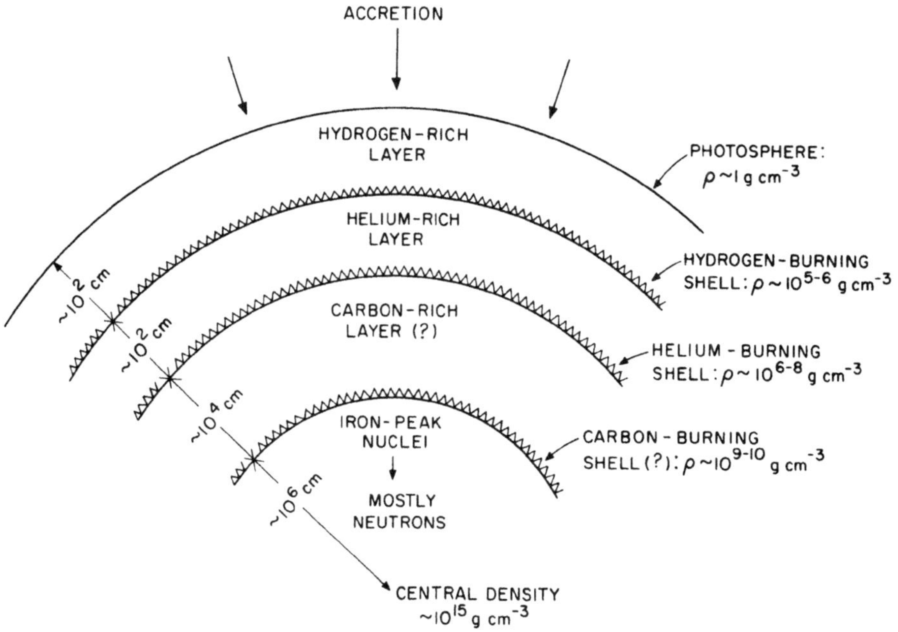

Figure 3. Schematic sketch of the surface layers of an accreting neutron star (from Joss 1979a).

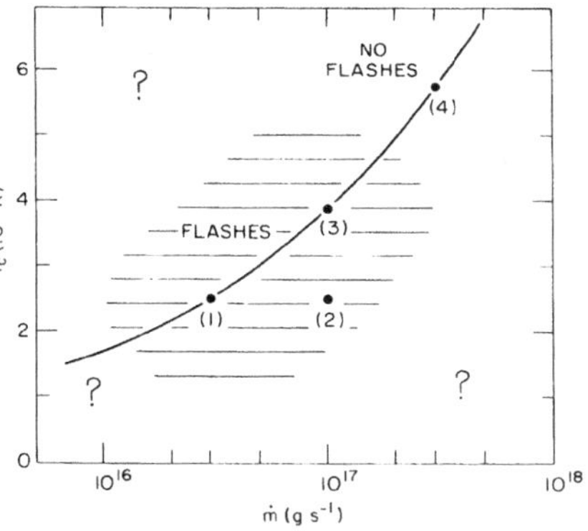

Figure 4. Mass accretion rates ($\dot{m}$) and core temperatures (T_c) for the evolutionary models of the helium-burning shell by Joss (1978). Points 1 through 4 denote the parameter values used in four models. All models assume a neutron-star mass of 1.4 $M_\odot$, a radius of 6.6 km, no magnetic field, and spherical symmetry. The solid curve passing through points 1, 3, and 4 is the estimated locus of parameter values for which the core of the neutron star is in thermal equilibrium (see text). Models 1, 2, and 3 all displayed thermonuclear flashes in the helium-burning shell; the hatched region denotes the range of parameter values for which flashes may be expected. The helium-burning shell in model 4 did not display flashes; the shell evidently becomes thermally stable at high values of T_c and $\dot{m}$, for reasons explained by Joss (1978). The behavior of the helium-burning shell at low values of T_c and $\dot{m}$ remains unexplored, but it is anticipated that flashing behavior will disappear at very low values of T_c and $\dot{m}$ ($T_c < 10^8$ K and $\dot{m} < 10^{15}$ g s^{-1}; Lamb and Lamb 1978). Also unexplored is the behavior of the helium-burning shell for parameter values far from the equilibrium curve (the upper left-hand and lower right-hand corners of the figure). (From Joss 1979a.)

in principle be unstable, but any runaways will be halted by the saturation effect before the release of a substantial amount of energy (see, however, the discussion of interacting hydrogen-helium shells in §IV below). The next shell inward is the helium-burning shell, which should be unstable over a wide range of conditions. It is uncertain whether there will be any other significant burning shells, as the matter might already burn to quite heavy elements in the helium shell (Taam and Picklum 1978; Joss 1978). However, if a carbon shell exists, it is very likely to be unstable also (Woolsey and Taam 1976; Taam and Picklum 1978).

Dimensional analysis (Joss 1977; Lamb and Lamb 1978) indicates that the helium-burning flashes should have the following properties: (1) They should occur after the accumulation of $\lesssim 10^{21}$ g of fuel and release total energies of $\lesssim 10^{39}$ ergs per flash. (2) For accretion rates comparable with those observed in X-ray pulsars ($\lesssim 10^{17}$ g s^{-1}), the time interval between flashes should be $\sim 10^4$ s, very roughly. (3) The transport of energy through the surface layers should result in the emission of bursts of electromagnetic radiation from the neutron-star photosphere with rise times of ~0.1 s, peak luminosities of $\sim 10^{38}$ ergs s^{-1}, decay time scales ~10 s, and peak blackbody temperatures of $\sim 3 \times 10^7$ K (if a full 10^{39} ergs of energy is indeed released in a single flash).

Carbon-burning flashes, if they exist, would occur much deeper beneath the neutron-star surface ($\sim 10^4$ cm, compared to $\sim 10^2$ cm for the helium shell) and would result in the release of substantially more energy. Hence, the duration of a "burst" resulting from a carbon flash should be much longer than for a helium flash (Joss 1977), unless dynamical effects are generated in the outermost surface layers.

IV. NUMERICAL MODELS

The above estimates, though very crude, suggest that thermonuclear flashes on accreting neutron stars could account for the observed properties of X-ray burst sources. With this encouragement, detailed numerical computations of the evolution of the surface layers of an accreting neutron star have been carried out.

Joss (1978) explored the evolution of the helium-burning shell (see also Hoshi 1980). In these calculations, the neutron star was chosen to have a mass of $M = 1.4\ M_\odot$ and a radius of $R = 6.6$ km. A simplified nuclear reaction network was used, incorporating the dominant reactions linking the nuclei from He^4 to Si^{28} and allowing the release of most of the available nuclear energy. The accretion was assumed to be spherical and the star was taken to be nonrotating and unmagnetized, so that spherical symmetry could be assumed throughout the calculations. The effects of hydrogen burning upon the structure of the surface layers was also neglected. The importance of these assumptions and approximations will be discussed below.

Joss' (1978) models contain two free parameters: the mass accretion rate, $\dot{m}$, and the core temperature of the neutron star, T_c. The values of $\dot{m}$ and T_c used in four models are indicated in Figure 4. If the core of the neutron star is in thermal equilibrium (i.e., if the heat flow into the core from the surface layers during thermonuclear flashes is just balanced by the heat lost from the core between flashes), then there is a unique relationship between $\dot{m}$ and T_c (Lamb and Lamb 1978); the estimated locus of these equilibrium values, as given by Joss (1978) (see also Joss and Li 1980), is shown in Figure 4. Three of the four models calculated by Joss (1978) displayed thermonuclear flashes in the helium-burning shell (see Figures 5 through 7). The properties of these flashes were in good agreement with those expected from dimensional analysis (Joss 1977; Lamb and Lamb 1978; see §III). More importantly, these calculations indicated that (1) a full $\sim 10^{21}$ g of matter accumulates on the neutron-star surface before each helium flash, (2) a flash consumes virtually all the available nuclear fuel and probably synthesizes mostly iron-peak elements, and (3) most of the energy of a flash is transported to the photosphere and lost as X-radiation, rather than carried inward to heat the interior of the star. These properties of the flashes had not been discerned prior to the performance of detailed evolutionary computations, at least in part because they depend upon the highly nonlinear characteristics of the flash growth and decay.

The behavior of the helium-burning shell was further explored by Joss and Li (1980), who investigated the sensitivity of the flash properties to the assumed mass and radius of the neutron star. They argued that since each helium-burning flash apparently consumes virtually all the available nuclear fuel, the nuclear physics essentially factors out of the problem and most of the mass- and radius-dependence of the flash properties follow from a few basic physical considerations. Thus, the recurrence interval τ_R between flashes is just the amount of time required for the base of the helium-rich layer to reach the critical temperature and density for a flash to commence, which is roughly proportional to the surface area ($A = 4\pi R^2$) of the neutron star and the scale height in its surface layers. The scale height, in turn, is roughly inversely proportional to the surface gravity $g = GM/R^2$. It then follows that

$$\tau_R \sim A\, g^{-1} \sim R^4\, M^{-1}. \qquad (1)$$

Similarly, the peak surface X-ray luminosity (L_{max}) following a flash just scales as the Eddington limit (Joss 1977), which is proportional to M but is independent of R; thus,

$$L_{max} \sim M. \qquad (2)$$

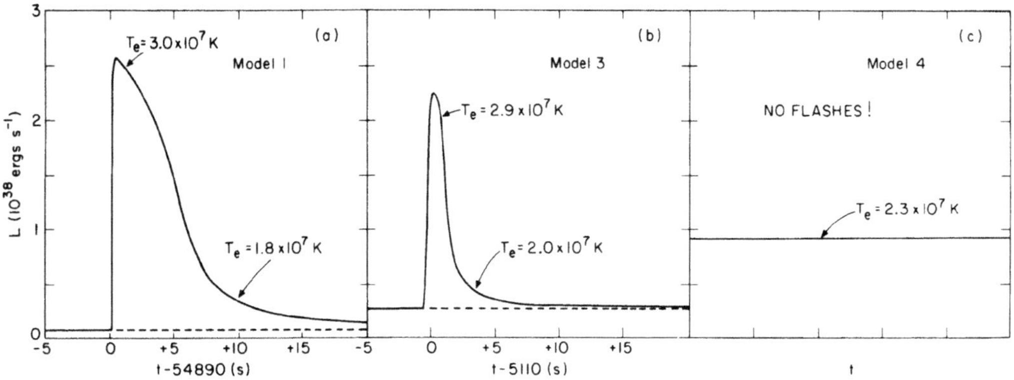

Figure 5. (a) The behavior of the surface luminosity L following a thermonuclear flash in model 1 by Joss (1978) (see Fig. 4). (b) Same for model 3. In each case, time t = 0 is at the start of accretion onto the neutron-star surface, the dashed line denotes the level of persistent accretion-driven luminosity, and the effective blackbody temperature (T_e) is indicated at a few points. The properties of these luminosity variations are in remarkably good agreement with the typical properties of observed X-ray bursts (see text). (c) The surface luminosity behavior for model 4. No flashes occur at the high core temperature and accretion rate of this model, so that the nuclear energy generation rate does not vary greatly and never produces more than a small perturbation on the accretion-driven luminosity. (From Joss 1979a.)

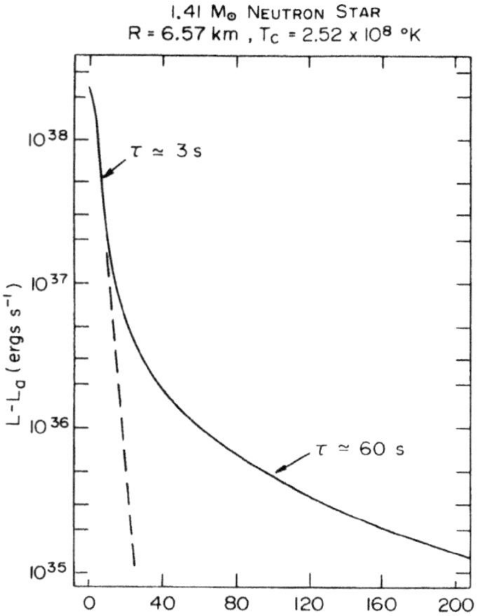

Figure 6. The decline from maximum X-ray luminosity in model 1 by Joss (1978), on time scales longer than those shown in Figure 5. L_a is the level of persistent accretion-driven luminosity from the neutron-star surface, so that ($L-L_a$) is the excess luminosity due to the thermonuclear flash. The initial decline is well fitted by an exponential decay (dashed line) with a time constant of $\tau \simeq 3$ s. However, for times greater than ~20 s after the burst peak, the decay time scale becomes much longer; 100 s after the peak, the local best-fit time constant is $\tau \simeq 60$ s. This "tail" of relatively soft X-rays (blackbody temperature $\simeq 1.3 \times 10^7$ K) contains ~10% of the total burst emission. These properties are in good agreement with those of soft X-ray "tails" in many observed X-ray burst sources. (From Joss 1980.)

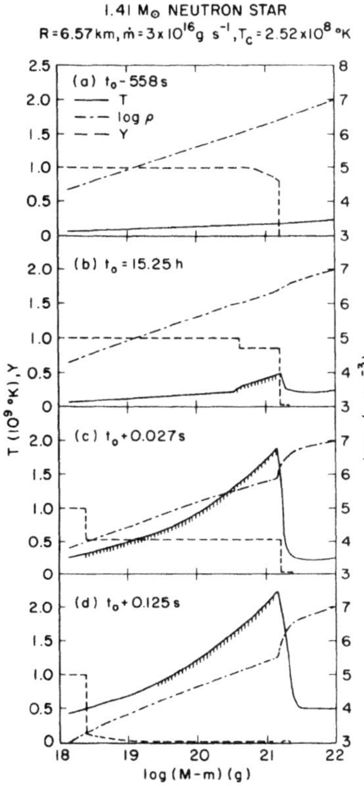

Figure 7. Structure of the surface layers of model 1 by Joss (1978) prior to and during the first helium-burning flash, which begins near time t_0. M is the total mass of the neutron star and m(r) is the mass enclosed within a sphere of radius r, with r = 0 at the stellar center; thus (M-m) is the total mass of the surface layers above level r. T is the temperature (left-hand scale), ρ the density (right-hand scale), and Y the fractional abundance of helium by mass (left-hand scale). The hatched regions indicate the extent of the convection zone generated by the flash. (a) Just prior to the flash; (b) near the start of the flash; (c) at the time when ~50% of the available fuel has been consumed; and (d) near the time of peak shell-burning temperature and peak surface luminosity. (From Joss 1978.)

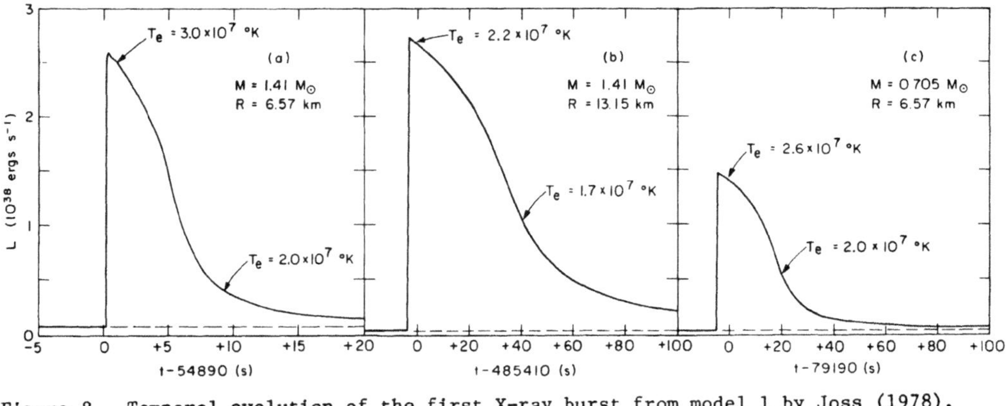

Figure 8. Temporal evolution of the first X-ray burst from model 1 by Joss (1978), with neutron-star mass M = 1.41 $M_\odot$ and radius R = 6.57 km. The notation is the same as in Figure 5. (b) Same as (a), but for a model by Joss and Li (1980) which has the same parameters as model 1 except that R = 13.15 km. (c) Same as (a), but for a model by Joss and Li (1980) which has the same parameters as model 1 except that M = 0.705 $M_\odot$. Note the difference in the scale of time for (b) and (c) compared to that of (a). The differences among these three models can be largely explained by simple scaling arguments (see text). (From Joss and Li 1980.)

Finally, the time scale (τ_D) of decline of surface X-ray emission following a flash is roughly directly proportional to the energy released in a flash, which is in turn directly proportional to τ_R and inversely proportional to the rate (L_{max}) at which that energy escapes the star; thus,

$$\tau_D \sim \tau_R/L_{max} \sim R^4\ M^{-2}. \qquad (3)$$

Joss and Li (1980) carried out numerical computations of the evolution of the helium-burning shell for neutron stars with a few different masses and radii (see Figure 8) and fitted their results for τ_R, L_{max}, and τ_B to power-law expressions in M and R. They obtained

$$\tau_R^{(fit)} \sim R^{3.1}\ M^{-0.5} \quad ;$$

$$L_{max}^{(fit)} \sim R^{0.1}\ M^{0.8} \quad ; \qquad (4)$$

$$\tau_D^{(fit)} \sim R^{3.1}\ M^{-1.4} \quad .$$

These expressions are in reasonably good agreement with relations (1)-(3). The existence of these simple scaling relations suggests the intriguing possibility that once the physics of neutron-star thermonuclear flashes is sufficiently well understood, it may be possible to deduce information on the masses and radii of neutron stars from the observed properties of X-ray bursts that result from such flashes. (As noted by Joss and Li, however, these scaling relations should not be applied to the observational data until the remaining major uncertainties in the theoretical calculations have been resolved.) The indirect measurement of general relativistic effects from the burst properties may be even more powerful in this regard (see §V below).

Joss and Li (1980) also investigated the evolution of the helium-burning shell in models wherein the effects of an intense surface magnetic field upon the surface layers were taken into account. They argued that if the magnetic field is sufficiently strong to funnel the accretion onto the magnetic polar caps of the neutron star, then the effective accretion rate in the polar cap regions is enhanced by a factor of $\sim 10^3$ (for a fixed total accretion rate $\dot{m}$) and the instability of the nuclear burning shells should be reduced (see also Taam and Picklum 1978 and Joss 1978). The surface magnetic field strength, B, required to funnel the accretion is not well determined, but available estimates (see, e.g., Arons and Lea 1980) yield $B \simeq 10^{12}$ G. Joss and Li (1980) found that surface magnetic fields in excess of $\sim 10^{12}$ G will also have some significant effects upon the heat transport properties of the surface layers, but none of these effects are likely to be as important as the funneling of the accretion flow.

It has become increasingly clear that hydrogen burning can have a major influence on the behavior of the helium-burning shell (Taam and Picklum 1978; Czerny and Jaroszyński 1979; Ergma and Tutukov 1980). The saturation of the CNO cycle by the finite lifetimes of the beta-unstable nuclei that participate in the cycle limits the hydrogen-burning rates to such an extent that, at higher accretion rates ($\gtrsim 1 \times 10^{16}$ g s^{-1}), the hydrogen-burning shell is forced inward until it overlaps the helium-burning shell. Taam and Picklum (1979) and Taam (1980) have carried out the first fully time-dependent computations of the evolution of the surface layers of a neutron star with a hydrogen-burning shell included and found that the hydrogen- and helium-burning shells can, indeed, interact in a complex way (see Figure 9). In fact, in their models the heating of the accreted material by the neutron-star core prior to a thermonuclear flash was insignificant compared to the heating that resulted from hydrogen burning. However, this effect appears to have been a consequence of the low core temperatures chosen for the neutron star ($<10^8$ K); such cores would generally not be in thermal equilibrium with the nuclear burning shells, but they might represent the properties of an old neutron star that has begun to accrete matter during the past $\sim 10^{2-3}$ yr.

Fujimoto, Hanawa, and Miyaji (1980) have also studied the interaction between the hydrogen- and helium-burning shells. They argued

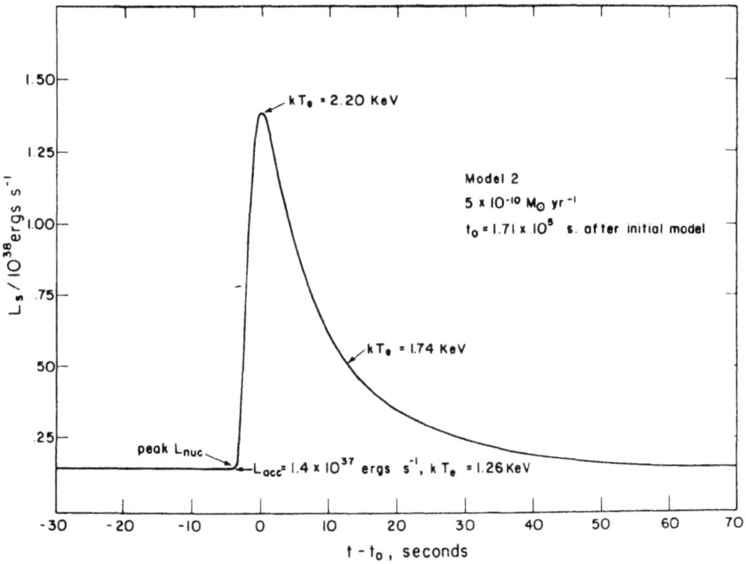

Figure 9. The behavior of the surface luminosity L_s following a thermonuclear flash in model 2 by Taam (1980). The parameters of the model are indicated in the figure. The accreting material is assumed to have an initial heavy-element abundance of 0.004 by mass. Time t = 0 is at the start of accretion onto the neutron-star surface, L_{acc} is the level of persistent accretion-driven luminosity, and the effective temperature (T_e) is indicated at a few points. The flash is driven by helium burning, but the entrainment of hydrogen into the flash has a strong effect upon its properties (see text). (From Taam 1980.)

that for neutron stars with low core temperatures there will be three modes of thermonuclear flashes: helium flashes followed by simultaneous helium burning and hydrogen burning when the two shells overlap at high accretion rates, pure helium flashes at intermediate accretion rates, and hydrogen flashes that ignite the helium-burning shell at low accretion rates. However, we note that due to the saturation of the CNO cycle, the rise in temperature ΔT due to a hydrogen flash is limited to

$$\Delta T \lesssim 10^8 \text{ K} \left(\frac{X_{CNO}}{10^{-2}} \right), \qquad (5)$$

where X_{CNO} is the fractional abundance by mass of CNO nuclei (Ayasli and Joss 1980). Hence, it seems that unless the abundances of the CNO nuclei in the accreting matter are substantially higher than their cosmic abundances, a hydrogen flash will usually be unable to fully ignite the helium shell and the third scenario described by Fujimoto _et al_. will not be realized. Nonetheless, it has become evident from all of the above work that the interactions between the hydrogen- and helium-burning shells may be very complex; it is highly likely that still further complications will be uncovered by future work.

Taam and Picklum (1978) studied the thermal evolution of a carbon-burning shell in an accreting neutron star but did not carry out their computations through a complete thermonuclear flash. Their models again assumed low core temperatures ($<10^8$ K) that would not be in thermal equilibrium with the surface nuclear burning shells.

V. GENERAL RELATIVISTIC EFFECTS

Utilizing the assumption that the spectra of X-ray bursts following peak luminosity could be represented by blackbodies, Van Paradijs (1978) demonstrated that in many cases the scale size of the X-ray emitting region is nearly constant; if the emitting region is a spherical surface, then its radius is ~7 km. This result lends compelling support to the idea that many X-ray bursts are thermal emission from the photospheres of neutron stars, and thus provides indirect evidence in favor of the thermonuclear flash model for such bursts.

However, it has become apparent that this argument is complicated by general relativistic corrections, such as the effect of gravitational redshift and time dilation upon the X-radiation emitted by the neutron-star photosphere (Goldman 1979; Van Paradijs 1979). The importance of general relativity can be seen by inspection of the parameter

$$\frac{2GM}{Rc^2} \simeq 0.60 \left(\frac{M}{1.4 \; M_\odot} \right) \left(\frac{R}{7 \text{ km}} \right)^{-1} . \qquad (6)$$

The left-hand side of equation (7) is just the ratio of the Schwarzschild radius of the neutron star ($2GM/c^2$) to its actual radius. For the indicated values of M and R, which have been used in many of the actual model calculations of thermonuclear flashes, it is evident that this parameter is not very much smaller than unity, so that general relativistic effects should be substantial.

It was shown by Goldman (1979) and Van Paradijs (1979) that when general relativistic corrections are included in determinations of the luminosities and effective blackbody temperatures of X-ray bursts, one obtains, at least in principle, strong constraints on the masses and radii of the underlying neutron stars. If taken at face value, these constraints would, in turn, severely constrain the equation of state of matter at densities in excess of nuclear-matter densities ($\rho > 2 \times 10^{14}$ g cm^{-3}). However, such constraints cannot yet be taken seriously, as other complications, including possible violations of spherical symmetry (see §VII) and deviations of the emitted spectrum from a simple blackbody (Swank, Eardley, and Serlemitsos 1979; Van Paradijs, Rybicki, and Lamb 1980), may turn out to be important. Once these additional complexities have been untangled, the observed properties of X-ray bursts may prove to be a powerful probe of the basic properties of neutron stars.

General relativistic corrections to the equations of stellar structure and evolution may also play a significant role in determining the behavior of the thermonuclear flashes themselves. Some of these corrections have already been included in some calculations (e.g., those by Taam and Picklum 1978, 1979; Czerny and Jaroszyński 1979; and Taam 1980). However, no fully time-dependent model computations have yet included all of the relevant general relativistic corrections.

VI. COMPARISON WITH OBSERVATIONAL PHENOMENA

A. X-Ray Bursts

The results of the numerical calculations of thermonuclear flashes (see Figures 4-9) strongly support the conjecture that the bursts from most observed X-ray burst sources result from such flashes. In particular, the typical burst rise times, decay time scales, peak luminosities, total emitted energies, spectral properties, low-energy "tails," and recurrence intervals (see Lewin and Clark 1980; Lewin and Joss 1980) are reproduced remarkably well by such flashes.

However, there are some difficulties with this model. It does not reproduce the observed complex burst structures and recurrence patterns, which vary from one burst source to another and often vary with time in a single source. It is quite possible that these complexities will be better understood when some of the approximations of the present model calculations are relaxed; we shall return to this point in §VII.

A particulary severe problem for the nuclear flash model is the ratios, α, of time-averaged persistent X-ray luminosity to time-averaged burst luminosity from the observed burst sources. In this model, α should just be the ratio of the gravitational energy released by accretion (more or less continuously) to the nuclear energy released in the flashes. For helium-burning flashes, the numerical value of α should thus be

$$\alpha \simeq 1 \times 10^2 \left(\frac{M}{M_\odot}\right)\left(\frac{R}{10\ \mathrm{km}}\right)^{-1} \quad (7)$$

(Joss 1977; Lamb and Lamb 1978). However, some burst sources have reported values of α significantly less than 10^2; one such case is 4U1608-52, which displayed two bursts separated by an interval of only ten minutes and a correspondingly small upper limit of $\alpha < 2.5$ during that interval (Murakami et al. 1980). Moreover, in some sources the recurrence intervals have been observed to increase (decrease) when the persistent luminosity increased (decreased), which is opposite to the trend expected from this model (see Figure 5).

If the nuclear flash model is correct, it is possible that the observed values of α are sometimes (perhaps always) reduced by the storage of nuclear fuel during burst-inactive phases (Lamb and Lamb 1978). Such a "battery" mechanism may be provided by large-scale violations of spherical symmetry, so that only a fraction of the surface of the neutron star participates in each flash, or from fluctuations resulting from the interaction between the hydrogen- and helium-burning shells (e.g., fluctuations in the amount of hydrogen entrained into the convection zone generated by a helium flash). Even in the absence of such fluctuations, the entrainment of hydrogen into a helium-burning flash could, in principle, reduce the value of α by up to a factor of ~5 (due to proton captures onto the heavier nuclei being synthesized and a concomitant increase in the energy yield per unit mass); the time-dependent calculations by Taam and Picklum (1979) and Taam (1980) do not indicate any reduction in α-values by this effect, but more complete hydrogen-burning reaction networks might yield different results (Wallace and Woosley 1980; Ergma and Kudrjashov 1980; Ayasli and Joss 1980). It is also possible that the accretion-driven luminosity is emitted anisotropically and/or largely shifted to photon energies $\lesssim$ 1 keV; Milgrom (1978) has proposed a specific mechanism (an accretion disk that extends all the way to the surface of a weakly magnetized neutron star) that could produce both these effects and thereby reduce the observed values of α by up to a factor of ~2.

B. The Rapid Burster

It is important to realize that the bursts from the "Rapid Burster," MXB1730-335, are at present unique and almost certainly cannot be the result of thermonuclear flashes. The recurrence intervals between bursts are $\sim 10^{1-3}$ s, and $\alpha \lesssim 2$ for this source (Lewin et al. 1976).

Hoffman, Marshall, and Lewin (1978) have described these bursts as "type II" and those from other sources as "type I". However, Hoffman, Marshall, and Lewin also found that the Rapid Burster occasionally emits "special" bursts whose properties much more closely resemble the type I bursts from other sources. Moreover, the ratio of time-averaged luminosity in the type II bursts from the Rapid Burster to that in its type I bursts is $\sim 10^2$ (Hoffman, Marshall, and Lewin 1978). Hoffman, Marshall, and Lewin made the intriguing speculation that the type I bursts from the Rapid Burster are the result of thermonuclear flashes on an accreting neutron star, while the type II bursts are the result of an unstable accretion flow onto the same object.

C. The Fast X-Ray Transients

The morphology of some of the "fast X-ray transients", which have durations of $\sim 10^{2-3}$ s, is suggestively similar to that of ordinary type I X-ray bursts (Hoffman et al. 1978). Joss (1979b) suggested that the fast transients may be the result of helium-burning flashes relatively deep within the surface layers of slowly accreting neutron stars ($\dot{m} \lesssim 10^{15}$ g s^{-1}). If this picture is correct, one would expect outbursts from the fast transients to recur, but only on time scales of weeks or longer. Some fast transients have been observed to have distinct precursors, consisting of relatively brief but intense X-ray emission just before the start of the main burst (Hoffman et al. 1978); a low accretion rate should result in a relatively large temperature contrast between the nuclear flashing shell and the outermost surface layers of the neutron star, so that the precursors might be the result of shock heating of the outer surface layers. However, detailed numerical computations of deep helium-burning flashes have yet to be carried out, and in the interim the above scenario must be regarded as highly speculative.

D. Relation to Binary X-Ray Pulsars and the Ages of Burst Sources

Let us accept, for the sake of discussion, that most observed X-ray burst sources can be understood as accreting neutron stars that are undergoing thermonuclear flashes. Since X-ray pulsars are also widely believed to be accreting neutron stars, it is then puzzling, at first sight, that these objects do not also display bursting behavior. However, the strong magnetic field ($\gtrsim 10^{12}$ G; see Arons and Lea 1980) that funnels the accretion onto the magnetic polar caps of an X-ray pulsar will also enhance the efficiency of radiative and conductive heat transport within and above its nuclear burning shells. Even more importantly, the heat released by accretion will have a much greater influence upon the inner burning shells if the freshly accreted matter is confined to the polar caps, rather than spread uniformly over the neutron-star surface. These effects should tend to reduce the instability of the nuclear burning shells of an X-ray pulsar against thermonuclear flashes (Taam and Picklum 1978; Joss 1978; Joss and Li 1980). Evolutionary models of the helium-burning shell in the presence of an intense magnetic field confirm the assertion that such fields reduce the instability of the shell (Joss and Li 1980).

With this picture, we can also understand why the persistent X-ray flux from type I X-ray burst sources is unpulsed: those neutron stars whose magnetic fields are too weak to funnel the accreting matter may be precisely those that can undergo thermonuclear flashes (Taam and Picklum 1978; Joss 1978; Joss and Li 1980). If the magnetic field was originally as strong as in an X-ray pulsar but has since decayed, then the neutron star must be fairly old (probably older than 10^7 yr; see Ruderman 1972 and references therein; Flowers and Ruderman 1977). The lack of X-ray eclipses in burst sources may also reflect membership in relatively old binary systems (Joss and Rappaport 1979) and may result from X-ray beaming effects that set in after the neutron-star magnetic field has decayed (see, e.g., Milgrom 1978). The concentration of X-ray burst sources in the direction of the galactic center and the identification of several of them with globular clusters (see Figure 1) may well be other manifestations of membership in an older galactic population than the X-ray pulsars, which are distributed through the disk of the galaxy and whose binary companion stars are often of early spectral type.

E. Gamma-Ray Burst Sources

Woosley and Taam (1976) suggested that carbon-burning flashes on accreting neutron stars result in cosmic γ-ray bursts. A more detailed look at the heat transport properties of the neutron-star surface layers demonstrates that the rising convective elements above any nuclear burning shell would cool to temperatures $\ll 10^{10}$ K before reaching the photosphere (Joss 1977). However, it is possible that under some circumstances, a thermonuclear flash generates dynamical effects in the outermost surface layers of a neutron star, resulting in the emission of a gamma-ray burst (see, e.g., Ruderman 1980).

VII. CONCLUDING REMARKS

We have just begun to grasp all of the intricacies of nuclear processes in the surface layers of accreting neutron stars. The theoretical problems are fascinating, not only as investigations in fundamental physics, but also for their probable applications to X-ray burst sources and other observational phenomena and as a potentially powerful tool for probing the basic properties of neutron stars.

The only complete evolutionary computations of neutron-star thermonuclear flashes that have been carried out to date (Joss 1978; Taam and Picklum 1979; Joss and Li 1980; Taam 1980) relied on a number of simplifying assumptions and approximations, such as the neglect of some general relativistic effects (see §V), the assumption of spherical symmetry, the neglect of possible dynamical effects, and, in some cases, the assumption that the neutron-star core is in thermal equilibrium. These approximations will have to be relaxed in future studies before this phenomenon and its observational implications can be more fully understood.

Small but significant violations of spherical symmetry might result from accretion through a relatively weak ($\lesssim 10^{11}$ G) magnetic field or the residual angular momentum of the accreting matter. If thermonuclear flashes result in X-ray bursts, such violations could be the key to the observed complexities in burst structure and recurrence patterns (see Lewin and Clark 1980; Lewin and Joss 1980). For example, a thermonuclear flash that ignites on one portion of the neutron-star surface may propagate around the star, in a pattern that varies from flash to flash and from one star to another (Joss 1978). A thorough investigation of such possibilities will eventually require two- or three-dimensional numerical computations, which will be much more difficult than the computations of spherically symmetric models that have been attempted to date.

Complexities in radiative transfer in the outer surface layers of the neutron star and possible mass ejection from the photosphere near the peak of a burst may also substantially complicate the observational properties of X-ray bursts and render their physical interpretation much more difficult. Some preliminary results on deviations from blackbody emission by a hot neutron-star atmosphere have been reported (Swank, Eardley, and Serlemitsos 1979; Van Paradijis, Rybicki, and Lamb 1980), but much work in these areas remains to be done.

The importance of thermal equilibrium of the neutron-star core has been only tentatively explored (see Figure 4). If there is a change in the average accretion rate, the thermal inertia of the core is sufficient to require the elapse of $\sim 10^{2-3}$ yr for thermal equilibrium to be reestablished (Lamb and Lamb 1978). Thus departures from thermal equilibrium are entirely possible, and they could have a substantial effect upon the behavior of the nuclear burning shells.

Many of these issues can be attacked by additional calculations in the immediate future. Further progress, and more surprises, are bound to be forthcoming.

ACKNOWLEDGEMENTS

I am grateful to S. Ayasli for helpful discussions. Earlier versions of some portions of this review were previously published in *Comments on Astrophysics, 8*, 109 (1979), *Annals of the New York Academy of Sciences, 336*, 479 (1980), and *Space Science Reviews* (in collaboration with W.H.G. Lewin; in press). This work was supported in part by the National Science Foundation under grant AST78-21993 and by the National Aeronautics and Space Administration under contract NAS5-24441 and grant NSG-7643.

REFERENCES

Arons, J., and Lea, S.M. 1980, *Astrophys. J.*, *235*, 1016.
Ayasli, S., and Joss, P.C. 1980, manuscript in preparation.
Barranco, M., Buchler, J.R., and Livio, M. 1980, preprint.
Cowie, L.L., Ostriker, J.P., and Stark, A.A. 1978, *Astrophys. J.*, *226*, 1041.
Czerny, M., and Jaroszyński, M. 1979, submitted to *Acta Astronomica*.
Ergma, E.V., and Kudrjashov, A.D. 1980, preprint.
Ergma, E.V., and Tutukov, A.V. 1980, *Astron. Astrophys.*, *84*, 123.
Flowers, E., and Ruderman, M. 1977, *Astrophys. J.*, *215*, 302.
Fujimoto, M., Hanawa, T., and Miyaji, S. 1980, preprint.
Giacconi, R., Gursky, H., Kellogg, E., Schreier, E., and Tananbaum, H. 1971, *Astrophys. J. (Letters)*, *167*, L67.
Giannone, P., and Weigert, A. 1967, *Zs. Astrophys.*, *67*, 41.
Goldman, Y. 1979, *Astron. Astrophys.*, *78*, L15.
Grindlay, J., Gursky, H., Schnopper, D., Parsignault, R., Heise, J., Brinkman, A.C., and Schrijver, J. 1976, *Astrophys. J. (Letters)*, *205*, L127.
Hansen, C.J., and Van Horn, H.M. 1975, *Astrophys. J.*, *195*, 735.
Hoffman, J.A., Lewin, W.H.G., Doty, J., Jernigan, G., Haney, M., and Richardson, J.A. 1978, *Astrophys. J. (Letters)*, *221*, L57.
Hoffman, J.A., Marshall, H.L., and Lewin, W.H.G. 1978, *Nature*, *271*, 630.
Hoshi, R. 1980, preprint.
Joss, P.C. 1977, *Nature*, *270*, 310.
Joss, P.C. 1978, *Astrophys. J. (Letters)*, *225*, L123.
Joss, P.C. 1979a, *Comments on Astrophys.*, *8*, 109.
Joss, P.C. 1979b, in "Compact Galactic X-ray Sources," ed. D. Pines and F. Lamb (Urbana, Illinois: Physics Dept., Univ. of Illinois), p. 89.
Joss, P.C. 1980, *Annals N.Y. Acad. Sci.*, *336*, 479.
Joss, P.C., and Li, F.K. 1980, *Astrophys. J.*, *238*, 287.
Joss, P.C., and Rappaport, S.A. 1979, *Astron. Astrophys.*, *71*, 217.
Lamb, D.Q., and Lamb, F.K. 1977, *Annals N.Y. Acad. Sci.*, *302*, 261.
Lamb, D.Q., and Lamb, F.K. 1978, *Astrophys. J.*, *220*, 291.
Lewin, W.H.G., and Clark, G.W. 1980, *Annals N.Y. Acad. Sci.*, *336*, 451.
Lewin, W.H.G., Doty, J., Clark, G.W., Rappaport, S.A., Bradt, H.V.D., Doxsey, R., Hearn, D.R., Hoffman, J.A., Jernigan, J.G., Li, F.K., Mayer, W., McClintock, J., Primini, F., and Richardson, J. 1976, *Astrophys. J. (Letters)*, *207*, L95.
Lewin, W.H.G., and Joss, P.C. 1977, *Nature*, *270*, 211.
Lewin, W.H.G., and Joss, P.C. 1980, *Space Sci. Revs.*, in press.
Lewin, W.H.G., McClintock, J.E., and Ricker, G.R. 1971, *Astrophys. J. (Letters)*, *169*, L17.
Maraschi, L., and Cavaliere, A. 1977, in "Highlights of Astronomy," Vol. 4, ed. E.A. Müller (Dordrecht: Reidel), Part I, p. 127.
Milgrom, M. 1978, *Astron. Astrophys.*, *67*, L25.
Murakami, T., Inoue, H., Koyama, K., Makishima, K., Matsuoka, M., Oda, M., Ogawara, Y., Ohashi, T., Shibazaki, N., Tanaka, Y., Tawara, Y., Hayakawa, S., Kunieda, H., Makino, F., Masai, K., Nagase, F., Miyamoto, S., Tsunemi, H., Yamashita, K., and Kondo, I. 1980, submitted to *Nature*.

Rosenbluth, M.N., Ruderman, M., Dyson, F., Bahcall, J.N., Shaham, J., and Ostriker, J. 1973, *Astrophys. J.*, *184*, 907.
Ruderman, M. 1972, *Ann. Rev. Astron. Astrophys.*, *10*, 427.
Ruderman, M. 1980, talk presented at the Erice School, to be published.
Schreier, E., Levinson, R., Gursky, H., Kellogg, E., Tananbaum, H., and Giacconi, R. 1972, *Astrophys. J. (Letters)*, *172*, L79.
Schwarzschild, M., and Härm, R. 1965, *Astrophys. J.*, *142*, 855.
Swank, J.H., Eardley, D.M., and Serlemitsos, P.J. 1979, preprint.
Taam, R.E. 1980, preprint.
Taam, R.E., and Picklum, R.E. 1978, *Astrophys. J.*, *224*, 210.
Taam, R.E., and Picklum, R.E. 1979, *Astrophys. J.*, *233*, 327.
Tananbaum, H., Gursky, H., Kellogg, E.M., Levinson, R., Schreier, E., and Giacconi, R. 1972, *Astrophys. J. (Letters)*, *174*, L143.
Van Horn, H.M., and Hansen, C.J. 1974, *Astrophys. J.*, *191*, 479.
Van Paradijs, J. 1978, *Nature*, *274*, 650.
Van Paradijs, J. 1979, *Astrophys. J.*, *234*, 609.
Van Paradijs, J., Rybicki, G., and Lamb, D.Q. 1980, talk presented at the Meeting of the High Energy Astrophysics Division of the American Astronomical Society, Cambridge, Massachusetts, January 1980, abstract in *Bull. Amer. Astron. Soc.*, *11*, 788 (1979).
Wallace, R.K., and Woosley, S.E. 1980, preprint.
Woosley, S.E., and Taam, R.E. 1976, *Nature*, *263*, 101.

DISCUSSION

Sugimoto: In the case of accreting white dwarfs, shell burning is ignited even when the core temperature is absolute zero at the onset of the accretion. What about the case of accreting neutron stars?

Joss: If the core temperature is less than about 10^8K, then the core is unimportant in heating the surface layers. However, as shown by the work of Taam and Picklum (1978), and Fujimoto, Hanawa, and Miyaji (1980), thermonuclear flashes will still occur even if the core is "cold" in this sense.

Tutukov: What heats the accretion shell between flashes, the hot neutron star or gravitational compression?

Joss: If the core of the neutron star is in thermal equilibrium with the nuclear burning shells, then under many, if not all, conditions of interest the heat stored in the core is more important than gravitational compression in heating the surface layers between flashes.

Sugimoto: You have shown a diagram of the core temperature against the accretion rate, on which there is only one line relating these two quantities. In the case of accreting white dwarfs, we can consider any point in that diagram, depending on the initial conditions. What determines that line in your case? If you claim it to be a limit cycle, what is the reason?

Joss: The equilibrium core temperature of an accreting neutron star was first estimated by Lamb and Lamb (1978); I incorporated a modified version of their estimate into my own model computations. More accurate calculations, currently underway, suggest that the earlier estimates of equilibrium core temperatures were too high, but not by a large factor.

Lamb: The recent investigation by Fujimoto, Miyaji, and Hanawa (1980) and the recent numerical calculations by Taam (1980) show unstable He burning at high accretion rates, in contrast to your own calculations (Joss 1978). This issue is important, as you point out, in trying to understand the distinction between the X-ray bursters and the pulsating X-ray sources such as Her X-1, Cen X-3, and SMC X-1, which apparently do not burst. What accounts for the difference between their results and yours, and which do you think represents more correctly the behavior of the burning at high accretion rates?

Joss: The difference, I believe, is a result of the high core temperature T_c, at high accretion rates that we assumed in our calculations, in contrast to the low values of T_c employed by Fujimoto et al. (1980). The values we adopted were based on the estimated conditions for thermal equilibrium of the core by Lamb and Lamb (1978), as modified by Joss (1978). As already remarked we are presently carrying out more accurate calculations of thermal equilibrium conditions and our preliminary results suggest that the original estimates of T_c were too high, but not by a very large factor. I therefore think that the burning will, indeed, be largely stabilized at high accretion rates, provided that the core is in thermal equilibrium.

Sugimoto: What are the important effects of general relativity other than "increasing" the gravity?

Joss: General relativistic effects enter the equations of stellar structure and evolution in a highly complex and nonlinear way (Thorne 1977 and references therein). As I indicated in my talk, certain results may be accounted for in terms of some particular general relativistic effect, such as time dilation or gravitational redshift. However, I know of no general way to isolate all of the possible effects of general relativity; one must simply incorporate the appropriate corrections into a detailed computation and see what emerges.

Schatzman: What kind of increased rate factor did you take for thermonuclear reactions in dense matter?

Joss: In our earliest calculations, we neglected screening corrections to the thermonuclear reaction rates. This is not very restrictive approximation since the screening corrections do not greatly change the temperature or density-dependence of the reaction rates and, as pointed out by Dr. Sugimoto in his talk earlier this morning, the absolute values of the reaction rates are much less important than their temperature dependences in determing the behavior of the nuclear burning shells. In our more recent calculations, we have incorporated the weak and strong screening corrections of Salpeter and Van Horn (1969).

Lamb: The helium flash calculations for nuclear burning on accreting neutron stars give values of α, the ratio of the background (accretion) luminosity to the time averaged burst luminosity in the range 200-800 (cf. Joss 1978, Joss and Li 1979, Taam and Picklum 1979, Taam 1980) and show all of the helium and hydrogen being consumed. However, early observations of MXB1743-28 by SAS-3 (Lewin 1977) indicated, and more recent observations of 1608-522 by Hakucho (Inoue et al. 1980) have shown clearly that some Type I burst sources exhibit $\alpha \lesssim 2$-3. Earlier, we had suggested on the basis of the SAS-3 observations that in some cases all the nuclear fuel must not be exhausted in a given flash, and that some type of "battery" model seems to be required (Lamb and Lamb 1978). Would you comment on this?

Joss: I fully agree with you that, in view of the clear observation by Inoue et al. of two Type I bursts from 1608-522 separated by only about 10 minutes, we must now take seriously the need for some of a "battery" idea to account for this phenomenon. However, I doubt that this will be a severe problem for the thermonuclear flash model. First, occurrences of Type I bursts in rapid succession seem to be quite rare and isolated, and their explanation may not play a central role in our understanding of the burst phenomenon. Second, the entrainment of hydrogen into the thermonuclear flashes, when fully taken into account, might already reduce the theoretical α-values to as low as ~10 even in the absence of a "battery" mechanism. As far as possible "battery" models are concerned, two thoughts come to mind: violations of spherical symmetry, so that less than the entire surface of the neutron star participates in a flash and some nuclear fuel is saved for the next flash; and fluctuations in the extent of the convection zone generated by each flash and in the resultant amount of nuclear fuel entrained in the flash. However, any such explanation must, of course, be regarded as highly speculative until it is supported by serious calculations.

SHELL-FLASHES ON ACCRETING NEUTRON STARS AND MODE PROFILES OF TYPE-I X-RAY BURSTS

Shigeki Miyaji[1], Masayuki Y. Fujimoto[2], and Tomoyuki Hanawa[3]

[1] Department of Earth Science and Astronomy, College of General Education, University of Tokyo
[2] Department of Astronomy, University of Illinois
[3] Department of Astronomy, University of Tokyo

Thermal stabilities of hydrogen and helium shell-burnings on accreting neutron stars are studied semi-analytically, the progress of nuclear reactions during the flash is followed numerically, and the mechanism which makes different modes of Type-I X-ray bursts is discussed.

The hydrogen shell-burning leads to the ignition of helium shell-flash through three different ways depending on the mass accretion rate dM/dt for the case of realistic core temperature; the stable hydrogen burning growing to the combined hydrogen and helium shell-flash with high dM/dt, the steady state hydrogen shell-burning and the pure helium shell-flash with intermediate dM/dt, and the hydrogen shell-flash developing into the combined hydrogen and helium shell-flash with low dM/dt.

The characteristics of pure helium flash is already described in the review of Joss in this symposium. These of combined hydrogen and helium flash is summarized to show the large variety of its burst profiles; the rise time is determined by the competition between the diffusion timescale and the nuclear timescale of 3α, (α,p), and (p,γ) reactions and the hardness ratio during the decay phase is governed by the β^+-decays of seed nuclei after the exhaustion of helium. Total released nuclear energy per unit mass during the flash also depends on the composition ratio.

These characteristics are consistent with observations of 1608-522 by HAKUCHO satellite both in 1979 and in 1980; short and long rise time ($\lesssim$ several seconds), fast and slow softening (e-folding time is upto ten seconds), and large variety of the ratio α (persistent X-ray luminosity / time averaged burst energy) from $\sim$500 to $\sim$30.

The possibility of partial (regional) shell-flash which grows independently at any part of accreting region is also pointed out.

D. Sugimoto, D. Q. Lamb, and D. N. Schramm (eds.), Fundamental Problems in the Theory of Stellar Evolution, 229–230.

DISCUSSION

Joss: In your case 3, where the hydrogen- and helium-burning shells are separated and the hydrogen undergoes a flash, the temperature probably must rise to a few times 10^8 K in order to ignite a helium flash. This requires that ~1% of the hydrogen must be consumed. However, each CNO nucleus can only capture a few protons, since the beta-decay timescales of the resultant proton-rich nuclei are longer than the local thermal diffusion timescale. Hence, it seems that one needs an enhanced CNO abundance, perhaps as high as 10% by mass, in order to ignite the helium. This is problematic, and especially so if the accreting matter has population II abundances. Can you see a way around this difficulty?

Miyaji: As you saw in my figure showing the stability curves, the burning is unstable to the right of the ignition line. Even when we take account of the saturation effect of the CNO cycle due to beta-decay, there we have $\varepsilon_n > \partial L_r/\partial M_r$, i.e. the nuclear timescale is shorter than the diffusion timescale so that the CNO-cycle can proceed many times.

THE CURRENT STATUS OF NEUTRON STAR COOLING THEORIES

S. Tsuruta*, T. Murai**, K. Nomoto***, N. Itoh****
* Department of Physics, Montana State University, Bozeman, Montana, U.S.A.
** Department of Physics, Nagoya University, Nagoya, Japan
*** NASA, Goddard Space Flight Center, Greenbelt, Maryland, U.S.A.
**** Department of Physics, Sophia University, Tokyo, Japan

There are serious discrepancies among some of the recent neutron star cooling calculations by various groups. We have been investigating the possible source of these discrepancies. In this paper, we report our findings. We also report the preliminary result of our most recent cooling calculations without assuming an isothermal stellar evolution code. In this work, we used the currently existing best energy transport theories, as well as general relativity, both in thermodynamics and hydrodynamics.

DISCUSSION

Lamb: While I appreciate being called a cool character by Sachiko, I would remark that the neutron star cooling curves computed by Ken van Riper and myself are not quite as cool as shown in her figure. Let me summarize our results as follows. At an age τ = 1000 yrs, we find for our soft star a luminosity $\cong$ 16 times lower (temperature $\cong$ 2 times lower) and for our stiff star a luminosity $\cong$ 40 times lower (temperature $\cong$ 2.5 times lower) than those found earlier by Tsuruta (1979). In the soft star, much of the difference appears to be due to our inclusion of general relativistic effects (particularly the gravitational redshift), although part appears to be due to our use of new radiative and conductive opacities. In the stiff star, general relativistic effects are smaller. We find, however, that it cools somewhat faster because superfluidity markedly reduces its heat capacity but does not affect its dominant energy loss processes, i.e. neutrino crust bremsstrahlung and photon emission from the stellar surface. Nevertheless, the luminosities of the soft and stiff stars differ very little during the neutrino cooling era. We can compare the recent results of Glen and Sutherland (1980) with ours only for the case of zero magnetic field, since we treated the effects of the field on the density structure of the outer layers of the

D. Sugimoto, D. Q. Lamb, and D. N. Schramm (eds.), Fundamental Problems in the Theory of Stellar Evolution, 231–232.

star whereas they did not. For zero magnetic field, our luminosities for the soft star agree (e.g. they are within a factor of $\cong$ 2 at τ = 1000 years) but our luminosities for the stiff star differ somewhat (ours is a factor $\cong$ 13 lower at τ = 1000 yrs). We are continuing our discussions with Peter Sutherland and expect to resolve our remaining differences soon.

Tsuruta: It is not correct to compare your results and my results in the reference as you indicate (Tsuruta 1979), because we are talking about two different things. Namely, I was talking about the local surface temperatures, while you are talking about the temperature observed at infinity (i.e. at the earth). Otherwise, your comment agrees qualitatively with my earlier remarks.

Schatzman: I am not clear about pion condensation and the rate of cooling that may result.

Tsuruta: If pion condensates occur in neutron stars, the cooling is so fast that there is no discrepancy with even the lowest observed lower limit to a neutron star surface temperature (e.g. SN1006, if a neutron star exists there). See, for example, S. Tsuruta, 1980, in X-Ray Astronomy, ed. by R. Giacconi and G. Setti, p. 73, Reidel, Dordrecht.

CYCLOTRON LINE EMISSION FROM ACCRETION ONTO A MAGNETIZED NEUTRON STAR

E.E. Salpeter
Physics and Astronomy Departments, Cornell University,
Ithaca, N.Y. U.S.A.

For material accreting along the magnetic field axis of a neutron star, electrons are quantized into Landau orbits. Collisional excitation of the first excited Landau level, followed by radiative decay, leads to the emission of a cyclotron line. The expected line is broad, because the optical depth is large, and its shape is difficult to calculate. Redshifts due to the recoil of a scattering electron and blueshifts due to scattering from the infalling accretion column are being calculated by I. Wasserman, as well as the proton stopping length in the presence of a magnetic field.

When the magnetic field is very strong, the intensity of the cyclotron line is small. When the magnetic field is of moderate strength, the line is broad (and may overlap with the next harmonic) but should still be observable from its affect on angular distribution ("phase" for an X-ray pulsar) and polarization. In principle, the gravitational potential ϕ at the neutron star surface can be obtained from detailed observations and analysis: The line intensity depends critically on $\phi\, m_e/E_{cyc}$, where m_e and E_{cyc} are the mass and the Landau excitation energy of an electron.

D. Sugimoto, D. Q. Lamb, and D. N. Schramm (eds.), Fundamental Problems in the Theory of Stellar Evolution, 233.

EFFECTS OF NUCLEAR BURNING ON X-RAY AND UV EMISSION FROM ACCRETING DEGENERATE DWARFS *

G.J. Weast*+, R.H. Durisen†, J.N. Imamura†, N.D. Kylafis+, and D.Q. Lamb*+
*Harvard-Smithsonian Center for Astrophysics
+Department of Physics, University of Illinois
†Department of Astronomy, Indiana University

The energy liberated by nuclear burning of matter accreting onto degenerate dwarfs can be more than an order of magnitude greater than that available from the release of gravitational potential energy. Nuclear burning therefore significantly alters the characteristics of X radiation from such stars. Here we report the results of two-fluid calculations in which steady burning occurs at various rates, and compare them with our earlier calculations which assumed no burning. If the star has a weak or no magnetic field, we find that nuclear burning enhances the soft X-ray flux emitted from the stellar surface, increases Compton cooling of the emission region and therefore reduces the hard X-ray luminosity and softens the hard X-ray spectrum. On the other hand, if the star has a strong magnetic field we find that nuclear burning enhances the soft X-ray flux emitted from the stellar surface but has little effect on the hard X-ray luminosity and spectrum. We apply the results of our calculations to the AM Her sources and to cataclysmic variables such as SS Cyg and U Gem, and discuss the evidence for and against nuclear burning of accreted material in these objects.

* This paper was presented by D.Q. Lamb

D. Sugimoto, D. Q. Lamb, and D. N. Schramm (eds.), Fundamental Problems in the Theory of Stellar Evolution, 234.

COMPTONIZED X-RAY EMISSION FROM THE ACCRETION DISK AROUND A MASSIVE BLACK HOLE

F. Takahara
Research Institute for Fundamental Physics, Kyoto University, Kyoto, Japan

We report the Monte Carlo simulation of the unsaturated Comptonization to probe the phyiscal state of the accreting plasma near a compact object. We calculate the emission spectrum and the Compton cooling rate of the plane-parallel hot plasma layer with the finite optical depth to the Thomson scattering, τ. The distribution of electrons is assumed to be a relativistic Maxwellian with the temperature T_e. The energy and angular distribution of the soft photons impinging on the plasma layer is taken to be Planckian with the temperature T_{ph}. We use the Klein-Nishina formula for the scattering between a photon and an electron.

For the two temperature disk model we have examined the cases $kT_e = m_e c^2$, $0.25\ m_e c^2$ and $0.0625\ m_e c^2$. The obtained spectrum turns out to be a power law if $y \equiv 4\tau^2 kT_e/m_e c^2 \lesssim 3$. For $y \gg 1$, it deviates from a power law and forms a bump near $h\nu \sim kT_e$. To fit the power law energy index α and the cooling rate per unit surface area, F_{compt}, to the observations of CygX-1 ($\alpha \sim 0.6$, $F_{compt} \sim 10^{22}$ erg cm^{-2}s^{-1}), we have found that the efficiency of Comptonization is rather low, i.e., $F_{compt}/F_{in} \approx 10 \sim 30$ and that an enormous flux of soft photons is needed to obtain the observed hard X-ray luminosity. It seems to be rather difficult to realize such a situation when one considers the geometry of the two-temperature disk.

As an alternative model we have also examined the hot corona model with $kT_e = m_e c^2$, $\tau \sim 0.03$, $kT_{ph} \sim 6 \times 10^{-4} m_e c^2$.

DISCUSSION

Schatzman: What are the two temperatures? What is the source of soft photons?

Takahara: "Two temperatures" means that the ion temperature and the electron temperature are different. For the two temperature disk model, the soft photon source is attributed to thermal emission from the cool middle of the disk.

D. Sugimoto, D. Q. Lamb, and D. N. Schramm (eds.), Fundamental Problems in the Theory of Stellar Evolution, 235.

ROTATION AND STELLAR EVOLUTION *

R. Kippenhahn, H.-C. Thomas
Max-Planck-Institut für Physik und Astrophysik, Institut für Astrophysik, Karl-Schwarzschild-Str. 1, 8046 Garching, W.-Germany

* This paper was presented by R. Kippenhahn.

Introduction

Does rotation influence stellar evolution? Does it cause observational effects other than line broadening? Can rotation be responsible for mixing of chemical elements throughout the star? Do evolved stars have rapidly rotating cores? This, for instance, is of interest if one wants to compute the details of supernova events. We are not sure whether rotation has really important effects on the life of a star. There might be no rapidly rotating cores. If we think that a fossile general magnetic field couples core and envelope of an evolved star, the core will always be slowed down by the big inertial momentum of the outer regions.

Indeed, there is observational evidence that rotation in the very interior of a star cannot be too important. White dwarfs seem to be rather slow rotators which indicates that they were slowly rotating when they still were cores of evolved stars. But we do not know too much about white dwarf rotation. They certainly do not rotate critically which would demand an equatorial velocity of 5000 km/sec. But the white dwarf in nova DQ Her with its rotational period of 142 sec indicates a rather rapid rotation with $\omega \approx 0.049$ sec^{-1} (compared to $\omega_{crit} \approx 0.32$ if we assume that the mass is 0.5 $M_\odot$). This dwarf rotates much faster than it would if it were coupled by a magnetic field to a red giant envelope. Then its angular velocity would only be $\approx 10^{-6}$. But we do not know whether accretion has sped up the rotation since the formation of the white dwarf out of an evolved red giant or a supergiant. But the high angular velocity can only have been obtained by accretion if the white dwarf has increased its mass by about 3%. This, on the other hand, seems to be a rather high amount of mass accreted, and therefore this system might give a hint that magnetic fields in evolved stars cannot couple completely cores and envelopes with respect to their rotation. There is also indication that the crab pulsar after the supernova event rotated faster than one would expect if it was formed out of a core which was in solid body rotation with a red supergiant envelope. Hardorp (1974) discussed the empirical

D. Sugimoto, D. Q. Lamb, and D. N. Schramm (eds.), Fundamental Problems in the Theory of Stellar Evolution, 237–256.

facts and came to the conclusion that there is some coupling between core and envelope, but that the rotation of the core is not completely slaved to that of the envelope.

But not only the question of rapidly rotating cores is important. Paczynski (1973) showed that the depletion of C relative to N in some early-type stars can be explained by Eddington-Vogt-circulation which mixes material (partially processed by the CNO cycle in the region near but outside the convective core) into the outer regions. Is the carbon depletion in these stars evidence for the existence of Eddington-Vogt circulation? Cottrell and Norris (1978) tried to explain the Bidelman-MacConnell weak g-band stars by circulation caused by rotation. Sweigart and Mengel (1979) used circulation to explain $^{12}C/^{13}C$ ratios and the weak g-band stars among evolved stars. They find a sufficiently big effect if they assume that ω is considerably bigger near the bottom of the convective zone than on the top. They give some arguments for that case, but since not very much is known about the rotation of convective regions, it is not clear that the stars these authors investigate really do have enough angular velocity to provide the mixing.

At the present moment, we are far from understanding how the angular velocity distribution of a star changes during its evolution. Even if one starts out with a rather simple angular velocity law at zero age main sequence, for instance assuming solid body rotation hoping that in the earlier Hayashi phase all differential rotation has been washed out, even then the future is rather unknown, even if one neglects magnetic effects. There is general acceptance that regions of varying molecular weight can create barriers which cannot be penetrated by circulation and which therefore insulate different regions in a star with respect to their rotation. But in chemically homogeneous radiative regions we are rather uncertain about the evolution of angular velocity distributions. Early attempts to predict the rotation of the interior of an evolved star have been made by Kippenhahn (1962), Kippenhahn, Thomas, Weigert (1965), Kippenhahn, Meyer-Hofmeister, Thomas (1969). Recently elaborate computations have been made by Endal and Sofia (1976), (1978) to follow the angular velocity distribution into the very advanced stages of stellar evolution. All these computations show that as long as there is no exchange of angular momentum throughout the μ-barriers rapidly rotating cores form which may become rotationally unstable. Whether these rapidly rotating cores really form and whether they are of importance for the fate of the star, is uncertain. But many people, ourselves included, would love to see that nature allows rapidly rotating cores in stars, because rapid rotation is much more interesting than slow rotation. The world would be less exciting without rapidly rotating stellar cores!

In the following we shall give some examples where the physics of rotation is not completely clear, or at least not correctly applied by some authors. We shall concentrate on three topics:

a) Can ω vary on equipotential surfaces?

b) What is the time scale of Goldreich-Schubert-Fricke instabilities?

c) How fast is circulation in surface-near regions?

We think we know the answer to the first two questions, while the third one might still be considered as open.

Can ω vary on equipotential surfaces?

Endal and Sofia (1978) assume that ω must be constant on equipotential surfaces. Law (1980) in her Yale thesis on differential rotation of low mass stars also assumed it. Papaloizou and Pringle (1979) believe that ω must be constant on equipotential surfaces and use it as an argument against the theory of accretion belts proposed by Kippenhahn and Thomas (1978). As long as one considers only axisymmetric perturbations the situation is clear. Along equipotential surfaces a compressible fluid behaves like an incompressible one because the equipotential surfaces are surfaces of constant density. Any exchange between elements on the same equipotential surface does not require compression or expansion. Therefore, along an equipotential surface the Rayleigh criterion is necessary and sufficient for stability:

$$\frac{ds^2\omega}{ds} \geq 0 \tag{1}$$

where s is the distance from the axis of rotation. This would indicate that there can be a variation of ω along equipotential surfaces as long as the condition (1) is fulfilled. It is well known that this criterion can easily be derived by computing the work which has been put into centrifugal force if one exchanges two tori of equal mass. Condition (1) then is equivalent with the condition that the net work against centrifugal acceleration is positive; one has to put energy in in order to make the exchange. It therefore does not occur spontaneously. The situation seems to be more complicated if non-axisymmetric perturbations are taken into account. At first sight it does not seem that conservation of angular momentum during the exchange is a good approximation, because then azimuthal pressure gradients occur, and by these angular momentum can, in principle, be transported. Indeed Cowling (1951) postulates "if more general (non axisymmetric) displacements are considered, azimuthal pressure gradients insure that the specific angular momentum does not remain constant". But in an inviscid fluid it is difficult to transport momentum even if there are azimuthal pressure gradients. A typical example is a rigid sphere carried horizontally through an inviscid fluid. It is well known that there is no drag. All the momentum lost on the front side of the sphere to the surrounding liquid is gained back on the rear side. The flow is a potential flow and no momentum is lost from the sphere. If this heuristic argument were correct, then Rayleigh's criterion (1) would also be valid for non-axisymmetric perturbations and ω could vary on equipotential surfaces as long as (1) is fulfilled. The general reluctance of many authors to believe this, might come from the feeling that differential rotation might cause shear instabilities. Also Zahn (1975) seems to be influenced by this argument when he tries to transform the Richardson number for plane parallel motion into one which can be used for rotational flow, where he replaced the plane parallel shear $\partial v/\partial z$ by $r d\omega/dr$ and concluded that shear instability (which for motion along equipotential surfaces cannot be stabilized by buoyancy) permits only ω = constant on equipotential surfaces. Cowling's and Zahn's arguments favour that in stars ω must be constant on equipotential surfaces and that this condition should be achieved within a dynamical time scale. But, as already mentioned, motions along equipotential surfaces in a stratified gas are very similar to motions in compressible fluids. Indeed Cowling's effect of azimuthal pressure components as well as Zahn's form of defining shear for a rotational flow should also hold for incompressible fluids. What, then do we know about the stability of flow? Chandrasekhar (1961) published a proof that the Rayleigh criterion (1) is necessary and sufficient for stability, even if one includes non-axisymmetric perturbations. This would indicate that an angular velocity distribution with $\omega \sim s^{-2}$ is marginal while $\omega \sim s^{-3/2}$, for instance, would be stable (the latter would be unstable according to Zahn's arguments). In order to make the confusion complete, in the 1970 print of Chandrasekhar's book the old statement is revised and the author finds that instabilities cannot be excluded even if condition (1) is fulfilled.

Fortunately, in the case of incompressible fluids experiment can give the answer. In Fig. 1 one can see that the marginal state is that of $d(s^2\omega)/ds = 0$. If one starts out with an angular velocity distribution which violates (1), then instability followed by a re-distribution of angular momentum will occur until the rotational state is marginally stable.

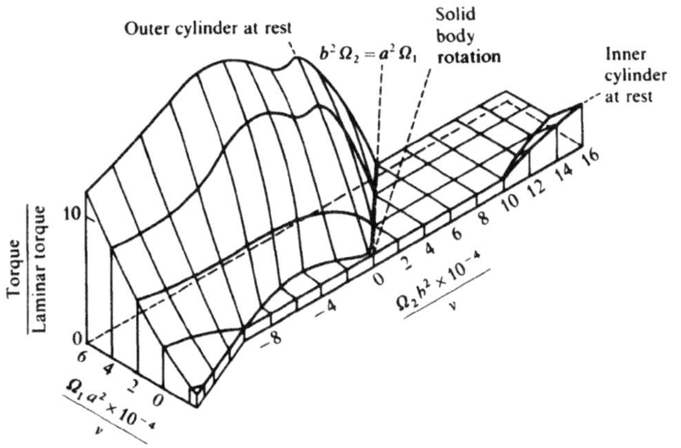

Figure 1 The stability of Couette flow (Joseph, 1976)

The torque in units of the laminar torque between two coaxial cylinders of radii a,b and of angular velocities Ω_1, Ω_2 is plotted. If its value is one the flow is laminar, the motion of the fluid is stable. If it is bigger than one the torque is enhanced by turbulence, the flow is unstable (the graph is based on experiments with rotating fluids). The coordinates in the Ω_1-Ω_2-plane are normalized with respect to the kinematic viscosity ν.

As one can see there is a line of marginal stability with $b^2\Omega_2 = a^2\Omega_1$, this is the line of constant specific angular momentum. The case of solid body rotation is stable. Between the two broken straight lines which indicate solid body rotation and constant specific angular momentum differential rotation is stable. There is also a wide area of differential rotation where the angular velocity increases outwards and the flow is stable. Only if the angular velocity increases outwards too rapidly instability sets in although the Rayleigh criterion predicts stability.

Solid body rotation, as one can see from Fig. 1, is not marginal, it is safe in the stable region. Only if condition (1) is fulfilled and the angular velocity gradient is very high, does a new instability set in. This instability can be compared with shear instability in the plane parallel case. Indeed, if the characteristic length scale for the variation of ω is small compared to the radius of curvature, the fluid behaves just like it does in the plane parallel case. But the instability on the left in Fig. 1 is well separated from that on the right by a large region of stability. The instability on the left can also be compared with the plane parallel shear instability, but one has to keep in mind that, for rotation, shear does not mean deviation from ω = constant but from constant specific angular momentum!

We therefore conclude that if one wants to compute the evolution of the angular velocity distribution in a star, one cannot put ω = const. on equipotential surfaces. The ω-distribution on equipotential surfaces depends on the history of the star. There is no effect which smoothes out differences in ω along equipotential surfaces.

Furthermore we want to emphasize that in the case of accretion belts (Kippenhahn, Thomas (1978)) there is no reason to assume that shear instabilities distribute the rapidly rotating accreted material from the equatorial region over the whole surface of the accreting star. Shear instability is sometimes claimed to be responsible for turbulence, and, therefore, for turbulent friction in accretion disks. But Kepler's law $\omega \sim s^{-3/2}$ is stable; there is no shear instability in accretion disks!

What is the time scale of the Goldreich-Schubert-Fricke instability?

When at least one of the conditions

$$\partial(s^2\omega)/\partial s \geq 0 \ , \quad \partial\omega/\partial z = 0 \tag{2}$$

is violated (s, θ, z being cylindrical polar coordinates about the rotation axis), the angular velocity distribution in a star is secularly unstable (Goldreich & Schubert, 1967, Fricke 1968). Goldreich and Schubert originally estimated the time scale in which this instability can redistribute the angular velocity distribution within the sun to be of the order of ten years. Later Colgate (1968) and Kippenhahn (1969) using different physical arguments estimated the time scale to be at least of the order of the Kelvin-Helmholtz time scale of the star, which, for the sun, is about 10^7 years. There are some hints that the Kelvin-Helmholtz time scale τ_{KH} is not very good as a lower bound. Since the instability must be driven by rotation, the time scale should depend on the angular velocity. Quickly rotating stars, being more unstable, should redistribute their angular momentum in a shorter time scale than slowly rotating stars. The simplest time scale which fulfills this condition is $\tau_{EV} \approx \tau_{KH}/\chi$, the Eddington-Vogt time scale which is the time scale in which meridional circulation caused by rotation moves throughout the star (it is about 10^{13} years for the sun). Here χ is a mean value over the star for the ratio of the absolute values of centrifugal to gravitational acceleration. Indeed, James and Kahn (1970, 1971) argued in favour of a time scale comparable to the Eddington-Vogt time scale of the star. But their theory has never been fully developed and it seems that it is based on different physical arguments than the estimate which has been given by Kippenhahn, Ruschenplatt, Thomas (1980a) and which we will now discuss in more detail.

In the paper mentioned above we give new arguments which favour a time scale comparable to that of the Eddington-Vogt circulation. Our estimates derived from following the motions due to the instability into the non-linear domain. Our arguments are similar to those which we used to estimate the time scale of a secularly unstable distribution of molecular weight in a star (Kippenhahn et al. 1980b). The simplest form of the argument is the case in which the condition $d(s^2\omega)/ds \geq 0$ is violated. Then any torus of matter which has expanded its big radius s_o by ℓ has - compared to its surroundings - an excess angular velocity

$$D\omega = -\frac{1}{s^2}\,\ell\,\frac{d\Theta}{ds}\,, \quad \Theta = s^2\omega\,. \tag{3}$$

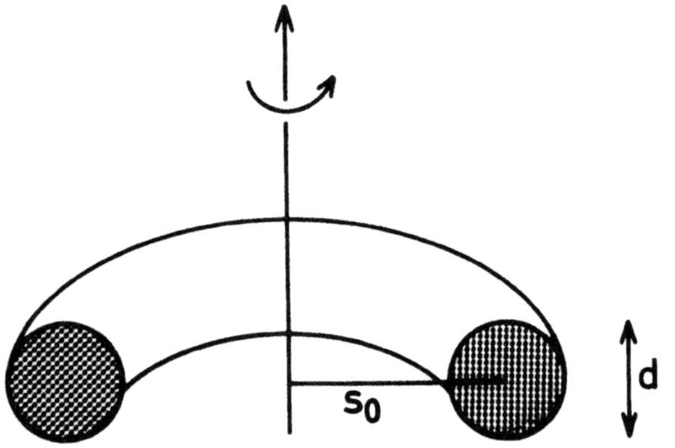

Figure 2 A torus of "big" radius s_o and "small" diameter d around the axis of rotation.

In order to have hydrostatic equilibrium during the expansion of the big radius it had to change its smaller diameter d in such a way that the density in the expanded torus is bigger so that the gravitational force and the centrifugal force compensate each other. Since there must be pressure equilibrium with the surrounding, the matter in the torus must have a temperature higher by DT from that of the surrounding:

$$\frac{DT}{T} = -2\,\chi\,\frac{D\omega}{\omega} < 0 \tag{4}$$

The torus, therefore, is not thermally adjusted. With the time scale τ^* of thermal adjustment it tries to heat up to the temperature of the surroundings. Consequently, it cannot remain steadily in hydrostatic equilibrium but must slowly increase its big radius again. An estimate of the velocity by which the big radius expands can be given (Kippenhahn 1969)

$$v_\omega = \frac{H_p}{(\nabla_{ad}-\nabla)\tau^*}\;2\,\chi\,\frac{D\omega}{\omega}\,,\quad \tau^* = \frac{3c_p\kappa\rho^2\zeta d^2}{8acT^3}\,, \tag{5}$$

where H_p is the pressure scale height. The other quantities have the usual meaning. Motions of this type by which torus-like mass elements expand their big radius and transport angular momentum constitute the mechanism by which the angular momentum is redistributed in order to get a stable or at least marginally stable angular velocity distribution. The question is, how effective this mixing is. Its effectivity depends on the velocity of the mass elements and their mean free path. The velocity as given by equ. (5) depends on the distance ℓ from the region where the mass element originated, as can be seen from equ. (3) and it therefore depends also on the mean free path. Whereas Kippenhahn, in his earlier estimate, assumed that the mean free path is limited by shear instabilities, we have now found a mechanism which destroys the mass element sooner and, therefore, determines the mean free path. This mechanism can be easily understood. While the torus is moving outwards, it is always cooler than the surrounding, its excess temperature follows from equs. (3), (4)

$$\frac{DT}{T} = 2\,\frac{\chi\ell}{\omega s^2}\,\frac{d\Theta}{ds} < 0\,, \tag{6}$$

and the temperature difference becomes bigger the further the mass element has moved from its original position. During its motion it is continuously receiving radiation from the neighbouring region.

It therefore acts as a heat sink for its surrounding and creates a small circulation pattern in its neighbourhood. The topology of the pattern is such that it tries to mix the matter of the mass elements with the surrounding (see Fig. 3).

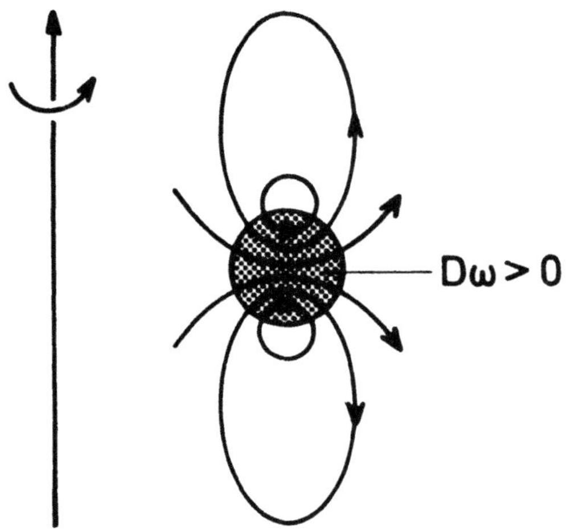

Figure 3 The velocity field in the neighbourhood of a torus with an excess velocity $D\omega > 0$. The matter in the torus (indicated by the dotted area) is cooler. In the area around the torus a circulation system is created which mixes the matter of the torus with its surrounding.

Kippenhahn et al. (1980a) have estimated that this "self-destruction" by mixing is effective within the time during which the torus expands its big radius by a length which corresponds to its small diameter. In equ. (6) we therefore have to replace ℓ by d and one finds

$$|v_\omega| = \frac{2\chi H_p d}{(\nabla_{ad}-\nabla)\tau^* H_\Theta} , \quad H_\Theta = |ds/d\ln\Theta| \tag{7}$$

The redistribution is therefore given by mass elements which move outwards or inwards and mix with the surrounding after they have moved along a distance comparable with their own size*). One can, therefore, define a diffusion coefficient:

$$D = v_\omega \ell = v_\omega d = \frac{2\chi H_p d^2}{(\nabla_{ad}-\nabla)\tau^* H_\Theta} . \tag{8}$$

It should be mentioned that this diffusion coefficient does not depend on the size of the elements since the thermal adjustment time scale τ^* is proportional to d^2. The corresponding diffusion time scale over a distance W is given by

*) Here we have assumed that the characteristic mass element has the form of a torus with small diameter d. The same would hold if the mass element had the topology of a sphere with mean diameter d. If these little "drops" move off or towards the axis of rotation, the motion is no longer axisymmetric. But, as long as friction can be neglected, drops which have an excess velocity can move through the matter as easily as tori, since in a frictionless fluid by azimuthal pressure gradients there is no exchange of angular momentum for reasons which we have already discussed in the foregoing section.

$$\tau_{diff} = \frac{H_\Theta (\nabla_{ad} - \nabla)}{H_p} \frac{\tau^W_{KH}}{2\chi\zeta} \tag{9}$$

where we have introduced the thermal adjustment time scale of the mass of a shell of thickness W, which corresponds to the thermal adjustment time scale of a mass element of thickness d according to the formula

$$\tau^W_{KH} = \frac{W^2}{d^2} \tau^* \tag{10}$$

If W becomes the radius of the star we find for the diffusion time scale

$$\tau_{diff} = \frac{H_\Theta (\nabla_{ad} - \nabla)}{H_p} \frac{\tau_{KH}}{2\chi\zeta} \tag{11}$$

which indeed is of the order of the Eddington-Vogt time scale.

Similar arguments can be used to show that this is also the time scale if the condition $\partial\omega/\partial z = 0$ is violated.

If, therefore, the angular velocity distribution in a star violates the Goldreich-Schubert-Fricke condition, then mass elements of all sizes will start a random motion and redistribute angular momentum. One can describe this random motion as a diffusion process and its effect on mixing angular momentum as kind of turbulent friction. In appendix A we give an estimate of the turbulent viscosity, which is caused by GSF instability.

How fast is circulation in surface-near regions?

Chemical anomalism in A-stars and correlation between pulsational variability and rotational velocity for A stars have provoked a series of papers in which the effect of a fractional sedimentation was used to explain these stars. For reference, see, for instance, the papers by Baglin (1972) and by Vauclair (1976). Although the outer convective regions of these stars are rather shallow and, therefore, convection does not contribute considerably to mixing, there is meridional circulation caused by the stellar rotation. Baglin (1972) takes this into account and uses shear instabilities caused by the circulation to explain why, in some stars, sedimentation seems to be effective but not in others.

In order to see the principles we assume an unevolved star of two solar masses with $R = 1.61\ R_\odot$, $X = 0.732$, $Y = 0.240$, $v_{equ} = 50$ km/sec. For this model we have $\chi = 0.011$ for surface-near equatorial regions. For our estimates we take the region with $\log P = 5.03$ as a representive layer, which is just below the stellar hydrogen convective zone. There we have

$H_p = 1.75\times10^8$, $1-\beta = 0.014$, $\nabla_{ad}-\nabla = 0.115$, $\ell_{opt} = (\kappa\rho)^{-1} = 1.23\times10^6$.

The classical Sweet formula (Sweet 1950) for this star would give a radial component of the circulation velocity in surface-near regions

$$v_r \approx 6.7\times 10^{-4}\chi = 7.4\times 10^{-6} \tag{12}$$

this velocity would not prevent sedimentation, which according to Baglin's formula (1972) is $v_{sed} = 1.4\times 10^{-4}$. But as Baker and Kippenhahn (1959) have shown, for non-uniform rotation, an additional term becomes important in surface-near regions. If there is no solid body rotation, their estimate gives

$$v_r \approx 6.7\times 10^{-4} \frac{\bar{\rho}}{\rho}\chi = 1.65\times 10^{2} \tag{13}$$

Here $\bar{\rho}$ is the mean density $3M/4\pi R^3$ of the star. But even if there is solid body rotation, another term appears if one takes into account second order effects in the small quantity χ:

$$v_r \approx 4.5\times 10^{-4} \frac{\bar{\rho}}{\rho}\chi^2 = 1.21 \tag{14}$$

This has been shown by Öpik (1951) and Mestel (1966).

As already mentioned, the Sweet term (12) does no harm to sedimentation, the other two would mix faster than separation by sedimentation. In the paper by Vauclair (1976) an argument by Osaki (1972) was used to ignore the dangerous $\bar{\rho}/\rho$-terms. In the following we show where the difficulties lie in getting the meridional circulation effects sufficiently small.

Osaki (1972) notes, correctly, that the estimates (12) - (14) are based on the assumption of a steady state. If time derivatives are taken into account in the energy equation, the situation is different. He considers the case where the terms containing time derivatives in the energy equation are large compared to the terms which describe the transport of energy by circulation. In this case one obtains an equation similar to the heat equation and one can assume that a steady state is reached after some time. But Osaki is aware that this steady state is GSF unstable and that, therefore, the angular velocity distribution will not remain in this state. We have a minor objection to the Osaki picture in the following sense: The equations from which Osaki derives his equations are identical with those from which the meridional motion caused by GSF instability is derived. Therefore, we do not think that first a steady state is reached and then instability sets in, but that Osaki's steady state is never reached. And, probably, this is what Osaki had in mind when he wrote that a "steady state may likely be established between the meridional circulation and the irregular motion due to this instability". Therefore if one wants to apply Osaki's reduction of the meridional circulation to the sedimentation problem, one must, at least, estimate how big the circulation is in the final state suggested by Osaki.

Recently we have rediscussed the problem and it seems that we arrive at a solution, which is probably equivalent to what Osaki has suggested. If one allows a small deviation ω^* from solid body rotation Ω, one can determine this deviation in such a way that the Baker-Kippenhahn term and the Öpik-Mestel term cancel each other. As is shown in detail in appendix B, ω^* is given by

$$\frac{\omega^*}{\omega} = \frac{4}{15}\frac{\Omega^2 r^3}{GM}\,[4\cot^2\theta(1-\cos^3\theta)-1] \tag{15}$$

Its θ-dependence is given in Fig. 4.

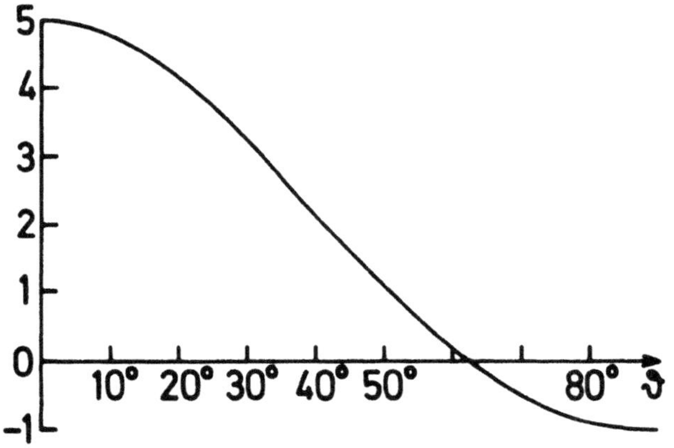

Figure 4 To demonstrate the θ-dependence of the angular velocity ω the function $4\cot^2\theta(1-\cos^3\theta)-1$ is plotted as a function of θ. It changes sign at $\theta = 62.^\circ 2$.

If, in the surface-near regions of a star, the angular velocity distribution is given by $\omega = \Omega + \omega^*$ then the meridional velocity is given solely by the Sweet estimate (12) up to second order terms in χ and cannot overcome sedimentation.

Unfortunately, the angular velocity distribution as given by equ. (15) is GSF unstable since ω varies on cylinders coaxial with the axis of rotation. As a consequence of this, we expect that turbulent friction will occur as given by equ. (A5). This friction would immediately cause deviations from the angular velocity distribution given by eq. (15), and, therefore, the compensation given by eq. (B23) would be distorted. If one demands a steady state the circulation must compensate the flux of angular momentum caused by friction:

$$s\rho(\underline{v}\,\underline{\nabla})(\omega s^2) = \underline{\nabla}\,(\eta_t s^3 \underline{\nabla}\omega). \tag{16}$$

The ω on the left-hand side of eq. (16) is roughly equal to Ω = const. while $\underline{\nabla}\omega$ on the right can be replaced by $\underline{\nabla}\omega^*$. Then an estimate for the right-hand side is given by

$$|\underline{\nabla}\ (\eta_t s^3 \underline{\nabla}\omega)| \approx \chi\Omega\ s^2\eta_t/H_p \tag{17}$$

where we have made use of the following estimates:

$$|\nabla\eta_t| \approx \eta_t/H_p\ , \quad s \approx r \approx R, \quad |\underline{\nabla}\omega| \approx \chi/R \tag{18}$$

and where we have assumed $H_p << r$. This gives with eq. (A5)

$$v_s \approx \frac{\chi\eta_t}{2\rho H_p} \approx \frac{16}{5}\frac{(1-\beta)\chi^3 c}{(\nabla_{ad}-\nabla)\zeta R\kappa\rho}\ . \tag{19}$$

Since $v_s \approx v_\Theta$ and $v_r \approx H_p v_\Theta/R$ we find

$$v_r \approx \frac{16}{5}\frac{(1-\beta)}{(\nabla_{ad}-\nabla)\zeta}\frac{H_p c}{\kappa\rho R^2}\chi^3. \tag{20}$$

For the stellar model used here we find with $\zeta = 1/6$ (spherical turbulent elements)

$$v_r \approx 1.4\times 10^{-3}. \tag{21}$$

The mixing due to circulation is about ten times more effective than separation by sedimentation. However our estimates are only approximate, and we think, therefore, that sedimentation as a mechanism for the separation of chemical elements in surface-near regions of stars with ineffective outer convective zones cannot be excluded completely. In our picture there are two effects which reduce the circulation velocity: the special angular velocity distribution which kills the $\bar{\rho}/\rho$-effects and the ineffectivness of the unstables modes in GSF unstable region due to self-destruction.

In the foregoing considerations we have shown only that there exists an angular velocity distribution which strongly reduces the circulation speed in surface near regions but we have not been able to show how out of a given initial angular velocity distribution such a special state of rotation could evolve.

We thank Drs. D. Galloway, L. Mestel, F. Meyer, Y. Osaki and R. Wegmann for stimulating discussions, Dr. J. Kirk corrected the English manuscript. Part of this work was supported by the Deutsche Forschungsgemeinschaft.

APPENDIX A: The turbulent viscosity induced by GSF instabilities

We define two characters with length scales:

$$H_\omega = |\partial z/\partial \ln \omega|$$

$$H_\theta = \begin{cases} -\partial s/\partial \ln\theta \ , & \text{if } \partial s/\partial \ln\theta < 0 \\ \infty \ , & \text{if } \partial s/\partial \ln\theta \geq 0 \ , \ \theta = s^2\omega . \end{cases} \qquad \text{(A1)}$$

They are measures for the degree of violation of the GSF instabilities. In case of stability they are infinite. The smaller they are the more unstable is the angular velocity distribution. In principle we should derive the turbulent viscosity separately for the two cases in which each of the two GSF instability conditions is violated. Instead of this we use the minimum of the two quantities

$$H = \mathrm{Min}(H_\omega, H_\theta). \qquad \text{(A2)}$$

Indeed as one can see from the detailed derivation in the paper by Kippenhahn et al. (1980a) the velocity of a turbulent element of size d which has moved along a distance of its size in both cases can be described by

$$v_t = \frac{H_p}{H} \frac{2\chi}{\nabla_{ad}-\nabla} \frac{d}{\tau^*} \qquad \text{(A3)}$$

If we then define an eddy viscosity by

$$\eta_t = \rho v_t d \qquad \text{(A4)}$$

we find

$$\eta_t = \frac{16}{3} \frac{\chi}{\nabla_{ad}-\nabla} \frac{H_p}{H} \frac{1}{\zeta} \frac{acT^3}{c_p \kappa \rho} \qquad \text{(A5)}$$

where again ζ is a dimensionless quantity of the order 1 which depends on the geometry of the elements. With this eddy viscosity one can define a timescale over which this viscosity can change a GSF unstable angular velocity distribution over a certain distance. This timescale then is of the order of the diffusion timescale already derived in eq. (9).

It is of interest to compare the eddy viscosity given in eq. (A5) with the radiative viscosity given by

$$\eta_R = \frac{2}{15} \frac{aT^4}{c\kappa\rho} . \qquad \text{(A6)}$$

The ratio of the two viscosities then is given by simple dimensionless factors

$$\frac{\eta_t}{\eta_R} = 40 \frac{\chi}{\nabla_{ad}-\nabla} \frac{H_p}{H} \frac{1}{\zeta} \frac{c^2}{v_c^2}$$

where v_c is the velocity of sound.

APPENDIX B: The angular velocity distribution with highly reduced circulation

We start from Eq. (7) of Baker & Kippenhahn (1959)

$$(\mathrm{div}\ \underline{F})^{(2)} = \frac{d}{d\psi}\left(- \frac{4acT^3}{3\kappa\rho} \frac{dT}{d\psi}\right) (\mathrm{grad}^2\psi)^{(2)} + \\ + \frac{4acT^3}{3\kappa\rho} \frac{dT}{d\psi} \left(\frac{1}{s} \frac{\partial(\omega^2 s^2)}{\partial s}\right)^{(2)} \tag{B1}$$

This formula holds for angular velocity distributions which have conservative centrifugal acceleration. ψ is the sum of gravitational and centrifugal potential. Any axisymmetric function A can be considered as the sum of a function $A^{(o)}$ which is constant on ψ-surfaces and a function $A^{(2)}$ which has vanishing mean values on all ψ-surfaces. Eq. (B1) is a relation between these latter types of functions. We apply eq. (B1) not only to the angular velocity distribution $\omega = \Omega$ = const. (which has a conservative centrifugal field) but also to distributions in the neighbourhood of that: $\omega = \Omega + \omega^*(s,z)$. As long as $|\omega^*/\Omega| < \chi$ the formula is still valid in second order in χ.

We want to show that the second order term of the first summand produced by solid body rotation (Öpik, 1951; Mestel, 1966), which does not depend on the density, can be compensated by a small correction ω to solid body rotation with Ω = const. via the second summand. For this we have to evaluate the first summand to second order in the rotation parameter $\chi = \Omega^2 r^3/GM$.

The luminosity is defined as

$$L = \int_\psi \underline{F}\, d\underline{\sigma} = - \frac{4acT^3}{3\kappa\rho} \frac{dT}{d\psi} \int_\psi |\nabla\psi| d\sigma. \tag{B2}$$

where $d\sigma$ is the surface element of an equipotential surface. With M_ψ defined as the total mass interior to the surface ψ = const. one has

$$\frac{\partial\psi}{\partial r} = \frac{GM_\psi}{r^2} - \Omega^2 r \sin^2\theta \tag{B3}$$

$$\frac{1}{r}\frac{\partial\psi}{\partial\theta} = - \Omega^2 r \sin\theta \cos\theta \tag{B4}$$

and therefore

$$d\sigma = 2\pi r^2 \sin\theta\, d\theta \sqrt{1 + \left(\frac{dr}{r d\theta}\right)^2}\ . \tag{B5}$$

Since $dr/rd\theta$ is of the order χ it can be neglected. And combining Eqs. (B3) and (B4) we obtain

$$|\nabla\psi| = \frac{GM_\psi}{r^2}(1 - \chi\sin^2\theta) + O(\chi^2). \tag{B6}$$

The integration then gives

$$\int_\psi |\nabla\psi| d\sigma = 4\pi GM_\psi (1 - \frac{2}{3}\chi_e) \tag{B7}$$

with $\chi_e = \Omega^2 r_e{}^3/GM_\psi$, r_e being the equatorial radius of the surface ψ = const. Therefore, we can write

$$-\frac{4acT^3}{3\kappa\rho}\frac{dT}{d\psi} = \frac{L}{4\pi GM_\psi}(1 + \frac{2}{3}\chi_e) \tag{B8}$$

and, to evaluate its derivative with respect to ψ, we need to compute $dM_\psi/d\psi$ and $d\chi_e/d\psi$. One has

$$dM_\psi = \rho dV_\psi = \rho d\psi \int_\psi \frac{d\sigma}{|\nabla\psi|} \tag{B9}$$

and

$$\int_\psi \frac{d\sigma}{|\nabla\psi|} = \frac{4\pi}{GM_\psi}\int_0^{\frac{\pi}{2}} d\theta \sin\theta\, r^4 (1 + \chi\sin^2\theta). \tag{B10}$$

Since this term depends on ρ we take only the lowest order, i.e. $r^4 = r_e{}^4$ = const. and obtain

$$\int_\psi \frac{d}{|\nabla\psi|} = \frac{4\pi r_e{}^4}{GM_\psi}. \tag{B11}$$

From the definition of χ_e one gets

$$\frac{d\chi_e}{d\psi} = 3\frac{\chi_e}{r_e}\frac{dr_e}{d\psi} \tag{B12}$$

with

$$\frac{dr_e}{d\psi} \approx \frac{r_e{}^2}{GM_\psi}. \tag{B13}$$

Differentiating Eq. (B8) leads to

$$\frac{d}{d\psi}\left(-\frac{4acT^3}{3\kappa\rho}\frac{dT}{d\psi}\right) = \frac{L}{4\pi GM_\psi}\left(-\frac{1}{M_\psi}\frac{dM_\psi}{d\psi} + \frac{2}{3}\frac{d\chi_e}{d\psi}\right) \tag{B14}$$

and inserting Eqs. (B9), (B11), (B12), and (B13) one has

$$\frac{d}{d\psi}\left(-\frac{4acT^3}{3\kappa\rho}\frac{dT}{d\psi}\right) = \frac{L}{4\pi GM_\psi}\left(-\frac{4\pi r_e^4\rho}{GM_\psi^2} + 2\chi_e\frac{r_e}{GM_\psi}\right). \tag{B15}$$

If we define a mean density by $M_\psi = \frac{4}{3}\pi r_e^3\bar{\rho}$, we obtain finally

$$\frac{d}{d\psi}\left(-\frac{4acT^3}{3}\frac{dT}{d\psi}\right) = -\frac{L\rho}{M_\psi}\frac{r_e^4}{G^2M_\psi^2}\left(1 - \frac{2}{3}\chi_e\frac{\bar{\rho}}{\rho}\right) \quad . \tag{B16}$$

To obtain $(\mathrm{grad}^2\psi)^{(2)}$ we have to split $(\nabla\psi)^2$ into one part which is constant on ψ-surfaces and another for which the mean value over a ψ-surface is zero. We do this at our level of approximation by substracting from $\mathrm{grad}^2\psi$ a term proportional to the Legendre polynomial P_2 such that a term is left which depends on ψ only. For this we need the function $r = r(\theta)$ for ψ = const. From Eqs. (B3) and (B4) one gets

$$\frac{dr}{rd\theta} = \frac{\Omega^2r^3}{GM_\psi}\sin\theta\cos\theta + O(\chi^2) \tag{B17}$$

and after integration

$$\frac{r_e^3}{r^3} = 1 + \frac{3}{2}\chi_e\cos^2\theta. \tag{B18}$$

For the gradient of ψ we have

$$(\nabla\psi)^2 = \frac{G^2M_\psi^2}{r^4}(1 - 2\chi_e\sin^2\theta) + O(\chi_e^2) \tag{B19}$$

and by inserting Eq. (B18) we obtain

$$(\nabla\psi)^2 = \frac{G^2M_\psi^2}{r_e^4}(1 + 2\chi_e - 4\chi_e\sin^2\theta) \tag{B20}$$

which we split into two terms according to the rule given above:

$$(\nabla\psi)^2 = \frac{G^2M_\psi^2}{r_e^4}\left(1 - \frac{2}{3}\chi_e\right) + \frac{8}{3}\chi_e\frac{G^2M_\psi^2}{r_e^4}\left(1 - \frac{3}{2}\sin^2\theta\right). \tag{B21}$$

So we now have

$$(\mathrm{grad}^2\psi)^{(2)} = \frac{8}{3}\chi_e\frac{G^2M_\psi^2}{r_e^4}\left(1 - \frac{3}{2}\sin^2\theta\right) \tag{B22}$$

and therefore can equate the two terms of Eq. (B1) which do not depend on the density:

$$\frac{4acT^3}{3\kappa\rho}\frac{dT}{d\psi}\left(\frac{1}{s}\frac{\partial(\omega^2 s^2)}{\partial s}\right)^{(2)} = -\frac{16}{9}\frac{L\bar{\rho}}{M_\psi}\chi_e^{\ 2}\left(1 - \frac{3}{2}\sin^2\theta\right). \tag{B23}$$

With the help of Eq. (B8) this transforms to

$$\frac{1}{\Omega^2}\left(\frac{1}{s}\frac{\partial(\omega^2 s^2)}{\partial s}\right)^{(2)} = \frac{16}{3}\frac{\Omega^2 r^3}{GM_\psi}\left(1 - \frac{3}{2}\sin^2\theta\right). \tag{B24}$$

Here we have replaced r_e by r because the θ-dependance of r is of higher order. We now write $\omega = \Omega+\omega^*(s,z)$ and obtain with the approximation $\omega^* \ll \Omega$

$$\frac{1}{s}\frac{\partial(\omega^2 s^2)}{\partial s} = 2\Omega^2 + 2\Omega\,\frac{1}{s}\frac{\partial(\omega^* s^2)}{\partial s} \tag{B25}$$

which, together with Eq. (B24) leads to

$$\frac{\partial(\omega^* s^2)}{\partial s} = \frac{8}{3}\frac{\Omega^3(s^2+z^2)^{3/2}s}{GM_\psi}\left(1 - \frac{3}{2}\frac{s^2}{s^2+z^2}\right). \tag{B26}$$

Integration over s results in

$$\omega^* = \frac{4}{3}\frac{\Omega^3}{GM}\frac{1}{s^2}\left[\frac{1}{5}(4z^2-s^2)(s^2+z^2)^{3/2} + f(z)\right], \tag{B27}$$

where we have replaced M_ψ by the total mass M, which corresponds to neglecting a term of the order $\rho/\bar{\rho}$. To determine the function f(z) we require ω^* to remain finite for $s \to 0$, so

$$f(z) = -\frac{4}{5}z^5 \tag{B28}$$

and we finally have

$$\omega^* = \frac{4}{15}\frac{\Omega^3}{GM}\frac{1}{s^2}\left[(4z^2-s^2)(s^2+z^2)^{3/2} - 4z^5\right], \tag{B29}$$

or, in spherical polar coordinates

$$\omega^* = \frac{4}{15}\frac{\Omega^3 r^3}{GM}\left[4\cot^2\theta(1-\cos^3\theta) - 1\right]. \tag{B30}$$

The angular velocity distribution $\Omega+\omega^*$ has to be distinguished from the so-called circulation-free angular velocity distributions as they are discussed by Schwarzschild (1942), Kippenhahn (1963), Roxburgh (1964). The latter describe states of rotation which up to first order in χ do not cause circulation. The velocity distribution determined in this appendix demands a circulation which up to second order in χ has no $\rho/\bar{\rho}$-term. Nevertheless it gives the normal Sweet type of circulation velocity in the first order. But this circulation is unimportant for the problem of sedimentation.

REFERENCES

Baglin, A.: 1972, Astron. Astrophys. 19, p. 45.
Baker, N., Kippenhahn, R.: 1959, Z. Astrophys. 48, p. 140.
Chandrasekhar, S.: 1961, in: "Hydrodynamic and Hydrogmagnetic Stability", Oxford Clarendon Press
Colgate, S.A.: 1968, Astrophys. J. (Letters) 153, p. L81.
Cowling, T.G.: 1951, Astrophys. J. 114, p. 272.
Cottrell, p.L., Norris, J.: 1978, Astrophys. J. 221, p. 896.
Endal. A.S., Sofia, S.: 1976, Astrophys. J. 210, p. 184.
Endal, A.S., Sofia, S.: 1978, Astrophys. J. 220, p. 279.
Fricke, K.: 1968, Z. Astrophys. 68, p. 317
Goldreich, P., Schubert, G.: 1967, Astrophys. J. 150, p. 571.
Hardorp, J.: 1974, Astron. Astrophys. 32, p. 133.
James, R.A., Kahn, F.D.: 1970, Astron. Astrophys. 5, p. 232.
James, R.A., Kahn, F.D.: 1971, Astron. Astrophys. 12, p. 332.
Joseph, D.D.: 1976, "Stability of Fluid Motions, Springer, Heidelberg.
Kippenhahn, R., Thomas, H.-C., Weigert, A.: 1965, Z. Astrophys. 61, p. 241.
Kippenhahn, R., Meyer-Hofmeister, E., Thomas, H.-C.: 1969, Astron. Astrophys. 5, p. 155.
Kippenhahn, R., Thomas, H.-C.: 1978, Astron. Astrophys. 63, p. 265.
Kippenhahn, R., Ruschenplatt, G., Thomas, H.-C.: 1980a, Astron. Astrophys. (in press).
Kippenhahn, R., Ruschenplatt, G., Thomas, H.-C.: 1980b, Astron. Astrophys. (in press).
Kippenhahn, R.: 1963, Proc. of the Varenna Conference "Star Evolution", ed. L. Gratton, p. 330.
Kippenhahn, R.: 1969, Astron. Astrophys. 2, p. 309.
Law, W.Y.: 1980, Yale Thesis.
Mestel, L.: 1966, Z. Astrophys. 63, p. 196.
Osaki, Y.: 1972, Publ. Astr. Soc. Japan 24, p. 509.
Paczynski, B.: 1973, Acta Astronomica 23, p. 191.
Papaloizou, J., Pringle, J.E.: 1979, Discussion remarks on the paper by Sparks and Kutter, IAU Coll. No. 53: "White Dwarfs and Variable Degenerate Stars", Ed. Van Horn and Weidemann, Rochester, p. 294.
Öpik, E.J.: 1951, M.N.R.A.S. 111, p. 278.
Roxburgh, I.W.: 1964, M.N.R.A.S. 128, p. 157.
Schwarzschild, M.: 1942, Astrophys. J. 95, p. 44.
Sweet, P.A.: 1950, M.N.R.A.S. 110, p. 548.
Sweigart, A.V., Mengel, J.G.: 1979, Astrophys. J. 229, p. 624.
Vauclair, G.: 1976, Astron. Astrophys. 50, p. 435.
Zahn, J.P.: 1975, Mém. Soc. Roy. des Sc. de Liège, 6e ser., tome VIII, p. 31.

DISCUSSION

Osaki: I would like to comment on the three points which Prof. Kippenhahn has raised. Firstly, I am happy to hear your argument that differential rotation on the equipotential surface is not sufficient to induce instability. I think that we must be more careful in discussing the viscosity and the stability of the accretion disk. Secondly, do you agree with me that the second term, $\partial s/\partial t$, in the energy equation which governs the meridional circulation velocity becomes important near the surface zone? How did you estimate the order of magnitude of the circulation velocity when $v \sim 10^{-2}$ cm/sec?

Kippenhahn: Concerning your second point, the term $\partial s/\partial t$ can become important, but I have difficulty understanding your conclusion that this term will lead to a circulation-free ω-distribution for the reasons I have given in our paper. We have estimated the circulation velocity by introducing an eddy viscosity caused by the Goldreich-Schubert-Fricke instability (using our value for the mean free path determined by self destruction). This viscosity gives a deviation $\tilde{\omega}$ from $\omega = \Omega + \omega^*$ which in a steady state demands an additional circulation velocity of the order I had mentioned.

Osaki: Thirdly, how sensitive is your result to your assumed form of perturbations? I ask this question because the perturbation you assumed does not conform to an eigenfunction of the linear mode. Did you consider the salt-finger type perturbation as well as the blob-type perturbation?

Kippenhahn: Concerning your third point, we have started our work with the problem of thermohaline mixing (the paper will appear in Astronomy and Astrophysics). There, as well as here, we have not investigated perturbations which correspond to the eigenfunctions of the linear theory. What we did is to investigate the fully developed nonlinear motion. We have made our estimate with a torus-like perturbation, but the mechanism of self destruction does not depend on the detailed geometry. All kinds of volumes that are in hydrostatic equilibrium with the surroundings, but have different angular velocities, will be destroyed by the circulation which they create.

Roxburgh: I understand that the most unstable modes in the Goldreich Schubert-Fricke unstable star are long thin modes, so I doubt the validity of an axially symmetric perturbation analysis. Since the stability analysis gives the growth rates, an estimate (or lower limit) on the diffusion rate can be obtained by determining the amplitude of the linear growing mode at which the nonlinear terms become important. Does this not give a reasonable estimate and has it been done?

Kippenhahn: The axisymmetry is not essential to our analysis. Self destruction appears in all kinds of perturbations you can make up. In the kind of analysis you suggest, you can learn how long it takes for the nonlinear terms to become important but you do not learn what will happen afterwards. You would not even find that a mechanism like self destruction exists.

Schatzman: Is the core of the sun rotating fast? Can you give an upper limit to the angular velocity of the solar core?

Kippenhahn: I do not know the upper limit by heart. However, it arises from the lack of an observed solar oblateness. An estimate has been given by D. Bortenverfer in a paper published in A & A about 6 years ago. You can probably find the latest value in one of the solar neutrino review papers.

MAGNETIC FIELDS AND STELLAR EVOLUTION

L. Mestel
Astronomy Centre, University of Sussex

1. INTRODUCTION

Magnetic fields are now observed or inferred in a wide variety of stellar objects. The class of early-type stars with strong large-scale fields extends from types B to F, with effective fields from 300 gauss up to several x 10^4 gauss (Borra and Landstreet 1980). Fields between 4×10^6 and 10^8 gauss have been inferred in a small percentage of white dwarfs, and of over 10^{12}gauss in neutron stars. Some Cepheids show measurable fields. Evidence has built up of solar-type activity in late-type stars. The pioneering work by Wilson (1978) on Ca activity has shown convincingly the occurrence of periodicity reminiscent of the solar cycle in a number of G, K and M stars. Ca II emission appears to be a good predictor of simultaneous X-ray emission from hot coronae around cool stars (Vaiana 1979, Mewe and Zwaan 1980). Fields of some 2×10^3 gauss have been reported in two late-type main sequence stars (Robinson, Worden and Harvey 1980).

The positive correlation between Ca activity and stellar rotation has been demonstrated both for main sequence stars (Wilson 1966, Kraft 1967) and more recently for giants (Middelkoop and Zwaan 1980). Perhaps most spectacularly, the RSCVn close binary systems show what appears to be greatly enhanced solar activity, with strong chromospheric, radio and X-ray emission, and provide evidence that a large fraction of their surfaces are covered by dark spots (Hall, D.S.in Fitch (ed.) 1976; papers in Plavec et al (eds.) 1980). Theoretical developments in dynamo theory, in which rotation (both absolute and differential) is crucial, encourage the treatment of all late-type stars with extensive outer convective zones as locations of solar-type dynamos, with the rotation as a fundamental parameter of magnitude determined by the star's history. From a complementary point-of-view, there have been a number of studies on the gross interaction between large-scale stellar magnetic fields and the thermal-gravitational-dynamical field, in both radiative and convective zones. Much of this work is independent of whether the field is being maintained by contemporary dynamo action, or is instead a "fossil" - a slowly-decaying relic of either the galactic

D. Sugimoto, D. Q. Lamb, and D. N. Schramm (eds.), Fundamental Problems in the Theory of Stellar Evolution, 257–272.

field pervading the gas from which the star formed, or of dynamo action in an earlier epoch in the star's life (e.g. the Hayashi phase). Dynamo action in the solar convection zone does not rule out a primeval field trapped within the solar radiative core, as postulated by several authors; nor do tentative arguments for a fossil theory for the observable fields of strongly magnetic early-type stars require the non-occurrence of dynamo action in their convective cores. When concerned with magnetic fields and stellar evolution, it is prudent at the present state of knowledge to keep options open.

2. STELLAR MAGNETIC FIELDS - GLOBAL CONSIDERATIONS

A strongly magnetic early-type star with a large-scale field of mean strength $\bar{B}$ over the radiative outer zone has a ratio of magnetic to gravitational energy of order

$$\varepsilon \equiv (\bar{B}^2/8\pi)(4\pi R^3/3)/(GM^2/R) \simeq F^2/(6\pi^2 GM^2), \qquad (1)$$

where $F = \pi\bar{B}R^2$ is an estimate for the total flux. The parameter ε is a convenient measure of whether the field in a magnetic star is globally "strong" or "weak". Stellar fields can be observed directly only if the field-strength is $\simeq 10^3$ gauss over a large part of the disk. The star with the strongest observed surface field B_s is HD 215441, with $B_s \simeq 3.4 \times 10^4$ gauss. If we extrapolate this inwards, adopting the largest-scale, slowest decaying Cowling mode (Cowling 1945; Wright 1974), we find an internal field $\bar{B} \simeq 10^6$ gauss, yielding $\varepsilon \simeq 10^{-5}$. This immediately raises the important theoretical question of the actual relation between the strength of the field that emerges through the photosphere to yield an observable Zeeman effect, and the mean field $\bar{B}$ deep within the star. We return to this below; for the moment we note that even for this exceptional star one needs $\bar{B}$ to be 30 times larger for ε to approach 10^{-2}.

Similar conclusions hold for magnetic white dwarfs and neutron stars, as follows if we assume that during contraction to form a degenerate star, the central regions of the parent star conserve their magnetic flux, so that $\bar{B}$ increases like $\bar{\rho}^{2/3}$, and the analogue of ε stays constant. Then as the density increases from a main-sequence central value of $\simeq 10$ gm/cm^3 to a typical mean white dwarf value of 10^6, $\bar{B}$ would increase from a hypothetical value of 10^6 to well over 10^8 - at the upper limit of the values inferred for the surface fields of white dwarfs. The analogous argument would predict a neutron star field of $10^{13} - 10^{15}$, again on the high side for both observation and theory. Thus there seems no <u>prima facie</u> reason to doubt that even the most strikingly magnetic stars are "weak". If anything, the flux-freezing argument suggests that 10^6 gauss would be exceptional for $\bar{B}$ within most main-sequence stars: $10^4 - 10^5$ would fit in better with the estimates for white dwarfs and pulsars.

The evidence is now rather strong that rapid flux leakage occurs during the molecular cloud phase of star formation. With field freezing re-established during the opaque phase, e.g. at a density of 10^9 particles/cm^3, the galactic field would be amplified to $\simeq 5 \times 10^4$ gauss on the main sequence, so there is no difficulty in producing magnetically "weak" Type I stars. However, Type II (and a fortiori Type III) stars may have formed from a gas of weakly polluted hydrogen and helium at $\simeq 10^4$ $^{\circ}$K (Hoyle 1953) which maintains a high level of ionization. Fragmentation can occur via preferential flow down the field-lines but would yield magnetically "strong" proto-stars (Mestel 1965a). It is known that even in sub-adiabatic stellar domains, magnetic fields of simple topology are subject to dynamical instabilities near their neutral points (Wright 1973; Markey and Tayler 1973, 1974); for under non-axisymmetric adiabatic motions that are nearly perpendicular to the gravitational field, the stabilizing effect of the sub-adiabaticity disappears, and magnetic energy is spontaneously released. To remove these obvious instabilities one needs to construct complex fields, with e.g. toroidal flux loops linking the poloidal loops. One wonders, however, whether this would be sufficient if the magnetic energy is high, or whether other dynamical instabilities inevitably arise unless the parameter ε is much below unity, whatever the structure of the field.

With dynamical stability supposed satisfied, there remains the question of secular stability. Even in a stable, sub-adiabatic radiative zone, individual flux-tubes tend to float to the surface via "magnetic buoyancy" (Parker 1979; Acheson 1979), at a rate determined by heat diffusion into the cool gas within the tube - an "Eddington-Sweet" effect (Sweet 1950). It may be that global fields with mutually linking poloidal and toroidal loops are subject to analogous instabilities, perhaps depending also on the finite resistivity which allows changes in field topology. Indefinite stabilization over a stellar lifetime may require a negative gradient of mean molecular weight μ.

The fossil theory of stellar magnetism is a theoretical possibility because a large-scale poloidal field $\underset{\sim}{B}_p$ is maintained by a toroidal current density $\underset{\sim}{j}_t = c\nabla \times \underset{\sim}{B}_p/4\pi$ flowing in the high conductivity interior, yielding a decay-time $\tau_p \simeq 4\pi\bar{\sigma}R^2/c^2$, where $\bar{\sigma}$ is a mean value. In the cooler outer regions the currents are much weaker, and the field in a large-scale Cowling mode is locally nearly curl-free. But the toroidal field $\underset{\sim}{B}_t$ linking $\underset{\sim}{B}_p$ (as required for stability) is maintained by poloidal currents $\underset{\sim}{j}_p$ that flow nearly parallel to $\underset{\sim}{B}_p$, since in a non-turbulent domain the total field must be nearly torque-free. Thus if a particular loop of $\underset{\sim}{B}_p$ passes through the surface regions, the currents flowing along it will suffer much greater dissipation, because the surface resistivity is some 10^3 higher than the mean. One therefore expects the surviving toroidal flux to be largely concentrated deep within the star, maintained by currents flowing primarily along poloidal loops that do not come too near the surface. Does this highly non-uniform distribution of $\underset{\sim}{B}_t$ have any serious effect on the dynamical stability of the field $\underset{\sim}{B}_p$? *Prima facie*, the stability near the neutral points should survive, as long as the toroidal flux is of the same order as the poloidal; but one would like reassurance that the effective Ohmic destruction time of the total field has not been substantially reduced.

The stability of stellar magnetic fields is clearly relevant to the question as to whether the fields of the strongly magnetic early-type stars are primeval, or require continual regeneration by a contemporary dynamo. The bewildering complexity of the observed parameters of the magnetic stars can be cited as an argument in favour of the flux being an extra parameter rather than one closely linked with the structure and the rotation of the star. Although our understanding of both kinematic and dynamical dynamos is very incomplete, and there remain the queries about instabilities possibly shortening the lifetimes of unregenerated fields, one can still claim that as yet there are no obvious astronomical advantages in insisting on a contemporary dynamo rather than a fossil explanation for the fields of the strongly magnetic stars; and likewise, a fossil field trapped within the radiative core of a late-type star remains a theoretical possibility.

3. MAGNETIC FIELDS AND STELLAR STRUCTURE

Since the ratio ε is apparently so small, the magnetic field should not have a great effect on the overall structure of the star. However, interesting effects can arise if stellar hydrodynamics leads to a local increase of field strength in regions of small scale, or into the low-density surface regions. And at least in non-convective domains, the field should be of paramount importance in its interaction with stellar rotation, as long as any mass motions have velocities well below the Alfvén speed. In particular, during the leisurely pace of normal stellar evolution, we can expect a contracting burnt-out core to be kept more or less corotating with an expanding envelope, as long as they remain magnetically linked. Thus there is no reason for surprise that even the most rapid pulsars are slow rotators at birth. Simple estimates show how essential is such redistribution of angular momentum. Suppose instead that in contracting from $\rho \sim 10$ to $\rho \sim 5 \times 10^{14} \mathrm{gm/cm^3}$, the central core of a main-sequence star with a rotation period of about 1 day had conserved its angular momentum: the period of a resulting neutron star would be 6×10^{-5} secs, with centrifugal forces eighty times gravity - clearly a *reductio ad absurdum*. Inverting the problem, we find that a pulsar will rotate at birth with centrifugal force just one percent of gravity if angular momentum conservation began when the core had reached a density near 10^4 gm/cm^3.

However, once contraction begins to be rapid, and approximate corotation is no longer maintained, important effects can follow from the conversion of the energy of non-uniform rotation - itself fed from the gravitational field - into toroidal magnetic energy. This local concentration of energy is crucial in some models of the dynamics of supernovae (LeBlanc and Wilson 1970; Kundt 1976). In the most recent study (Müller and Hillebrandt 1980), the rebound of the core at nuclear densities generates a shock wave which delays infall of the mantle, so giving the rotational shear time to generate a toroidal magnetic pressure near the core that is close to the thermal pressure. The resulting second hydromagnetic shock can lead to mass ejection with high energy, leaving a central neutron star.

Even a weak magnetic field radically alters the problem of constructing a self-consistent model of a rapidly rotating stellar radiative envelope. Over the bulk of the zone, the advection of angular momentum by the slow, centrifugally-driven Eddington-Sweet circulation is easily off-set by the field, since the speed v_p is much less than any likely Alfvén speed (Mestel 1961). The approximation of ignoring the magnetic as compared with the centrifugal disturbance to hydrostatic and thermal equilibrium is consistent over the bulk of the zone, since

$$\rho v_p^2 << B_p^2/4\pi << \rho\Omega^2 r^2. \tag{2}$$

But in the low-density surface regions the magnetic forces need not be negligible; also, a general perturbing force yields circulation speeds that become large like $\bar{\rho}/\rho$ (Baker and Kippenhahn 1959), so the whole order of approximation could break down.

A limited number of steady self-consistent models have been constructed. The simplest class suppose that the star has achieved radiative equilibrium (with no circulation) - the magnetic forces are significant not only in the surface regions but over the whole radiative envelope. In models with the rotation and magnetic axes parallel, there is as a consequence a general tendency for the fraction of the prescribed total flux that appears above the surface to decrease with increasing rotation rate (Davies 1968; Wright 1969; Moss 1973, 1975). In the appropriate parameter range, the ratio $\bar{B}/B_s$ is $\simeq$ 1500 - much higher than in a Cowling mode, implying $\bar{B} \simeq 10^6$ or more even when B_s is no more than 10^3. Mestel and Moss (1977) have constructed approximate models with a non-vanishing circulation field that reduces deep down to the Eddington-Sweet flow. It is found that the magnetic forces in the surface regions adjust themselves so as to kill off the $\bar{\rho}/\rho$ terms, leaving both $\underset{\sim}{v}_p$ and $\underset{\sim}{B}$ small in the surface regions, and again with the flux strongly concentrated into the deep, high density regions.

It should be emphasized that in much of this work there are severe mathematical difficulties, which force the premature truncation of expansions in orthogonal polynomials. One suspects that this is largely responsible for failure to cover all the physically allowed parameter range. The attempts on analogous non-axisymmetric models have had markedly less success, probably for similar reasons (Monaghan 1973; Moss 1977b). However, the work may give some qualitative understanding of the respective properties of magnetic and non-magnetic early-type stars. It is known that there is a gross anti-correlation between rotation rate and the appearance of magnetic flux above the stellar surface, or of the abundance peculiarities that are often a tracer of such external flux. There is also some tentative evidence that within the class of magnetic Ap stars, the observed fields are systematically weaker in the more rapid rotators (Borra and Landstreet 1980). More effective magnetic braking by a stronger external field may account for this in part, but the tendency for rotationally-driven circulation to concentrate the flux deep in the star may also be playing a role.

All this work has assumed no μ-gradients in the radiative zone. In particular, the model with the Eddington circulation deep down assumes there is a "μ-barrier" at the surface of the convective zone, where the vertical velocity is zero, and where the simple theory predicts large horizontal velocities by continuity. Thus the μ-barrier region is again singular; if the field is at least approximately frozen into the fluid, the associated horizontal component becomes large enough again to contribute to hydrostatic and thermal balance. It is reasonable to look for a self-consistent steady state in which the combined effects of centrifugal and magnetic forces reduce the thermally-driven vertical speed to zero at the barrier. Rough estimates predict that the magnetic contribution should be significant in a layer of thickness $D \simeq r(\bar{B}^2/8\pi\bar{\rho}\Omega^2 r^2)^{1/3}$. However, the attempted construction of such a model has run into even more severe truncation difficulties than in the surface regions. More seriously, it is not clear that one can satisfy the conditions of zero slip and zero stress at the barrier that are strictly required, even though the viscosity of stellar material is so small. It may be that the picture of a μ-barrier separating the domain of a steady increase in μ from an envelope with zero μ-gradient should be questioned (cf. the non-magnetic study of Huppert and Spiegel 1977). Perhaps instead a μ-gradient extends itself steadily through the radiative zone, so that the Eddington-Sweet circulation suffers from "creeping paralysis" (cf. Mestel 1953, 1965b). If so, then the tendency of the magnetic field to be concentrated into high-density regions will be at most temporary; as the circulation dies out, the field over the bulk of the radiative envelope would diffuse back into a Cowling mode. However, residual motions near the surface could still be important for the local field structure, and in particular for the amount of observable flux.

The strength of sub-surface magnetic fields is of particular interest for the theory of Cepheid variables. The observations of magnetic fields in some Cepheids have prompted Stothers (1979) to see what effect a magnetic pressure would have on the pulsational properties of Cepheids. He postulates a small-scale field, with the ratio ν of magnetic to thermal pressure an adjustable parameter. He is able to remove the discrepancy between the pulsational mass and the evolutionary mass if ν has a uniform value of 0.8 in a surface layer comprising 10^{-3} of the stellar mass, implying a field strength of $\simeq 10^4$ gauss at the base of the layer. This model is certainly somewhat ad hoc, but it is worth noting that a frozen-in poloidal magnetic field B_p, consistent with a circulation field v_p, satisfies $B_p \propto \rho v_p$; hence if the circulation speeds were to retain something like the $\bar{\rho}/\rho$ dependence, the distorted magnetic field could have a structure with B only weakly dependent on ρ. A value of 10^4 gauss just below the surface would then not be hydromagnetically inconsistent with a similar value deep down. Certainly one feels that the whole problem of self-consistent, thermally-driven, hydromagnetic flow in the outer layers of a star is still ill understood. Further work should probably pay more attention to the detailed physical properties of the surface regions, especially to the local convective zones.

4. THE OBLIQUE ROTATOR

The case for this simplest model for the magnetic Ap stars (Deutsch 1958; Preston 1971) gains support from the plot of the spectroscopically measured Vsini against the period P of magnetic, spectral and luminosity variations, which shows the points lying below a well-defined hyperbolic envelope PV = constant, to be expected if these stars do not deviate too much from the main-sequence A-star region. Even if a magnetic field is not measured directly, one is always tempted to interpret periodicity in some other feature as the consequence of azimuth-dependent structure with a magnetic field as a possible cause. Equally, any global consequences of non-axisymmetry may be only weakly dependent on the amount of flux that penetrates the photosphere, and so may affect the evolution of observably "non-magnetic" stars.

For definiteness, we suppose the magnetic field symmetric about an axis $\underset{\sim}{p}$ inclined at an angle χ to the rotation axis k. An oblique rotator has some properties in common with the classical problem of a body with three unequal axes of inertia (Spitzer 1958). In slowly-rotating, weakly magnetic stars, the departures from spherical symmetry of the density-pressure field are the superposition of a part symmetric about $\underset{\sim}{k}$ and a part symmetric about $\underset{\sim}{p}$. The motions within such a body can be most simply analyzed into (1) the basic rotation $\Omega \underset{\sim}{k}$; (2) the Eulerian nutation with a frequency ω about $\underset{\sim}{p}$; and (3) a field of "$\underset{\sim}{\xi}$-motions" with frequency ω that ensure that the star remains in hydrostatic equilibrium (Mestel and Takhar 1972; Mestel et al 1980; Nittmann and Wood 1980). To order of magnitude

$$\omega \sim \Omega\varepsilon\ , \qquad \xi \sim \ell\ (\Omega^2 r^3/Gm(r)) \tag{3}$$

where ℓ is the local scale-height. Thus the nutation period $2\pi/\omega$ >> the rotation period $2\pi/\Omega$, but can easily be much less than the stellar lifetime or the Kelvin-Helmholtz time. In slow rotators ξ/ℓ is small, but the appropriate generalization to rapid rotators suggests that the motions would not then be trivial: they could yield mixing of matter between part of a stellar envelope and a convective core, so modifying tracks in the H-R diagram; and in fact Nittmann and Wood (1980) have suggested $\underset{\sim}{\xi}$-motion mixing in rapid rotators as an explanation of the blue straggler phenomenon. Also, as already noted, a μ-gradient in a radiative zone can effect indirectly the distribution of magnetic flux through the star.

However, as pointed out by Spitzer, the $\underset{\sim}{\xi}$-motions will be subject to dissipation, which acts as a drain on the rotational kinetic energy of the star. This has the form $h^2/2I$, where h is the angular momentum and I the moment of inertia about the instantaneous axis of rotation. Thus if dissipation is fairly efficient, and nothing else intervenes to affect the flux-distribution, the magnetic axis should rotate in space until the star is rotating about its maximum moment of inertia, and the $\underset{\sim}{\xi}$-motions cease. Following the report (Preston 1971) that magnetic obliquities seemed to be concentrated either at χ large or χ small, the theory was tentatively linked with the requirement that stable magnetic fields must have linked poloidal and toroidal flux of comparable magnitudes. The suggestion was that cases with dominant poloidal flux would be dynamically oblate about the field axis, and so would tend to approach $\chi = 0$, while those with dominant toroidal flux may be prolate and so approach $\chi = \pi/2$. The observational evidence is now more obscure: Hensberge et al (1979) argue that there is no non-randomness in χ, while according to Borra and Landstreet (1980) there is marginal evidence for non-randomness. The immediate applicability of the argument thus depends on the time-scale for dissipation. In Mestel et al (1980) an upper limit is found by constructing the distortions $\underset{\sim}{B}'$ to the magnetic field due to the $\underset{\sim}{\xi}$-motions and then computing the volume integral of the Ohmic dissipation $(c\nabla \times \underset{\sim}{B}'/4\pi)^2/\sigma$. Not surprisingly, the dissipation is strongly peaked at the cooler surface regions. The total dissipation rate is sensitive to the nutation frequency ω and so to ε; in particular, if the field is not as centrally condensed as in the models of Wright, Moss, Mestel and Moss etc., but is closer to Cowling's slowest decaying mode, then the time for Ohmic dissipation of the $\underset{\sim}{\xi}$-motions is too long. One could then imagine an initial gross obliquity and so also the associated $\underset{\sim}{\xi}$-motions lasting through the star's lifetime and so affecting stellar evolution.

It is well, however, to remember that there are other processes that can affect the apparent obliquity, e.g. the precessional torque associated with magnetic braking (Mestel and Selley 1970), or just the slow kinematic effect of horizontal surface motions (Moss 1977a). There may be more powerful dissipative processes affecting the $\underset{\sim}{\xi}$-motions. With more complicated field structures than those studied, there may not be an unambiguous relation between the state with rotation about the maximum moment of inertia and the superficial flux distribution. And if there is some interaction between the convective core and the field in the envelope, e.g. by dynamo generation of new flux, then the theoretical uncertainties are multiplied.

Any non-axisymmetric feature on a stellar surface is a potential probe of a rotation. Thus Dicke (1979) interprets his most recent solar oblateness measurements, which pick up a 12.2 day period, in terms of the rotation of a perpendicular magnetic rotator in the radiative core. His particular model is questionable on stability grounds, as he postulates a toroidal field with an energy far greater than the poloidal. But if the Ap star fields are primeval, then it is tempting to argue that late-type stars also contain such flux, which is however largely prevented from appearing at the surface by the powerful outer convective zone; in which case, all the questions of stability, obliquity, coupling with the convective zone etc. remain relevant for the central regions.

The problems are frustrating, because of the difficulty in making an unambiguous link with observation. The discovery that some Cepheids have magnetic fields may offer more fruitful scope. For many years it has been known that RR Lyrae shows a 41-day cycle in its light and velocity curves (e.g. Detre and Szeidl 1973). The earlier report that the star has an observable magnetic field has not been confirmed; however, this need not mean that magnetic forces are unimportant in the observed low-density regions. It is well worth trying to construct the back-reaction on the density-pressure field of the forces exerted by a magnetic field that is periodically distorted by an essentially radial pulsation. If significant effects are found, and the basic magnetic field is oblique, then the theory should predict variations with the rotation period.

5. STELLAR CONVECTION ZONES: DYNAMO ACTION

Since the turbulent velocities deep in a convective zone are highly subsonic, one might expect significant magnetic interference with convection for magnetic energy densities much below the thermal. For example, with $\rho \sim 1$ and $v_t \sim 10^4$ cm/sec, $B^2/8\pi \sim \rho\, v_t^2$ requires $B \sim 5 \times 10^4$ gauss. Nearer the surface the turbulent velocities are higher but the densities much lower, so that $B \sim 10^2$ gauss is sufficient. However, for a magnetic field to interfere globally with convective heat transport and so affect seriously the structure and evolution of the star, the field would need to be much stronger: one must compare the field energy with the energy that the turbulence would develop if the temperature gradient retained the strongly superadiabatic value it has if radiative equilibrium were maintained, rather than relaxing to just a fraction 10^{-6} above the adiabatic value, as in the classical Biermann-Cowling estimate. Detailed stability studies by Gough, Moss and Tayler (1966, 1969) show that complete suppression of convection requires a field energy comparable with the thermal energy. Quite apart from conflict with observation, such a strong field could itself be spontaneously unstable. However, a field anchored in a deeper radiative zone may very well stabilize the weak convective regions in the surface regions of early-type stars, so enabling element diffusion to occur and yield abundance anomalies.

A series of laminar flow studies within the Boussinesq approximation (e.g. Galloway, Proctor and Weiss 1977, 1978) have demonstrated how in the presence of a magnetic field an unstable zone maximizes the efficiency of heat transport by assembling the imposed magnetic flux into isolated tubes, so that unimpeded convection can occur in the rest of the zone. A similar picture is suggested (Galloway and Weiss 1979) for a turbulent convective zone. A significant conclusion is that local field-strengths are well above the value given by equipartition with the convective energy, as is indeed observed in sunspots. The upper limit is clearly given by $B \sim \sqrt{8\pi p}$, implying zero thermal or turbulent energy within a flux-tube. One can argue that an external magnetic field may not be totally expelled from a neighbouring strongly convective zone, but rather that some flux penetrates and is concentrated by the turbulence into ropes. The field should thus ensure some dynamical coupling between contiguous stable and turbulent zones.

Accepting that the observed solar fields are generated and destroyed as part of a periodic dynamo, with the solar rotation as a basic parameter, one is led to ask whether magnetic activity could ever rise to a level when it makes a substantial difference to stellar luminosity. We recall the old problem of the "missing heat flux" from a sunspot: a bright ring was expected around a spot, to compensate for the reduced flux into the spot (according to the Biermann picture - though this has been challenged by Parker (1979)). This difficulty was removed by Spruit (1977), who showed that the area of this ring should be too large to yield an observable temperature excess. More recently, Foukal and Vernazza (1979) have found a weak dependence of solar luminosity on magnetic activity, at the level of 7×10^{-4} of the continuum, with periods of 28 days, and consistent with changes in the area of magnetic faculae and sunspots. They argue that this is merely a redistribution of heat flow and not a sign of magnetic influence on the steady heat flow to the photosphere. But would such an effect be completely negligible in a rapidly rotating late-type star - e.g. in the synchronized members of RSCVn close binaries, which as noted appear to show greatly exaggerated solar activity?

Whatever the answer, one can predict that these stars will continue to offer scope for the application of hydromagnetic ideas. Thus Shore and Hall (in Plavec *et al* 1980) have applied a model analogous to Babcock's solar dynamo, with the two stars in near synchronous rotation, but with the differential rotation less than in the sun. There is also the possibility that the binary structure introduces new effects. For example, Dolginov and Urpin (1979) have discussed the possibility of Herzenberg-type dynamo action in binary systems that are not yet synchronized, and with rotation axes inclined towards the orbital plane.

The rapid development in dynamo theory and the corroboration from observation of a strong dependence of magnetic activity on stellar rotation are very impressive, but one should note that there remain theoretical difficulties. Layzer *et al* (1979) have criticized the mathematical and physical basis of the "mean field" dynamo equations. They argue for a return to the torsional oscillation model for the solar cycle, which they suggest is not a genuine self-maintained dynamo, but requires a continuous supply of flux from a non-regenerated, irregular field, largely confined to the radiative core. The occurrence of solar-type activity in late-type stars that are probably fully convective is perhaps a difficulty for attempts to replace stellar dynamos by "amplifiers" dependent on an externally anchored source of flux. Dynamo models based more on the individual flux-rope picture face the problem that magnetic buoyancy in a superadiabatic zone can be embarrassingly efficient. A toroidal tube of the strength required to account for the observed solar surface flux and in temperature equilibrium with the non-magnetic surroundings will rise in about a month, as compared with the solar time-scale of years. Several proposals exist for increasing the rise-time. Some workers have argued that the tube will be similar to a tube in a radiative zone - at the same density as its surroundings, but with a lower temperature, so that it rises at the rate fixed by the inward leakage of heat. However, Spruit (1980) has noted that this would not resolve the difficulty, as the tube would be unstable against buckling in vertical planes. Perhaps more plausibly, Zwaan (1978) has proposed that the difference in turbulent pressures maintains the balance between a flux tube with the same density as its surroundings.

We have seen that there is strong observational evidence that magnetic activity in late-type stars increases with rotation. However, this is by no means an obvious consequence of all dynamo theories. For example, Moffat (1970, 1972) has produced a model in which the increasing rotation makes the turbulence more nearly two-dimensional, so reducing the "α-effect". It cannot be ruled out that the fields of early-type magnetic stars are not fossils but are generated in their convective cores by steady dynamos of this type. Moss (1980) has suggested alternatively that rapid rotators produce oscillatory dynamos within their cores, which fail to yield observable fields at the surface, since the fields propagate to the surface as strongly damped waves.

The expected rotation law within convective zones remains ill-understood. Deep within a nearly adiabatic convective envelope, the rotation would be expected to approximate to constancy on cylindrical surfaces, with only minor departures due to locally strong magnetic forces, or to circulation driven e.g. by the Biermann-Kippenhahn anisotropic viscosity (Kippenhahn 1963). It appears that only in a comparatively thin surface layer will the anisotropy and the small scale-height enforce something closer to an $\Omega(r)$ law (Galloway and Mestel, in preparation), and this layer gets thinner in more rapid rotators. How this will affect dynamo action is as yet unclear.

6. MAGNETIC BRAKING OF STELLAR ROTATION

The braking process most studied involves a magnetically-controlled stellar wind (due either to the pressure of a hot corona or to the input of momentum by line absorption from early-type stars). The simple theory yields a flow of angular momentum equivalent to effective co-rotation out to the Alfvénic surface S_A where the wind speed $v = v_A \equiv B/(4\pi\rho)^{1/2}$:

$$-\frac{d}{dt}(k^2R^2M\Omega) = -\left(\frac{dM}{dt}\right)(\Omega R_A^2) \tag{4}$$

where kR is the radius of gyration of the star and R_A a measure of the distance to the Alfvénic surface (Mestel 1966; Weber and Davis 1967). There is in fact something a little paradoxical about what is being demanded. We want the magnetic stresses to be able to control the rotation, and this is satisfied within S_A, where $v^2 < B^2/4\pi\rho$. However, one expects the wind energy density to be comparable with the energy density driving it - e.g. a thermally-driven wind to be supersonic but not hypersonic; in which case, a magnetic energy density that dominates over the thermal should seriously interfere with the outflow of gas. The simplest argument for the solar wind is that a corona heated to $10^{6\circ}$ cannot be held in by the pressure of the interstellar medium; but magnetic field-lines anchored in the solar convection zone can act as a "lid", preventing the expansion of gas with too little thermal energy. In fact, evidence has accumulated that it is the weaker field regions on the sun which are pulled out by the solar wind to become the interplanetary field. The point to note is that (4) shows how the field can increase the <u>efficiency</u> of angular momentum loss for a given mass-loss rate $-\dot{M}$; but a crucial question is how $-\dot{M}$ itself varies with changes in the external magnetic flux (e.g. due to the rotation-dependent dynamo). The maximum braking occurs if the field is supposed pulled out to be nearly radial, so that all the field-lines partake in the braking process (Weber and Davis 1967). A plausible minimum results by supposing the field to be virtually curl-free out to S_A, with extensive dead zones; this yields a braking rate which is only weakly dependent on the strength of the surface field (Mestel 1968). The real value is between these limits, but is difficult to calculate (cf. Okamoto 1974).

The problem is important for understanding the rotational history of the sun and other late-type stars. Anticipating that the external flux and so also R_A increase with Ω, we still need to know to what extent $-\dot{M}$ is reduced as the flux increases. Fortunately, we have observational evidence of the variation of rotation period with age (Kraft 1967) and of the decline in Ca activity with age (Wilson and Woolley 1970). It appears that the specific rate of braking does increase with Ω, implying that the increase of R_A^2 is more important than the interference of the increased magnetic field with $-\dot{M}$. One can plausibly parametrize the net effect by writing $-\dot{M}R_A^2$ in (4) as a simple increasing function of Ω; there results an algebraic rather than an uncomfortable exponential law of variation of Ω with time (Spiegel 1968; Skumanich 1972). On extrapolating back to the zero-age main sequence via the inferred rotations of late-type stars in the young Hyades and Pleiades clusters, it is estimated that the sun began there with only 10 - 20 times its present very slow rotation (Ostriker 1972). This puts a constraint on the later stages of the formation of the sun and similar stars. Even the most efficient processes for angular momentum removal during the pre-opaque phases of star formation are very unlikely to produce proto-stars that can contract all the way to the main sequence without running into centrifugal trouble. It is more plausible that proto-stars reach the pre-main sequence phase as rapid rotators, and subsequently redistribute their remaining angular momentum during the final contraction. The evidence cited implies that late-type stars become slow rotators during this phase. Magnetic effects may again be crucial, e.g. via coupling to a strongly enhanced stellar wind during the Hayashi phase, or between a central condensation and a nascent planetary system.

For stars with cool coronas (of temperatures below $10^{5\circ}$) and so without thermally-driven winds, magnetic braking can occur via an accretion process. Gravitationally inflowing gas will compress the field until the magnetic stresses halt the inflow at the Alfvénic surface defined by $B^2/8\pi \sim \rho GM/r_A$. This state is Rayleigh-Taylor unstable: gas can slide into troughs between magnetic planes, and so pick up angular momentum from the star via magnetic pressure gradients. If $\Omega^2 r_A > GM/r_A^2$, the gas is then driven outwards under the centrifugal slingshot, again carrying off the angular momentum of approximate co-rotation at r_A. The process clearly stops when $\Omega^2 r_A^3 = GM$. Rough estimates suggest that the process depends only weakly on the accretion rate; it has a typical time-scale of about 10^7 years (Mestel 1975), and yields a period of a few days at the cut-off. This is certainly of the right order for the majority of the strongly magnetic stars, for which one does not expect strong winds, since they lack outer convection zones, and are not hot enough for radiation-driven winds.

Magnetic coupling between members of a close binary system can interchange spin and orbital angular momentum, and so can be significant in the later stages of star formation. The classical synchronization process - tidal friction - can be efficient for late-type stars with a strong turbulent friction, but it decreases strongly with decreasing ratio of radius to mutual separation. As in the wind braking process, the effectiveness of magnetic synchronization is likely to depend rather critically on the field structure, in particular on the amount of flux coupling the two stars.

A combination of synchronization and efficient loss of angular momentum could be crucial for the evolution of RSCVn and other close binary systems. Break-up into a binary system is one way that a rapidly rotating proto-star can resolve its angular momentum problem; it would be amusing if subsequent magnetic braking led the two stars ultimately to coalesce.

REFERENCES

Acheson, D. J. 1979, Solar Phys., 62, 23.
Baker, N. and Kippenhahn, R. 1959, Z. f. Ap., 48, 140.
Borra, E. F. and Landstreet, J. D. 1980, Ap. J. Suppl. 42, 421.
Cowling, T. G. 1945, M.N.R.A.S., 105, 166.
Davies, G. F. 1968, Aust. J. Phys., 21, 294.
Detre, L. and Szeidl, B. 1973, in Fernie, J.D. (ed.), Variable Stars in Globular Clusters, Reidel.
Deutsch, A. J. 1958, in Lehnert, B. (ed.), Electromagnetic Phenomena in Cosmical Physics, C.U.P.
Dicke, R. H. 1979, Ap. J., 228, 898.
Dolginov, A. Z. and Urpin, V. A. 1979, Astron. Astrophys., 79, 60.
Fitch, W. S. (ed.) 1976, Multiple Periodic Variable Stars, Reidel.
Foukal, P. and Vernazza, J. 1979, Ap. J., 234, 707.
Galloway, D. J. and Mestel, L. 1980 (in preparation).
Galloway, D. J., Proctor, M. R. E. and Weiss, N. O. 1977, Nature, 266, 686.
Galloway, D. J., Proctor, M. R. E. and Weiss, N. O. 1978, J. F. M., 87, 243.
Galloway, D. J. and Weiss, N. O. 1979 (preprint).
Gough, D. O. and Tayler, R. J. 1966, M.N.R.A.S., 133, 85.
Hensberge, H., van Rensbergen, W., Gossens, M. and Deridder, G. 1979, Astron. Astrophys., 75, 83.
Hoyle, F. 1953, Ap. J., 118, 513.
Huppert, H. and Spiegel, E. A. 1977, Ap. J., 213, 157.
Kippenhahn, R. 1963, Ap. J., 137, 664.
Kraft, R. P. 1967, Ap. J., 150, 551.
Kundt, W. 1976, Nature, 261, 673.
Layzer, D., Rosner, R. and Doyle, H. T. 1979, Ap. J., 229, 1126.
LeBlanc, J. M. and Wilson, J. R. 1970, Ap. J., 161, 541.
Markey, P. and Tayler, R. J. 1973, M.N.R.A.S., 163, 77.
Markey, P. and Tayler, R. J. 1974, M.N.R.A.S., 168, 505.
Mestel, L. 1953, M.N.R.A.S., 113, 716.
Mestel, L. 1961, M.N.R.A.S., 122, 473.
Mestel, L. 1965a, Q.J.R.A.S., 6, 265.
Mestel, L. 1965b, in Aller, L.H. and McLaughlin, D.B. (eds.) Stellar Structure, Chicago.
Mestel, L. 1966, Liège Colloquium, 1966.
Mestel, L. 1968, M.N.R.A.S., 138, 359.
Mestel, L. 1975, Mem. S.R. Sci. Liège, 6 série, 8, 79.
Mestel, L. and Moss, D. L. 1977, M.N.R.A.S., 178, 27.
Mestel, L., Nittmann, J., Wood, W. P. and Wright, G. A. E. 1980 (preprint).
Mestel, L. and Selley, C. S. 1970, M.N.R.A.S., 149, 197.
Mestel, L. and Takhar, H. S. 1972, M.N.R.A.S., 156, 419.
Mewe, R. and Zwaan, C. 1980 (preprint).
Middelkoop, F. and Zwaan, C. 1980 (in preparation).
Moffatt, H. K. 1970, J. F. M., 44, 705.
Moffatt, H. K. 1972, J. F. M., 53, 385.

Monaghan, J. J. 1973, M.N.R.A.S., 163, 423.
Moss, D. L. 1973, M.N.R.A.S., 164, 33.
Moss, D. L. 1975, M.N.R.A.S., 173, 141.
Moss, D. L. 1977a, M.N.R.A.S., 178, 61.
Moss, D. L. 1977b, M.N.R.A.S., 181, 747.
Moss, D. L. 1980 (preprint).
Moss, D. L. and Tayler, R. J. 1969, M.N.R.A.S., 145, 217.
Müller, E. and Hillebrandt, W. 1980, Astron. Astrophys. , 80, 147.
Nittmann, J. and Wood, W. P. 1980 (preprint).
Okamoto, I. 1974, M.N.R.A.S., 166, 683.
Ostriker, J. P. 1972, in Reeves, H. (ed.), On the Origin of the Solar System, CNRS.
Parker, E. N. 1979, Cosmical Magnetic Fields, Oxford.
Plavec, M. J., Popper, D. M. and Ulrich, R. K. (eds.) 1980, IAU Symposium 88, Close Binary Stars, Reidel.
Preston, G. W. 1971, P.A.S.P., 83, 571.
Robinson, R. D., Worden, S. P. and Harvey, J. W. 1980 (in preparation).
Skumanich, A. 1972, Ap. J., 171, 565.
Spiegel, E. A. 1968, Highlights of Astronomy 1, Reidel, Dordrecht, p. 261.
Spitzer Jr., L. 1958, in Lehnert, B. (ed.), Electromagnetic Processes in Cosmical Physics, C.U.P.
Spruit, H. C. 1977, Solar Phys., 55, 3.
Spruit, H. C. 1980 (preprint).
Stothers, R. 1979, Ap. J., 234, 257.
Sweet, P. A. 1950, M.N.R.A.S., 110, 548.
Vaiana, G. S. 1979, Highlights of Astronomy 5, Reidel, Dordrecht, p. 419.
Weber, E. J. and Davis Jr., L. 1967, Ap. J., 148, 217.
Wilson, O. C. 1966, Ap. J., 144, 695.
Wilson, O. C. 1978, Ap. J., 226, 379.
Wilson, O. C. and Woolley, R. 1970, M.N.R.A.S., 148, 463.
Wright, G. A. E. 1969, M.N.R.A.S., 146, 197.
Wright, G. A. E. 1973, M.N.R.A.S., 162, 339.
Wright, G. A. E. 1974, M.N.R.A.S., 167, 527.
Zwaan, C. 1978, Solar Phys., 60, 213.

DISCUSSION

Schatzman: I am a theoretician but I like a theory better when it is in agreement with observations. A simple-minded view suggests that microscopic diffusion (under gravity or under radiation pressure) can take place where the magnetic field stabilizes the convection. However in Ap stars, the regions of strong magnetic field do not coincide with the regions where the elements separated by the Michaud mechanism show up.

Similarly, diffusion explains the general spectroscopic features of Am stars, however it fails to explain in detail any specific case.

Mestel: My comment on the relation between theory and observation was just that I feel defeated if one has to appeal to observation to resolve difficulties within a well-defined theoretical problem, instead of just solving the problem completely and then compare its predictions with observation.

I remarked that a weak convective zone in the outermost regions of an early-type star can be suppressed by a magnetic field of the strength observed, and that this is presumably a necessary condition for Michaud diffusion to occur. I did not claim that any particular theory of abundances is in good agreement with observation.

Roxburgh: Would we not expect all stars to have at least a weak magnetic field? Calculations of rotating or close binary stars which reglect the effect of magnetism may not have much relevance to the real world.

Mestel: That tends to be my view, at least for processes that have long timescales. However, while we are still uncertain as to what happens to primeval field in the Hayashi phase, and dynamo theory is still far from complete, studies of the strictly non-magnetic problems retain their interest and possible relevance.

Vilhu: What ideas do you have about how the braking depends on Ω, e.g. if the sun were to rotate more and more rapidly, up to say ~200 km/s?

Mestel: I recall that the Ω^3 law suggested by Skumanich and Spiegel seems to be consistent with the limited number of observations we have.

CONCAVE HAMBURGER EQUILIBRIUM OF ROTATING BODIES

T. Fukushima[1], Y. Eriguchi[2], D. Sugimoto[2], and G.S. Bisnovatyi-Kogan[3]
1 Astronomical Division, Hydrographic Department
2 Department of Earth Science and Astronomy, College of General Education, University of Tokyo
3 Space Research Institute, Moscow, USSR

Equilibria of rigidly rotating polytropic gas with small compressibilities are computed in order to investigate the relation between the incompressible and compressible equilibria. The equilibrium figure varies from a spheroid-like shape to a concave hamburger as the angular velocity increases. This result is supported by the fact that a concave hamburger equilibrium is obtained even in the complete incompressible case. Thus the Maclaurin spheroid does not represent the incompressible limit of the rotating polytropic gas because of its restriction of the figure. The computed sequence of equilibria clarifies the relation between the Maclaurin spheroid and the Dyson-Wong toroid. Moreover it is the sequence of minimum-energy configuration. These results suggest that our solutions are more physical and probably stabler than any other equilibrium of incompressible fluids.

DISCUSSION

Roxburgh: If you have a sequence corresponding to increasing angular momentum, do you find a ring of increasing diameter and decreasing angular velocity? Is there not another sequence with a Maclaurin spheroid inside a ring?

Fukushima: The method used here is not applicable to bodies whose central density is exactly zero and/or which have disconnected regions. So a nearly ring structure is obtained for high angular momentum, but the exact ring or Maclaurin-spheroid-inside-a-ring structure is not computed.

Durisen: What are the values of $T/|W|$ at which the hamburger, and then the toroidal figures first appear along one of your sequences and how do they compare with other stability limits and bifurcations?

Fukushima: When the value of $T/|W|$ reaches about 0.48, the concave hamburger structure appears. We have not carried out a stability analysis, but the total energy of the concave hamburger is less than that of spheroids and toroids.

D. Sugimoto, D. Q. Lamb, and D. N. Schramm (eds.), Fundamental Problems in the Theory of Stellar Evolution, 273.

RAPIDLY ROTATING AND FULLY GENERAL RELATIVISTIC POLYTROPES

Yoshiharu Eriguchi
Department of Earth Science and Astronomy, College of General Education, University of Tokyo

A technique used in the numerical computation of Newtonian rotating polytropes has been generalized and applied to the rapidly rotating polytropes in general relativity.

The star is assumed to rotate uniformly and to be axially and equatorially symmetric. The polytropic relation

$$P=K\varepsilon^{1+1/N}$$

is also assumed. The full equations for the rotating polytrope in general relativity have been numerically integrated without any approximation. Strength of relativity is measured by $\mu = P_c / \varepsilon_c c^2$ where P_c, ε_c, and c are the central pressure, the central energy density, and the light velocity, respectively. Rotating sequences with $\mu = 0.001$ (almost Newtonian case), 0.25 (mildly relativistic case), and 0.5 (highly relativistic case) have been computed for the polytropic index N=1.5 .

For the same value of the central energy density and the same value of K, the rest mass increases as the angular momentum increases but the increase for the large value of μ case is rather mild one.

For the same value of the rest mass the gravitational mass always increases as the angular momentum increases. The rotation does not act as stabilizing the highly relativistic models.

D. Sugimoto, D. Q. Lamb, and D. N. Schramm (eds.), Fundamental Problems in the Theory of Stellar Evolution, 274.

THE INFLUENCE OF ROTATION ON HORIZONTAL BRANCH STARS AND ON RR LYRAE PULSATIONAL PROPERTIES.

V.Castellani
Istituto Astrofisica Spaziale, Frascati
and
University of Rome

The occurrence of rotati on in Globular Cluster stars has been suggested (see e.g. Renzini 1977) as a mechanism producing the observed colour spread in actual Horizontal Branches. If this is the case, canonical results on evolutionary properties of HB stars have to be revisited in order to account for rotation-driven structural variations: faster Main Sequence rotators delay the He flash increasing the mass -size M_c of the He core at the flash and loosing a greater amount of mass during the Red Giant stage.
Both these effects act in the sense of increasing the surface temperature of HB stars: the increase in M_c also drives an increase in the luminosity of the Zero Age HB locus. As a consequence canonical results are relased as for as the close correlation between the original helium content Y and the luminosity of HB stars is concerned. It follows that the mean period of RR Lyrae pulsators is no more a "bona fide" function of Y, neither the ratio R between the number of HB and RG stars (Iben, 1968) will remain a completely reliable Y indicator (Castellani et al 1980).
By constructing "sinthetic" Horizontal Branches under wide assumptions on the original chemical composition, we find (Castellani and Tornambè 1980) that observations can be watched with theory if galactic HB_s are populated by stars evolving along typical low-He ($Y \sim 0.22$) evolutionary tracks but overluminous will respect to the expectation from the canonical evolutionary frame.
If MS rotation is causing the quoted HB morphology, one can derive that galactic globular clusters are fitted by the values $Y \sim 0.22$, $t \sim 12\text{-}10^{9}$ yrs, $\omega_o = 3.6\text{-}10^{-4}$ provided that the efficiency of mass loss is supposed to sensitively increase in increasing the metal content Z.

References

Castellani,V., Ponte,G., Tornambè,A.: Astrophys.Space Sci (in publ.)
Castellani,V., Tornambè,A.: 1980, Astron.& Astrophys. (in publ.)
Iben,I.Jr: 1968, Nature 220, 143
Renzini,A.: 1977, Proceedings of "Saas-Fee", 7th Advanced Course.

D. Sugimoto, D. Q. Lamb, and D. N. Schramm (eds.), Fundamental Problems in the Theory of Stellar Evolution, 275.

CAN STELLAR ATMOSPHERES BE IN QUASI-STATIC EQUILIBRIUM IN THE PRESENCE OF MAGNETIC FIELDS ?

Kyoji Nariai
Tokyo Astronomical Observatory, University of Tokyo
Mitaka Tokyo 181, Japan

Among the observational astronomers it is quite customary to assume potential magnetic fields in calculating the strength and the direction of the magnetic fields on the stellar surface. People use this assumption because

1. it is easy to calculate,
2. it is possible to fit the observations with the calculations to some extent using the parameters inherent to the model,
3. at a first glance, there is no difficulty in assuming both potential magnetic fields and static stellar atmosphere because they do not interact each other (except at the pole).

However, the basis of this assumption is physically unsound because we have to introduce a current system around the magnetic pole in order to reproduce the magnetic fields in the outer region, and with this current system it is impossible to maintain the plasma in a static state near the pole and to keep the electromagnetic quantities time-independent at the same time.

It should also be kept in mind that the energy of plasma integrated in a flux tube is by far larger than the magnetic energy integrated in the same flux tube if we assume potential magnetic fields.

Once we admit the interaction between the electromagnetic fields and the stellar plasma, it follows immediately that the stellar atmosphere cannot be in equilibrium in the presence of magnetic fields.

DISCUSSION

Mestel: I am not sure if I have followed your arguments. Certainly a field that is curl-free near the surface must be maintained by a current deeper down. One can construct hydrostatic models of magnetic stars in which the current density falls off near the surface, e.g., like the

D. Sugimoto, D. Q. Lamb, and D. N. Schramm (eds.), Fundamental Problems in the Theory of Stellar Evolution, 276–277.

density, so that near the surface the field approximates very closely to the curl-free state. Concerning one of your equations, Biermann showed many years ago that if ρ is not a function of $\underline{P}$, one can get the equivalent of a "battery" generating a magnetic flux. But the timescale of the process is very long, and there is no contradiction in assuming that the star remains in hydrostatic equilibrium.

Nariai: The treatments of stellar structure in the presence and in the absence of a magnetic field are quite different from each other, although they are concerned with the same equations. I agree with you that one can construct models of magnetic stars in which the current density falls off near the surface if you talk about the magnetic field from the standpoint of an astrophysicist concerned with the stellar interior. I am saying that we have to attack the problem of the surface field from the viewpoint of specialists in stellar atmosphere, to whom the current density is finite although it might be "zero" from the viewpoint of those who are concerned with the magnetic field in the interior.
Concerning the time scale of the process, I think that the problem is worth study, with special attention given to the region $\tau = 1$ because this is the region where γ varies greatly because of radiative transfer, and because ρ is almost zero from the viewpoint of specialists in stellar structure.

INFLUENCE OF CONVECTION ON THE PULSATIONAL STABILITY OF STARS

Y. Osaki* and G. Gonczi
Observatoire de Nice, Nice, France
*Department of Astronomy, University of Tokyo, Tokyo, Japan

Influence of convection on the pulsational stability of stars is examined based on Unno's (1967) theory of time-dependent convection. This theory is an extension of Vitense's local mixing-length theory of convection to the time-dependent problem. The equations of linear non-adiabatic pulsations, which include the thermal interaction between convection and pulsation, have been solved numerically on a series of Cepheid models. It is found that the thermal eigenfunctions (of $\delta T/T$ and $\delta L/L$) exhibit spatial oscillations with short wavelength in cool stellar models where the energy is mostly transported by convection. This phenomenon of spatial oscillations is interpreted, and it is shown that they arise because of the phase lag of the convective flux to pulsation and they are related to the local character of the convection theory. It is found that the spatial oscillations play a determinant role in the pulsational stability in the region of the red edge of the Cepheid instability strip, as far as the thermal convection-pulsation coupling is concerned. The details of this study have been pulbished in Astronomy and Astrophysics.

REFERENCES

Gonczi, G. and Osaki, Y. : 1980, Astronomy and Astrophysics, 87, pp304-310.
Unno, W. : 1967, Publ. Astron. Soc. Japan, 19, pp140-153.

DISCUSSION

Wheeler: Bob Deupree of Los Alamos has used a time-dependent non-local two-dimensional numerical treatment of convection to address the question of the red edge of the instability strip. My impression was that he had provided a satisfactory quantitative solution to the red edge problem. Can you comment on Deupree's work?

D. Sugimoto, D. Q. Lamb, and D. N. Schramm (eds.), Fundamental Problems in the Theory of Stellar Evolution, 278–279.

Osaki: I think the two-dimensional numerical treatment of convection by Deupree is not satisfactory for this problem since convection is essentially a three-dimensional phenomenon.

VISCOSITY, THERMAL AND ELECTRICAL CONDUCTIVITIES IN TURBULENT CONVECTION

Wasaburo Unno, Tohru Nakano*, and Masa-aki Kondo**
Dept. of Astronomy, Faculty of Science, Univ. of Tokyo
*Dept. of Physics, Chuō Univ.
**Dept. of Earth Science and Astronomy, College of General Education, Univ. of Tokyo

Abstract

Turbulent diffusivities are often used for representing nonlinear interactions of turbulent elements on the motion of a larger scale. In turbulent convection, the average life of a representative element is substantially lengthened by bouyancy. Taking this effect into account, we calculate turbulent viscosities, thermal and electrical conductivities for Boussinesq fluids on the basis of a spectral theory of turbulent convection (Nakano, Fukushima, Unno, and Kondo, 1979). The effect of bouyancy results in the increase of turbulent diffusivities, compared with the case without bouyancy. We also propose the generalization of the method such that a stellar convection zone can be theoretically constructed without recourse to the mixing length.

Reference

Nakano, T., Fukushima, T., Unno, W., and Kondo, M. 1979, Publ. Astron. Soc. Japan, 31, 713.

D. Sugimoto, D. Q. Lamb, and D. N. Schramm (eds.), Fundamental Problems in the Theory of Stellar Evolution, 280.

ON THE PRESENCE OF TURBULENT DIFFUSION IN STARS AND ITS EFFECT ON STELLAR EVOLUTION

E. SCHATZMAN
Observatoire de Nice
B.P. 252 - Nice-Cedex 06007 - France

The presence of a mild turbulence inside rotating stars has been considered by Howard et al (1967) and by Bretherton et al (1968) in connection with the problem of the solar flattening, raised by Dicke (1970). The possibility of using abundance determination to test the existence of that turbulence has been considered by Schatzman (1969).

It is only in 1977 that the interpretation by Schatzman (1977) of the abundance determination of lithium in giants by Alschuler (1975) has brought a reasonable proof of the existence of a mild turbulence deep in stars. The presence of a mild turbulence at greater depths was established by Genova and Schatzman (1979) by a consideration of the (12C/13C) ratio in giants.

Further proof is given by the consideration of the solar neutrino flux. Schatzman and Maeder (1980) have shown that turbulent diffusion brings H and 3He in the solar core. The effect is to reduce the central temperature of the Sun, with the consequence of a drastic decrease of the neutrino flux. It leads also to an explanation of the surface 3He abundance and fullfills the constraint on the solar luminosity.

For all these effects, the turbulent diffusion, supposed to be due to some marginal Reynolds instability, induces a turbulent diffusion coefficient $D_T = Re^{*} \nu$. A rough estimate by Schatzman (1977) for a Kolmogoroff spectrum leads to Re^{*} = 100 to 200. If Re^{*} is considered as a purely phenomenological parameter, the various astrophysical estimates agree with a value Re^{*} = 100 to 200.

Evolution of a 1 $M_{\odot}$ star has been computed by Maeder (1980) with turbulent diffusion mixing. The star does not evolve towards the first ascending branch of the giants, but becomes a blue straggler. An inhibition of the turbulence has to be introduced to provide the evolution towards the giant branch. This is probably due to the stabilization by the μ-gradient.

These results suggest important modification to stellar evolution patterns.

D. Sugimoto, D. Q. Lamb, and D. N. Schramm (eds.), Fundamental Problems in the Theory of Stellar Evolution, 281–283.

DISCUSSION

Roxburgh: I can see a problem in trying to explain the solar neutrino problem by such diffusion. In order to lower the neutrino flux, the diffusion has to be strong enough to keep the sun more or less homogeneous. This then raises a problem in explaining the agreement between inhomogeneously evolved models and observations of globular clusters.

Schatzman: This raises problem of the blue stragglers. If nothing happens, and if diffusion is not turned off, a star would evolve more or less in the direction of the blue stragglers and would never reach the giant branch. It is necessary for most of the stars to turn off turbulent mixing at some proper time. It seems that the assumption that turbulent diffusion stops when the μ-gradient exceeds a certain critical value, which might be related to the angular velocity, is quite reasonable. There would be at least two classes of stars: those which keep experiencing mixing, and those for which mixing stops at some time.

Roxburgh: What mechanism do you think drives the weak turbulence? I would like to suggest that one attractive possibility for driving diffusion is the instability driven by the build-up of ^{3}He away from the centre. Diffusion could then keep the sun on the verge of instability until such time as the chemical inhomogeneity stabilised the sun and suppressed the weak turbulence.

Schatzman: I am interested in this possibility, which might prevent the ^{3}He peak from becoming too large and thus would limit the amount of ^{3}He which is driven by turbulent diffusion to the surface. On the other hand, I have been thinking that magnetic braking is followed by a continuous redistribution of angular momentum inside the sun and that, as supposed by Spiegel and co-workers, some sort of turbulence takes care of the process.

Castellani: I am a bit surprised by your statement about the influence of central ^{3}He on the decrease of the central temperature in Sun. It seems to me that in the central region the ^{3}He-abundance is determined only by the equilibrium conditions. This is just because the characteristic time scale to reach ^{3}He equilibrium is much shorter than the time scale for turbulent diffusion. So I suggest that the decrease in temperature is driven by the increase in H, and that a (small) increase in ^{3}He is only the consequence of such a decrease.

Schatzman: In order of magnitude, the excess of ^{3}He can be written:

$$\frac{\Delta X_3}{X_3} = \frac{1}{2}\,\frac{3D_t (X_3)_{peak}}{R_c^2}\,\frac{1}{K_{11} X_1^2 A_3 m_H} ,$$

with $(X_3)_{peak} = 2.4 \times 10^{-3}$, $(R_c/R_\odot) = 0.3$, and $D_t = 100$. We obtain

$$\frac{\Delta X_3}{X_3} \cong 0.085 .$$

The direct calculation gives

$$(X_3/X_1)_{balance} = 3.44 \times 10^{-5}$$

$$(X_3/X_1)_{diffusion} = 3.7 \times 10^{-5} \quad (D_t = 100)$$

where $(X_3/X_1)_{balance}$ has been calculated for the central temperature of the diffusion model. It is not the diffusion time which is important, but the flow, which depends on both D_t and ∇X_3.

Massevitch: Have you considered the possibility that the use of evolutionary sequences other than Iben's, with slightly different chemical compositions and/or other opacity tables, may change the estimated Li-deficiencies so that they coincide better with the observed values?

Schatzman: Without diffusion, all models for the giants lead to a plateau followed by sudden drop of the Lithium abundance as soon as the convective zone has become deep enough to reach the level at which Li is burnt. This is complete disagreement with the observations, which show as a general trend a continuous decrease of the Lithium abundance as a function of spectral type.

ON THE NUMBER RATIO OF HORIZONTAL BRANCH STARS TO RED GIANT STARS IN GLOBULAR CLUSTERS

N. Arimoto* and M. Simoda**
* Astronomical Institute, Tohoku University, Sendai, Japan
**Department of Astronomy and Earth Sciences, Tokyo Gakugei University, Tokyo, Japan

The number ratios of horizontal branch stars to red giant stars were obtained for globular clusters and Draco dwarf galaxy and the helium abundance was estimated using model results without semiconvection zone (SCZ) and with fully developed one. The analysis was confined to the four clusters (M4, M5, M13, and 47 Tuc) and the Draco galaxy, for which fairly precise star counts had been carried out. The effect of the difference in radial distribution between horizontal and red giant branch stars were taken into account, if necessary. The statistically significant difference in R exists among these objects. The cause may be the difference in the helium abundance and/or in the development of the SCZ. In the case of the fully developed SCZ, the helium abundance for M5 and Draco is appreciably smaller than the value given by the big-bang cosmology. It may be taken as an evidence against the full development of the SCZ for the horizontal branch stars in these objects.

D. Sugimoto, D. Q. Lamb, and D. N. Schramm (eds.), Fundamental Problems in the Theory of Stellar Evolution, 284.

SUPERNOVAE: PROGENITOR STARS AND MECHANISMS

J. Craig Wheeler
Department of Astronomy
University of Texas at Austin

ABSTRACT

Type II supernovae probably arise predominantly in stars of 8-15 $M_\odot$ which leave neutron star remnants but accomplish little in the way of nucleosynthesis. Stars in the mass range $\sim$ 15-70 $M_\odot$ may either explode or collapse. Their evolution and final outcome, including their contribution to nucleosynthesis, may depend strongly on processes of mass loss. Type I supernovae probably involve a deflagrative explosion in a carbon-oxygen core surrounded by a distended helium envelope. The evolutionary origin of such a configuration is obscure.

INTRODUCTION

The purpose of this review is to give an overview of the present theory and basic observations which pertain to the question of which stars explode and why. The following presentations by Nomoto and Mazurek will go into more detail concerning specific models. The developments presented here are given in more detail in reviews by Sugimoto and Nomoto (1980) and by Wheeler (1981).

TYPE II SUPERNOVAE

The rates and kinematics (Tammann 1978; Maza and van den Bergh 1976) light curves (Falk and Arnett 1977; Chevalier 1976; Weaver and Woosley 1980) and spectra (Kirshner et al. 1973) of Type II supernovae (SN II) are all consistently understood in terms of the explosion of a normal massive star. The explosion takes place within an extended red-giant hydrogen envelope. Recent progress has been made toward refining the mass range in which SN II occur and determining the final evolution. Much work is currently being done to understand the explosion mechanism.

One way of setting the lower limit to the mass of stars which become SN II is to determine the mass below which stars leave white

D. Sugimoto, D. Q. Lamb, and D. N. Schramm (eds.), Fundamental Problems in the Theory of Stellar Evolution, 285–294.

dwarf remnants, and die a quiet death. This can be done by counting the number of white dwarfs in young open clusters and assigning that number of stars to the main sequence above the cluster turnoff in accordance with the cluster mass function. This exercise has been redone recently by Romanishin and Angel (1980) with the result that all stars below $M \sim 8\ M_\odot$ seem to leave white dwarfs, and to not make supernovae. Koester and Weidemann (1980) study the observed mass distribution of white dwarfs and reach the same conclusion.

Recent theoretical developments strengthen the conviction that $8\ M_\odot$ represents the lower limit to supernovae for normal single star evolution. Although the result depends somewhat on the theory of convection, there has for some time been general agreement that the dividing line between stars which form degenerate carbon/oxygen cores and those which burn carbon in a non-degenerate manner is at $M \sim 8\ M_\odot$ (Sugimoto 1971; Paczyński 1970; Becker and Iben 1979). Work by Barkat and collaborators (Tuchman Sack and Barkat 1979) has given special significance to this dividing line. They find that as the stars with degenerate carbon/oxygen cores evolve and brighten, they become catastrophically pulsationally unstable and eject their envelopes, leaving the cores to cool into white dwarfs.

Thus the pulsational calculations say that all stars which develop carbon/oxygen cores eject their envelopes. Evolutionary calculations say that stars with $M \lesssim 8\ M_\odot$ form such cores and hence should eject their envelopes. These combined theoretical statements are in good accord with the observations that stars with $M \lesssim 8\ M_\odot$ indeed do eject their envelopes and leave white dwarfs. The steep mass function is such that if stars much above $8\ M_\odot$ also failed to explode there would be a difficulty in accounting for the rates of SN II. Stars much above $\sim 15\ M_\odot$ may not make SN II, as discussed below, so the range for the progenitors of most SN II is ~ 8-$15\ M_\odot$.

The evolution of stars in this mass range is very complex, involving semi-dynamical shell flashes and electron capture. Progress in exploring the final evolution in this mass range has recently been made by Nomoto and collaborators (see his contribution) and by Weaver and Woosley (1980; see also Woosley, Weaver, and Taam 1980). Although the details change very rapidly with mass there seems little question that electron captures and/or photodisintegration will cause the cores of these stars to collapse to form neutron stars.

The crucial question is then the mechanism by which the process of neutron star formation causes an explosion. A great deal of work is currently being done on this question. The realization that neutrinos would become degenerate and trapped in the collapsing core (Mazurek 1974; Sato 1975) led to the conclusion that the collapse would be nearly adiabatic and proceed to greater than nuclear densities before the equation of state would stiffen, halting the collapse (Mazurek 1977; Arnett 1977; Lamb et al. 1978; Bethe et al. 1979). In the basic one-dimensional calculations (see Mazurek herein) a cold homologous core of $\sim 0.7\ M_\odot$ forms and collapses to neutron star densities. The bounce of this core and subsequent infall generates an outward moving shock. Early calculations with parametrized equations of state (Van Riper and Arnett 1978; Lichtenstadt, Sack and Bludman 1980) indicated that with a proper choice of core mass near the limiting neutron star mass for a given equation of state the core bounce could give a shock which ejected the envelope with supernova-like energy $\sim 10^{51}$ ergs. Refinements have led to the shock petering out, partly due to the loss of energy in photodisintegrating the infalling material, although numerical problems can not be ruled out.

The outcome may depend on core mass. With their schematic equation of state Van Riper and Arnett found that for very large cores, $M \gtrsim 2.5\ M_\odot$, a hot core bounced and formed an explosion but later collapsed. In principle at least, a star could both explode and leave a black hole although in most cases the result is an explosion with a neutron star remnant, or total collapse with no explosion.

Addition of rotation may cause the core to halt collapse at lower densities (Tohline, Schombert and Bass 1980) and perhaps alter the dynamics appreciably. Convective overturn of the core initiated by rotational deformaties or by alteration of the composition by neutrino losses (Epstein 1979) may enhance the neutrino losses (Livio, Buchler and Colgate 1980) or promote PdV pumping of the core into the envelope (Colgate and Petschek 1980).

The masses and the evolution of the progenitor stars of SN II are being steadily refined. They probably come from 8-15 $M_\odot$ and leave neutron star remnants. There is little material between the collapsing core and the envelope in this mass range so the bulk of SN II can contribute only little to the synthesis of the heavy elements.

VERY MASSIVE STARS

The final stages of evolution of more massive stars $\sim$15-70 $M_\odot$ is more tractable than in somewhat lower masses and has been more thoroughly explored (e.g. Weaver, Woosley, and Zimmerman 1976; Sparks and Endal 1980). These stars form iron cores which collapse to make neutron stars because of photodisintegration of the iron. There are probably too few of these stars to contribute appreciably to the rate of SN II. Even if they explode many of these massive stars may not contribute to the observed rate of optical supernovae. The reason is that these stars are prone to strong mass loss on the main sequence and even late in the core helium burning, Wolf-Rayet, phase. This mass loss probably prevents the formation of an extended envelope. If the envelope has a radius appreciably less than 10^{13} cm the energy deposited by the shock of the explosion will be dissipated in adiabatic expansion before the envelope becomes thin enough to radiate. In such a case there is very little optical display.

There is indirect evidence that many of these stars explode. Detailed calculations of the evolution, dynamics (assuming an explosion is triggered, e. g. by core bounce) and nucleosynthesis show that a solar distribution of basic elements is created if all stars in this mass range explode (Weaver and Woosley 1979).

This conclusion is important both for supernova theory and for nucleosynthesis so some of the caveats should be presented. Studies of the abundances in old field stars show that the oxygen abundance was high by a factor ~ 3 in the past compared to carbon and iron (Sneden et al. 1980) as if the oxygen were produced more rapidly in the past. This would not happen if the basic nucleosynthetic source of these elements were a fixed ensemble of stars as normally assumed in the theoretical calculations which match the present abundances. One possibility is that the mass distribution of stars is not constant but was weighted more heavily in the past to more massive stars which naturally produce an oxygen excess. Alternatively oxygen, carbon, and iron could have their source in disparate objects, for instance carbon in carbon stars and iron in Type I supernovae as discussed by Tinsley (1979). In this case, the reproduction of the solar distribution by the massive star calculations could be fortuitous.

Another problem is that stellar winds can strongly affect the yield of heavy elements from massive stars. Both main sequence winds (Chiosi and Caimmi 1979) and those during the helium burning Wolf-Rayet phase (Vanbeveren and Olson 1980) can serve to reduce the mass of the core compared to a constant mass star, and hence the ultimate yield. There are many uncertainties concerning the stellar evolution, mass function and birthrates; thus while there are encouraging results the conclusion that all these massive stars do explode in the manner envisaged remains unsure.

Two astronomical objects illustrate the possible divergent fates of massive stars. The supernova remnant Cassiopeia A clearly resulted from an explosion of a massive star which ejected freshly synthesized material, predominantly the products of explosive oxygen burning (Chevalier and Kirshner 1979). There was no optical outburst of the magnitude of a supernova (even the 6th magnitude star reported by Ashworth (1980) is an unsure association) which is consistent with the loss of the envelope to form the nitrogen-enhanced low velocity filaments (Lamb 1978). Thus Cas A could be the result of a typical exploding massive star. No neutron star is observed in Cas A in X-rays (Murray et al. 1980). This may be because the neutron star has cooled rapidly (Glen, and Sutherland 1980; Van Riper and Lamb 1980). Alternatively, the progenitor of Cas A may have been totally disrupted, perhaps by the oxygen deflagration which occurs in very massive stars ($M \gtrsim 70\ M_\odot$).

The second illustrative object is Cygnus X-1. If this X-ray source is a black hole of $M \sim 10\ M_\odot$ it probably represents the core of a star of $\sim 30\ M_\odot$ which lost its envelope in a stellar wind. If

30 $M_\odot$ stars leave massive black holes, they can not at the same time be the major contributors to the synthesis of heavy elements as indicated by the nucleosynthesis calculations.

Some stars in the mass range 15-70 $M_\odot$ may explode, others may collapse totally depending on the mass and details of the final evolution. These stars may generate a significant portion of the heavy elements but they probably make little if any contribution to classical SN II.

TYPE I SUPERNOVAE

Type I supernovae (SN I) are more difficult to understand than SN II. They are hydrogen deficient and display the famous exponential decay which has defied explanation. The last year has seen a resurgence of interest in the idea that the exponential decay is produced by the radioactive decay sequence $^{56}Ni(6^d.1,\gamma)\rightarrow{}^{56}Co(77^d,\gamma,e^+)\rightarrow{}^{56}Fe$. The principle impetus has been detailed calculations of the γ-ray and positron energy deposition by Colgate, Petschek, and Kreise (1980) and of the deposition and resulting late-time spectra by Axelrod (1980). These and many related ideas are contained in the proceedings of the Austin SN I Workshop (Wheeler 1980).

Colgate _et al_. argue that increasing transparency of the expanding envelope to the positrons from Co decay causes the modulation of the intrinsic 77^d half-life into the observed, more rapid decline. Their particular model ejects 1/4 $M_\odot$ of Ni and 1/4 $M_\odot$ of inert material and requires a rise time of $\sim 6^d$. This is less than one-half the observed rise time so more Ni and more total mass is probably required to fit the observations. The synthetic spectra of Axelrod also require more Ni to be ejected $\gtrsim .5$ $M_\odot$. In Axelrod's model the positrons are all trapped by a putative magnetic field but a progressively larger fraction of the energy goes into the infrared at the expense of the optical. This picture seems to violate the observations that the IR flux drops more rapidly than the optical (Kirshner _et al._ 1973).

Another unresolved question concerns the nature of the initial peak. Models in which the peak and the exponential tail come solely from radioactive decay have not yet proven totally self-consistent. The model of Colgate _et al._ has too short a rise time. A model based on the same deposition theory by Chevalier (1980) ejects 1.4 $M_\odot$ to get the proper rise time and 1 $M_\odot$ of Ni to get the peak (nearly) bright enough but as a result produces too much light in the exponential tail. An alternative is to create most of the peak light with the energy of the initial shock. This requires a large envelope to avoid adiabatic losses and implies the progenitor is not a white dwarf despite the paucity of hydrogen.

The light curves and late-time spectra indicate SN I eject $\sim$ 1/2-1 $M_\odot$ of ^{56}Ni. This is very difficult to do in any model in which a neutron star forms and the Ni is produced by silicon burning in the explosion; only $\sim 0.1\ M_{\odot}$ of Ni can be produced this way. The only models which have been calculated which produce $\sim 1/2\ M_{\odot}$ of Ni proceed from a thermonuclear detonation or deflagration and completely disrupt the star leaving no neutron star. Relevant to this conclusion are the observations of historical remnants of SN I by the Einstein X-ray satellite. As for Cas A no neutron stars are observed in the remnants of Tycho's or Kepler's supernovae or SN 1006 (Helfand, Chanan and Novick 1980). Again the neutron stars may have cooled very quickly, but the observations are quite consistent with models which predict no neutron star at all.

The spectra at peak light show no H, Ni or Co (Branch 1980). This implies the existence of a blanketing envelope which is probably mostly He, perhaps enriched in Si, Ca and Fe. Furthermore the absorption lines are observed, even at late times, to have velocity $v > 8000$ km s^{-1}. This feature implies that the progenitor did not have a monotonically decreasing density gradient as would be expected for a white dwarf. If this were the case, the velocity of the material at the photosphere should decrease monotonically in time; there would be no cutoff at 8000 km s^{-1}. Rather this cutoff implies a non-monotonic density gradient as would obtain with a distended envelope which could be ejected as a shell. Whether such an envelope is necessary and whether it is large enough that some shock energy can be contributed to the peak remains to be seen.

The best one can currently say in terms of progenitors is what they are not. They are probably not degenerate He dwarfs because in order to generate enough Ni for the light curve and spectra the resulting velocity would be too high. They are not the result of rapid mass transfer onto a carbon/oxygen white dwarf because a hydrogen rich extended envelope would form. They are not due to moderate mass transfer onto a carbon/oxygen dwarf because despite the fascinating new calculations by Nomoto (1980) and Woosley, Weaver and Taam (1980) that such stars ignite an off-center double detonation wave in the degenerate helium shell, the result is complete burning to Ni with no He blanket remaining. They may result from slow accretion onto a carbon/oxygen dwarf ($\dot{M} \lesssim 4\times10^{-10}\ M_{\odot}$ yr) because such stars can burn carbon subsonically, and disrupt totally, shoving off an unburned He shell. Unfortunately that shell, once a part of the dwarf structure, will probably not satisfy the kinematic constraint of $v > 8000$ km s^{-1}. In any case the most successful binary dwarf model must be old, triggered as it is by <u>slow</u> mass transfer.

SN I are probably not the explosion of He cores produced by the loss of the envelope from a star of $\sim 8\text{-}10\ M_{\odot}$. These He stars could form distended helium envelopes but, as mentoned in the first section, such stars will develop collapsing cores and probably cannot eject sufficient Ni. In addition such stars may be required to concentrate in spiral arms whereas SN I are not observed to do so (Maza and van den Bergh 1976). In any case this picture for the origin of SN I requires them to be quite young.

An important lesson here is that to construct a model which can be either young or old by the twitch of a parameter is decidely non-trivial. Changing the age of a model, e.g. by changing $\dot{M}$ in the binary picture, qualitatively alters the nature of the model. This is in violation of the observation that SN I are a very homogeneous class particularly in terms of the early spectra which determine the composition and kinematics of the envelope.

To even discuss the age of SN I as a possible parameter is a relatively new proposition. Classically the association of SN I with elliptical galaxies has been interpreted in terms of old progenitor stars. Oemler and Tinsley (1979) have argued strongly, however, that SN I are connected with regions of recent star formation. The question of the age of SN I progenitors is thus an active one, and far from settled.

In a sense the current status of SN I complements that for SN II. For SN II the evolution is beginning to be relatively well understood although the actual explosion mechanism still defies proper explanation. For SN I, there is a growing feeling that the most plausible mechanism is a carbon deflagration in a degenerate core, itself buried in a partially, if not fully, distended helium envelope. The evolutionary origin of such a configuration is still obscure.

REFERENCES

Arnett, W. D.: 1977, Astrophys. J., 218, pp.815.
Ashworth, W.: 1980, B.A.A.S., 11, pp.660.
Axelrod, T.: 1980, in Supernovae Spectra-Atomic and Spectroscopic Data Needs, ed. R. E. Meyerott and G. H. Gillespie, in press.
Becker, S. A., and Iben, I. Jr.: 1979, Astrophys. J., 232, pp.831.
Bethe, H. A., Brown, G. E., Applegate, Jr. and Lattimer, J. M.: 1979, Nuclear Phys. A., 324, pp.487.
Branch, D.: 1980, in Proceedings of the Texas Workshop on Type I Supernovae, ed. J. C. Wheeler (Univ. of Texas, Austin).
Chevalier, R. A.: 1976, Astrophys. J., 207, pp.872.
Chevalier, R. A.: 1980, preprint.
Chevalier, R. A., and Kirshner, R. P.: 1979, Astrophys. J., 233, pp. 154.
Chiosi, C., and Caimmi, C.: 1979, Astron. and Astrophys., 74, pp.62.
Colgate, S. A., and Petschek, A. G.: 1980, Astrophys. J. (Letters), 236, pp.L115.
Colgate, S. A., Petschek, A. G., and Kriese, J. T.: 1980, Astrophys. J. Letters, 237, pp.L81.
Epstein, R. I.: 1979, Monthly Notices Roy. Astron. Soc., 188, pp.305.
Falk, S. W., and Arnett, W. D.: 1977, Astrophys. J. Suppl., 33, pp.515.
Glen, G., and Sutherland, P.: 1980, Astrophys. J., 239, pp.000.
Helfand, D. J., Chanan, G. A., and Novick, R.: 1980, Nature, 238,p.337.

Kirshner, R. P., Oke, J. B., Penston, M. V., and Searle, S.: 1973a, Astrophys. J., 185, p.303.
Kirshner, R. P., Willner, S. P., Becklin, E.E., Neugebauer, G., and Oke, J. B.: 1973, Astrophys. J. (Letters), 180, p.L97.
Koester, D., and Weidemann, V.: 1980, Astron. and Astrophys., 81,p.145.
Lamb, S. A.: 1978, Astrophys. J., 220, p.186.
Lamb, D. Q., Lattimer, J. M., Pethick, C. J., and Ravenhall, D. G.: 1978, Physics Rev. Letters, 41, p.1623.
Lichtenstadt, I., Sack, N., and Bludman, S. A.: 1980, Astrophys. J., 237, p.903.
Livio, M., Buchler, J. R., and Colgate, S. A.: 1980, Astrophys. J. (Letters), 238, p.L139.
Maza, J., and van den Bergh S.: 1976, Astrophys. J., 204, p.519.
Mazurek, T. J.: 1974, Nature, 252, p.287.
Mazurek, T. J.: 1977, Comments on Astrophys., 7, p.77.
Murray, S. S., Fabbiano, G., Fabian, A. C., Epstein, A., and Giacconi, R.: 1979, Astrophys. J. (Letters), 234, p.L69.
Nomoto, K.: 1980, in Proceedings of the Texas Workshop on Type I Supernovae, ed. J. C. Wheeler (Univ. of Texas, Austin).
Oemler, A., and Tinsley, B. M.: 1979, Astron. J., 84, p.985.
Romanishin, W., and Angel, J. R. P.: 1980, Astrophys. J., 235, p.992.
Sato, K.: 1975, Prog. Theor. Phys., 53, p.595.
Sneden, C., Lambert, D. L., and Whitaker, R. W.: 1979, Astrophys. J. 234, p.964.
Sparks, W. M., and Endal, E. S.: 1980, Astrophys. J., 237, p.130.
Sugimoto, D.: 1971, Prog. Theor. Phys., Kyoto, 45, p.761.
Sugimoto, D., and Nomoto, K.: 1980, Space Science Review, 25, p.155.
Tammann, G. A.: 1978, Mem. Soc. Astron. It., 49, p.315.
Tinsley, B. M.: 1979, Astrophys. J., 229, p.1046.
Tohlin, J. E., Schombert, J. M., and Boss, A. q.: 1980, IAU Colloquium No. 58, Stellar Hydrodynamics Los Alamos.
Tuchman, Y., Sack, N., and Barkat, Z.: 1979, Astrophys. J., 234, p.217.
Vanbeveren, D., and Olson, G. L.: 1980, Astron. and Astrophys., 81,p.228.
Van Riper, K. A., and Arnett, W. D.: 1978, Astrophys. J. (Letters), 225, p.L129.
Van Riper, K. A., and Lamb, D. Q.: 1980, Astrophys. J. (Letters), 000, p.000.
Weaver, T. A., and Woosley, S. E.: 1980, B.A.A.S., 11, p.724.
Weaver, T. A., and Woosley, S. E.: 1980, Ann. N.Y. Acad. Sci., 336, p.335.
Weaver, T. A., Zimmerman, G. B., and Woosley, S. E.: 1978, Astrophys. J. 225, p.1021.
Wheeler, J. C.: 1980, Proceedings of the Texas Workshop on Type I Supernovae, ed. J. C. Wheeler, (Univ. of Texas, Austin).
Wheeler, J. C.: 1981, Reports on Progress in Physics, in press.
Woosley, S. E., Weaver, T. A., and Taam, R. E.: 1980, Proceedings of The Texas Workshop on Type I Supernovae, ed. J. C. Wheeler (Univ. of Texas, Austin).

DISCUSSION

Joss: In the old carbon-detonation scenario that you described, the degenerate cores of stars with masses in the range 4-8 $M_\odot$ ingnited carbon before the core mass reached the Chandrasekhar limit and before the stellar envelope was ejected. However, according to the new results that you reported, for stellar masses up to 8 $M_\odot$ the envelope becomes unstable and is ejected prior to core carbon ignition. Is there a physical reason for the apparent coincidence between the upper mass limits for carbon detonation and prior envelope ejection, or is this merely a conspiracy of nature?

Wheeler: As I understand Barkat the growing core mass sets the growing luminosity, which in turn determines the pulsational properties of the envelope. If his calculations are correct, there is a direct physical connection between the growing carbon core, the double burning shells and the ultimate ejection. Whatever the mass is below which degenerate carbon cores form, that is the mass below which envelope ejection occurs. The formation of a degenerate carbon core automatically forces the envelope ejection in Barkat's picture.

Sugimoto: Evolution of stars in the mass range 4-8 $M_\odot$ toward carbon deflagration is considered also to be the origin of carbon stars. If all of such stars lose their envelope by dynamical instability, the number of carbon stars should be significantly decreased. What do you think about this matter? I am asking this question, because the dynamical instability of the envelope should depend sensitively on the assumed value of the ratio of the convective mixing length to the scale height.

Wheeler: I cannot speak in too much detail concerning the calculations of Barkat et al. They are attempting to reproduce the observations of Mira variables with some success. The question of whether they can produce carbon stars depends on whether envelope ejection proceeds or follows any carbon enrichment by dredge-up. I am unsure what their calculations say on this point. They are now actively exploring the sensitivity of their results to the assumed mixing length.

Vanbeveren: I want to make a remark concerning the frequency of stars with masses larger than ~15 $M_\odot$ (corresponding to O-types). A set of 200 O-type stars has been extensively studied by P. Conti and collaborators; special attention was given to spectral types and luminosity classes. In a paper that will appear soon in A & A, I have tried to determine the IMF for O-type stars using the former set of 200 stars. Very surprisingly, this IMF goes like M^{-1} which predicts much more massive stars than have been thought in the past (IMF ~M^{-2} as proposed by Dr. Lequeux in 1979), thus enhancing nucleosynthesis yields from massive stars.

Wheeler: Certainly, the mass function of O-type stars is uncertain. I simply wanted to illustrate that the uncertainties are such that our present understanding does not require that massive iron-core stars carry the burden of galactic nucleosynthesis, despite the plausibility of the picture.

Tayler: In the process which may produce either a neutron star or a black hole, does the maximum mass of a neutron star play any crucial role or do cores, which could in principle become neutron stars, become black holes?

Wheeler: The initial suggestion was that the dividing line between neutron star formation and black hole formation was precisely at a core mass equal to the stability limit. Subsequent studies of the equation of state and dynamics suggest that this interpretation may have been somewhat simplistic.

Lamb: From the calculations of Van Riper and Arnett (1979), I would construct the diagram you have shown somewhat differently. At low values of the mass M_h of the homologous core (that part of the initial configuration that falls inward as one piece), there is a "dud". The rest of the star will then fall in, accrete onto the homologous core, and form a massive black hole. At values of M_h less than but close to M_{max}, the maximum mass of a stable zero temperature neutron star, there is a violent explosion and the formation of a neutron star. But at values of M_h slightly more than M_{max}, there is a violent explosion and the formation of a hot neutron star which is partially supported by thermal pressure. As this neutron star cools, it will become unstable and collapse to form a black hole. Thus one will have a two-stage formation process for the black hole, and its formation will be accompanied by a violent explosion. Finally, if $M_h >> M_{max}$, a black hole is formed without an explosion. I would like to emphasize that the case in which M_h is slightly greater than M_{max} may well be important in explaining the apparent lack of neutron stars in Cas A and other young supernova remnants indicated by the Einstein observations.

Wheeler: I believe the case you make is plausible. I may have oversimplified the behavior near the point of maximum energy. On the other hand, whether the two-stage process you described actually works will depend on the details of the dynamics. If the homologous core is near the mass limit, then the picture may be as you say. If, however, a smaller homologous core forms and the shock depends on the nature of the subsequent accretion, then the outcome does not depend on how near the core is to the mass limit, and your reasoning may not apply. I suppose a black hole and an explosion might have occured in Cas A, but I find that is an implausible explanation for the failure to detect a neutron star in SN 1006, Kepler or Tycho.

SUPERNOVA EXPLOSIONS IN DEGENERATE STARS
---DETONATION, DEFLAGRATION, AND ELECTRON CAPTURE---

Ken'ichi Nomoto*
Laboratory for Astronomy and Solar Physics
NASA/Goddard Space Flight Center
Greenbelt, MD 20771, USA

ABSTRACT

Presupernova evolution and the hydrodynamic behavior of supernova explosions in stars having electron-degenerate cores are summarized. Carbon deflagration supernovae in C+O cores disrupt the star completely. On the other hand, in electron capture supernovae, O+Ne+Mg cores collapse to form neutron stars despite the competing oxygen deflagration.

Also discussed are white dwarf models for Type I supernovae (SN I). Supernova explosions in accreting white dwarfs are either the detonation or deflagration type depending mainly on the accretion rate. The carbon deflagration model reproduces many of the observed features of SN I.

1. INTRODUCTION

Electron degeneracy in the stellar interior has a strong influence on the structure and evolution of small and intermediate mass stars, which has been investigated as a fundamental problem since Chandrasekhar (1939). In the final stages of evolution, electron degeneracy provides several triggering mechanisms of supernova explosions. Supernovae in normal red giant stars forming degenerate cores probably contribute to a major fraction of the total rate of Type II supernovae (SN II) and neutron star formation in the Galaxy. Also, it has been suggested that Type I supernovae (SN I) are related to degenerate stars with no hydrogen-rich envelope. However, we have not yet reached agreement about either the final fates of such stars (total disruption or neutron star formation) or the origin of SN I.

In this review, we summarize the recent efforts and achievements in the above problems; the following types of supernovae are discussed.
(1) A thermonuclear explosion in a degenerate core was proposed as one of the triggering mechanisms of supernovae by Hoyle and Fowler (1960).

---
*On leave from Department of Physics, Ibaraki University, Japan

D. Sugimoto, D. Q. Lamb, and D. N. Schramm (eds.), Fundamental Problems in the Theory of Stellar Evolution, 295–315.

Many computations have shown that this type of explosion occurs as a carbon deflagration supernova in stars of mass $6 \pm 2 < M/M_\odot < 8 \pm 1$ (Arnett 1969; Nomoto et al. 1976).

(2) Electron captures have also been suggested to trigger the collapse of degenerate cores (Rakavy et al. 1967). However, little has been known about the presupernova evolution and the hydrodynamic behavior of this type of supernovae until the recent computations by Miyaji et al. (1980) and Nomoto (1980a). They showed that stars of 8 to 10 $M_\odot$ evolve into electron cpature supernovae in which the degenerate O+Ne+Mg core collapses to form neutron stars.

(3) White dwarfs have been thought to be candidates for SN I because of the absence of hydrogen lines in their spectrum and their appearance in elliptical galaxies (Finzi and Wolf 1967). Recent progress in the radioactive decay model for the light curve of SN I (Colgate et al. 1980) has renewed the interest in the detonation/deflagration type supernovae in the accreting white dwarfs.

2. EVOLUTION OF DEGENERATE CORES

In the advanced phases of evolution, electrons become degenerate in a core of He, C+O, and O+Ne+Mg for the stars of mass smaller than about 2 $M_\odot$, 8 $M_\odot$, and 10 $M_\odot$, respectively. Since the pressure depends only slightly on temperature in the equation of state of degenerate electrons, onset of electron degeneracy changes the global thermodynamics of the core as a self-gravitating system, i.e., the sign of the gravothermal specific heat changes from negative to positive (e.g. Kippenhahn 1970; Hachisu and Sugimoto 1978). As a result, the star evolves in the following way, which differs greatly from the more massive stars evolving into gravitational collapse.

2.1 Critical Core Mass for Non-Degenerate Ignition

During the gravitational contraction of the partially degenerate core, the temperature attains its maximum and then decreases by emitting photons and neutrinos. Such a maximum temperature is obtained for the core of a fixed mass, and its value is lower if the core mass is smaller. This implies that there exists a certain critical core mass for nuclear burning to be initiated under non-degenerate or partially-degenerate condition. Such critical core masses are 0.08 $M_\odot$ (Kumar 1963), 0.31 $M_\odot$ (Cox and Salpeter 1964), 1.06 $M_\odot$ (Murai et al. 1968), and 1.37 $M_\odot$ (Boozer et al. 1973) for the ignition of hydrogen, helium, carbon, and neon, respectively.

In the actual evolution, the core mass increases because materials in the envelope flow into the core through the nuclear burning shell. As an example, let us discuss the C+O core. (Hereafter M_H, M_{He}, and M_C, denote the core mass contained interior to the burning shell of hydrogen, helium and carbon, respectively.) In a star of mass $M > (8 \pm 1)M_\odot$, the mass of the C+O core, M_{He}, reaches the critical core mass 1.06 $M_\odot$ before the electron degeneracy sets in so that carbon burning proceeds under non-degenerate condition. For a smaller mass star, on the other hand, M_{He} is smaller than 1.06 $M_\odot$ when the maximum temperature is attained. Accordingly non-degenerate ignition does not occur and the cool electron-degenerate C+O core is formed.

2.2 Development of Degenerate C+O Core

The evolutionary path of such a degenerate C+O core in the central density ρ_c and temperature T_c plane is shown in Figure 1, which was computed from the helium burning through the carbon ignition for the helium core of initial mass $M_H^{(0)} = 1.5\ M_\odot$ in the 7 $M_\odot$ star (Sugimoto and Nomoto 1975; see also Rose 1969, Paczyński 1970). As is well-known, evolution of the intermediate mass stars converges to the same track in this plane irrespective of their total mass (Paczyński 1970), becuase it is determined only by the competition between the growth rate of M_{He} ($\simeq M_H$) due to the H/He double shell-burnings and the neutrino cooling rate. Carbon burning is ignited at $\rho_c \simeq 2.5 \times 10^9$ g cm^{-3} and grows into the flash because of the positive specific heat of the degenerate matter. This is the trigger of the carbon deflagration supernova.

Details of such an evolution are seen in the recent reviews by Sugimoto and Nomoto (1980) and Mazurek and Wheeler (1980). It should be noted that the mass range of the star which evolves through the carbon deflagration would be roughly $(6 \pm 2 - 8 \pm 1)M_\odot$ but is very uncertain because it depends on the mass loss in the red giant stages (see e.g. Wheeler 1978b) and also on chemical composition (Alcock and Paczyński 1978; Becker and Iben 1979).

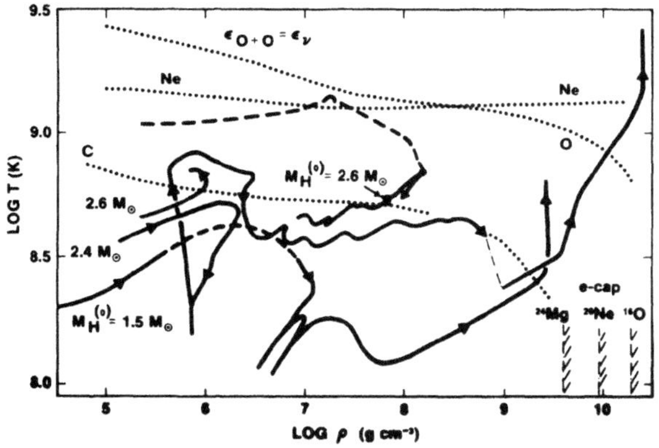

Figure 1. Evolution of the central density and temperature for the helium core of initial mass $M_H^{(0)}$ = 1.5 $M_\odot$, 2.4 $M_\odot$, and 2.6 $M_\odot$ (solid). Dashed line for the 2.6 $M_\odot$ core shows a structure from the center toward the core edge at the off-center Ne ignition. Dotted are ignition lines for C, Ne, and O, and dot-dashed are the threshold densities for electron captures.

2.3 Formation of Degenerate O+Ne+Mg Core

In stars of $M > (8 \pm 1)M_\odot$, O+Ne+Mg core is formed after the stable carbon burning. If we apply the same discussion in section 2.1 to this core, we can expect that there exists a mass range for which a degenerate

O+Ne+Mg core is formed. Although the stars in this mass range would be the progenitors of most of neutron stars (Barkat et al. 1974; Wheeler 1978b), little attention has been paid to the evolution beyond the carbon burning until quite recently.

This expectation was recently confirmed by Nomoto (1980a) and Miyaji et al. (1980). Nomoto (1980a) computed the evolution of helium cores of initial masses $M_H^{(0)} = 2.4\ M_\odot$ and $2.6\ M_\odot$ starting at the helium burning stage. These cores correspond to the stellar mass of $(10 \pm 1) M_\odot$ depending on the mass loss and chemical compositions. Evolutionary paths of the central density and temperature for both cases and chemical evolution of the $2.4\ M_\odot$ core are shown in Figures 1 and 2, respectively. The core of $M_H^{(0)} = 2.4\ M_\odot$ evolves through the following stages.

(1) Carbon burning in non-degenerate core: Carbon is ignited off-center, and then the burning layer shifts inward due to heat conduction and reaches the center as seen in Figure 2. This is similar to the result by Boozer et al. (1973) and Ergma and Vilhu (1978). Afterwards carbon burning proceed in non-degenerate condition at $\rho_c \sim 10^6 g\ cm^{-3}$ (Figure 1).

(2) Growth of O+Ne+Mg core: The resultant O+Ne+Mg core of mass M_C grows through the phases of several carbon-shell flashes (Figure 2). These shell flashes are mild; the peak energy generation rate attained is only $L_{C+C} = 2 \times 10^7\ L_\odot$. In this core, a temperature inversion appears. After a maximum temperature of $1.13 \times 10^9 K$ is attained in the shell of $M_r = 1.26\ M_\odot$, the temperature begins to decrease because of the neutrino loss. Neon ignition does not occur (see the ignition line of $\varepsilon_{Ne} = \varepsilon_\nu$ in Figure 1) because $M_{He} = 1.343\ M_\odot$ is smaller than the critical mass of $1.37\ M_\odot$ for the neon ignition (Boozer et al. 1973). The electron degeneracy becomes stronger as M_C increases.

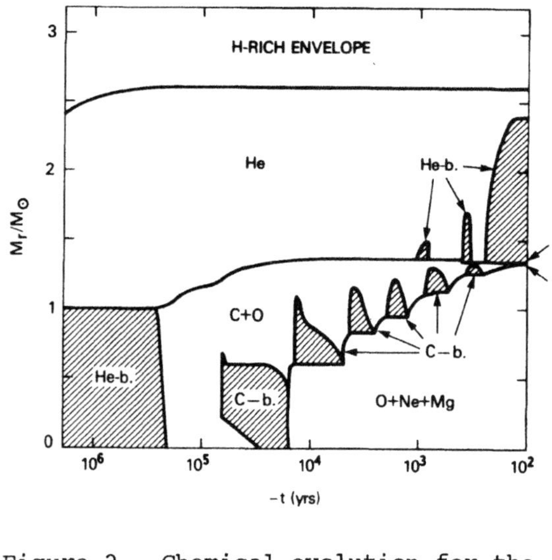

Figure 2. Chemical evolution for the star of $M_H^{(0)} = 2.4\ M_\odot$.

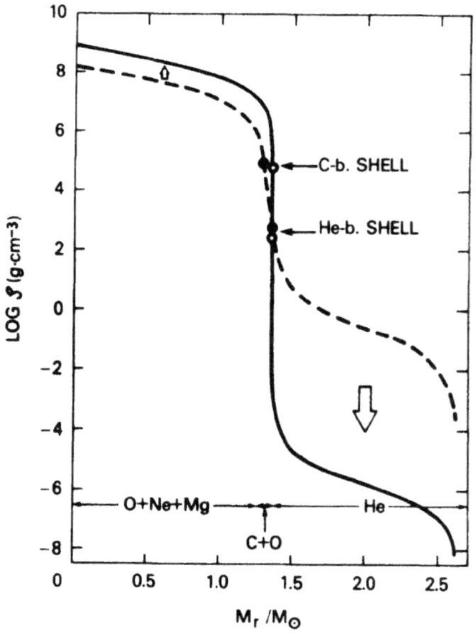

Figure 3. Expansion of the helium zone at the end stage in Figure 2.

(3) Expansion of helium zone and penetration of the surface convection zone: When M_C reaches 1.339 $M_{\odot}$ which is very close to M_{He} (=1.343 $M_{\odot}$), the helium zone expands greatly as seen in Figure 3. This is the same phenomenon as occurs in the 2 $M_{\odot}$ helium star having the degenerate C+O core (Paczyński 1971). The mechanism of such expansion is the same as when a main-sequence star becomes a red giant star (cf. Sugimoto and Nomoto 1980). The temperature at the core edge drops to 10^4K which implies that the surface convection zone would penetrate into the helium zone if the hydrogen-rich envelope were taken into account in the computation (cf. Becker and Iben 1979). The mass M_H will be reduced down to $M_H \simeq 1.343\ M_{\odot}$, and the star will have very thin helium and carbon zones.
(4) Core growth by triple shell-burning: After that the O+Ne+Mg core will grow toward the Chandrasekhar limit through the phases of triple shell-burning of hydrogen, helium, and carbon, which would be almost the same as the growth of the C+O core by the double shell-burnings. Therefore evolution of 8 to 10 $M_{\odot}$ stars in ρ_C - T_C Plane will converge to the same track. The ultimate fate of such stars is an electron capture supernova (Miyaji et al. 1980) as will be discussed in section 4.

2.4 Off-Center Neon Burning

In the core of $M_H^{(0)} = 2.6\ M_{\odot}$, carbon is ignited in the center. Subsequent evolution of the growing O+Ne+Mg core through the phases of carbon shell-flashes are qualitatively the same as in the core of $M_H^{(0)} = 2.4\ M_{\odot}$. However, M_{He} reaches 1.45 $M_{\odot}$ exceeding the critical mass of 1.37 $M_{\odot}$, so that neon is ignited off-center as seen from the structure line (dashed) in Figure 1. Peak energy generation rate reaches $L_{Ne} = 3 \times 10^{13} L_{\odot}$ but no dynamical effect is induced.

The inward shift of such neon/oxygen burning shell was computed for the 8 $M_{\odot}$ star having core of $M_{He} \simeq 1.4\ M_{\odot}$ (Barkat et al. 1974) and for the C+O star of $M_{He} = 1.5\ M_{\odot}$ (Ikeuchi et al. 1972). The result depends on M_{He}: A degenerate O+Ne+Mg core of 0.3 $M_{\odot}$ remains unburned in the former case, while neon burning-shell reaches the center and a degenerate Si core is formed in the latter.

Recently, the evolution of a 10 $M_{\odot}$ star was computed up to the supernova stage by Weaver and Woosley (1979). In their model, $M_{He} \simeq 1.55\ M_{\odot}$ so that neon is ignited and a deflagration wave develops. Helium and hydrogen layers are then ejected and the resultant 1.55 $M_{\odot}$ core evolves into the phase of iron core collapse to form a neutron star.

These results show that the evolution in the mass range of (10 ± 1) to (12 ± 1)$M_{\odot}$ is quite sensitive to stellar mass because the C+O core mass, M_{He}, varies from 1.37 $M_{\odot}$ to 1.6 $M_{\odot}$ crossing the Chandrasekhar limit. In other words, such a mass range corresponds to the transition from degenerate to non-degenerate stars.

3. CARBON DEFLAGRATION SUPERNOVAE

The ultimate fate of the stars having degenerate C+O core has been extensively investigated since Arnett (1969) proposed a model of carbon detonation supernova. We discuss some basic problems in the detonation/deflagration models which can also be applied to the white dwarf models for SN I (section 5). (See details in Sugimoto and Nomoto 1980, and Mazurek and Wheeler 1980).

3.1. Carbon Detonation Assumption

In the carbon detonation model, ignition of carbon grows into a thermonuclear explosion as follows.
(1) Thermal runaway into deflagration: The central temperature T_c rises developing a super-adiabatic convective core and reaches a point where the dynamical effect is essential. This critical point is defined by $\tau_n = \tau_{ff}$, where $\tau_n \equiv c_p T/\varepsilon_n$ and $\tau_{ff} \equiv (24\pi G\rho)^{-1/2}$ are the nuclear timescale for a rise in temperature and the free fall timescale, respectively; the corresponding temperature is denoted as the deflagration temperature T_{def}. (The deflagration temperatures for the helium, carbon, neon, and oxygen burnings will be shown in Figure 6 and denoted by T_{def}(He), T_{def} (C), etc.) The convective URCA process (Paczyński 1972) could suppress the runaway of carbon burning initially (Iben 1978) and delay the occurence of the deflagration to somewhat higher density (Couch and Arnett 1975). Afterwards matter is incinerated into nuclear statistical equilibrium (NSE) abundances in the very central region within the free fall timescale. The temperature reaches 8×10^9K, which produces the overpressure. This is a formation of the deflagration front.
(2) Carbon detonation assumption: In the hydrodynamical model of carbon detonation (Arnett 1969; Bruenn 1971), it was assumed that the deflagration front grows into the detonation wave, i.e., the shock wave is so strong as to deflagrate the material ahead of the deflagration front. Once formed, the detonation wave propagates self-consistently (Buchler et al. 1971) while incinerating almost all materials exclusively into iron peak elements. The star is totally disrupted with the explosion energy of 1.6×10^{51}erg. It leaves no neutron star remnant and ejects 1.4 $M_\odot$ iron peak elements.

However, the assumption of the formation of a detonation wave was found to be wrong by numerical computations (Ivanova et al. 1974; Buchler and Mazurek 1975; Nomoto et al. 1976) and by semi-analytical investigation (Mazurek et al. 1977). The reason is as follows: Mazurek et al. (1977) investigated the necessary condition to initiate a Chapman-Jouguet (CJ) detonation by the shock-tube analysis. In order to form a CJ detonation at densities higher than 10^9g cm^{-3}, it is required that the nuclear energy, q, released by the deflagration is as much as the initial specific internal energy, u_0, of the degenerate matter, i.e., $q \simeq u_0$. (Hereafter the subscript 0 denotes the initial condition before the nuclear runaway occurs.) Moreover, the spherical geometry in the central region, which was not taken into account by Mazurek et al. (1977), has a large damping effect on the out-going shock wave (Ōno 1961; Lee 1972), so that a larger value of q/u_0 is required. On the other hand, the actual energy release for the establishment of NSE state at $\rho_c = 2.5 \times 10^9 g\ cm^{-3}$ is $q = 3 \times 10^{17} erg/g$ which is only 20 percent of u_0. Therefore the resultant thermal overpressure is too small to initiate a detonation. Although q/u_0 is larger at lower densities, the carbon detonation wave is unlikely to form even at the density as low as $5 \times 10^7 g\ cm^{-3}$ (Mazurek et al. 1977).

3.2 Nucleosynthesis in Carbon Deflagration Supernovae

Without growing into the detonation, the deflagration wave propagates at a speed of v_{def} by convective energy transport across the front. Such a model was computed by Buchler and Mazurek (1975) only through a relatively early stage of the propagation and by Nomoto et al. (1976) through the explosion of the star. Nomoto et al. (1976) presented the following model of carbon deflagration supernova. The propagation velocity, v_{def}, was treated as a parameter by applying the mixing length theory of convection. Since v_{def} is slower than the sound velocity, v_s, the core expands appreciably during the subsonic propagation of the deflagration wave. Such expansion makes the deflagration weaker and, at last, quenches the nuclear burnings. In Figure 4, change in the temperature profile for the model of $v_{def} \simeq 0.2\ v_s$ is shown; the temperature decreases appreciably at late times of the propagation of the deflagration wave. Despite this quenching, the total nuclear energy release exceeds the initial binding energy, so that the star is totally disrupted.

Nucleosynthesis in the deflagration wave ($v_{def} \simeq 0.2\ v_s$) proceeds as follows.

(1) Core materials of $M_r \leq 1.03\ M_\odot$ are incinerated into NSE compositions. In the region of $1.03 < M_r \leq 1.31\ M_\odot$, partial burnings of silicon, oxygen, neon, and carbon synthesize Ca, S, Si, Ne, O, etc., in the decaying

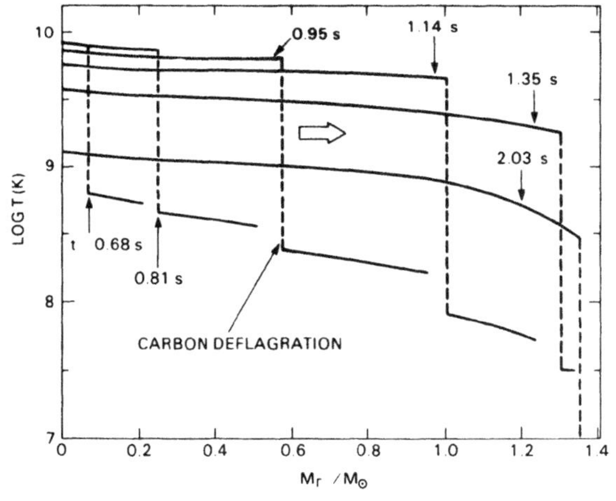

Figure 4. Propagation of the deflagration wave and change in the temperature profile.

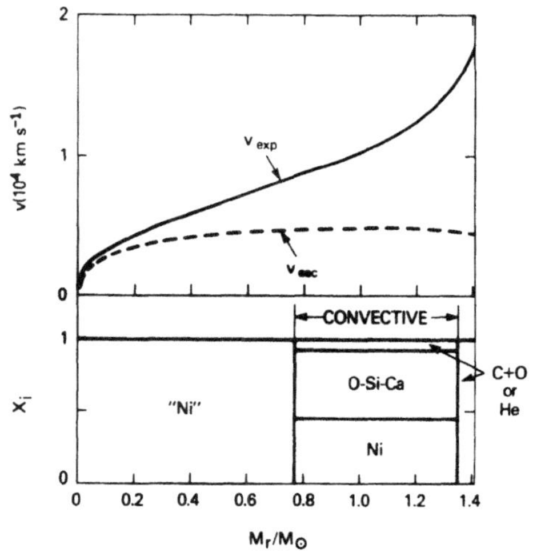

Figure 5. Chemical composition Xi, expansion velocity v_{exp}, and escape velocity v_{esc} of the carbon deflagration supernova at t = 2.03 sec.

deflagration wave. The rest of 0.1 $M_\odot$ C+O remains unburned near the core edge.
(2) These elements are partly mixed by convection. The incinerated NSE region is convectively stable initially, i.e., entropy increases outward because energy release, q, increases along the decreasing initial density ρ_0. On the other hand, in the partially burned region of Ca-Si-O, entropy decreases outward because q decreases as the deflagration wave decays. Then the convective zone mixes the material. The convective shell expands both inward and outward so that some amount of ^{56}Ni and unburned C+O are mixed with Ca-Si-O. In Figure 5, the abundance distribution is shown at t = 2.03 s after the initiation of the deflagration. At this time, materials in the region of $0.77\ M_\odot \lesssim M_r \lesssim 1.35\ M_\odot$ are mixed. Also shown is the expansion velocity, v_{exp}, compared with the escape velocity, v_{esc}. The explosion energy in this model is $E_{expl} = 1.3 \times 10^{51}$ erg.

The abundance in the ejecta and E_{expl} depend on the assumption of the value of v_{def}. For a model of $v_{def} \simeq 0.01\ v_s$, $E_{expl} = 5 \times 10^{49}$ erg and masses or iron peak elements, partially burned Ca-Si-O, and unburned C+O are 0.16 $M_\odot$, 0.23 $M_\odot$, and 1.02 $M_\odot$, respectively.

The above features (1) and (2) are quite different from the detonation model in which materials are almost exclusively incinerated into iron peak elements. Existence of appreciable amount of Ca-Si-O mixed with iron peak elements in the carbon deflagration model is important to interpret the chemical composition of SN I by white dwarf models as will be discussed in section 5.

3.3 Type II Supernovae with Exponential Tail or Faint Supernovae

The carbon deflagration supernova with $v_{def} \simeq 0.2\ v_s$ in the intermediate mass stars would be observed as a Type II supernova (SN II) because it has an extended hydrogen-rich envelope. Moreover, it could be identified by the exponential tail in its light curve at late times because a substantial amount of ^{56}Ni provides radioactive decay energy; there exists such a kind of SN II (Weaver and Woosley 1980; Arnett 1980). If the deflagration velocity is as slow as $v_{def} \simeq 0.01\ v_s$, the explosion would be fainter than normal supernovae because its explosion energy is as small as 5×10^{49} erg.

From the point of view of nucleosynthesis, the carbon deflagration supernovae could contribute to the synthesis of iron peak elements, Ca, Si, S, O, and C in the Galaxy (cf. Wheeler et al. 1980). Their amount, however, depends on the deflagration velocity and on the lower mass limit for this mass range. This implies also that the so-called over-production difficulty of iron-peak elements in the carbon detonation model could be avoided.

4. ELECTRON CAPTURE SUPERNOVAE

4.1 Effects of Electron Capture

Electron degeneracy provides another triggering mechanism of supernova explosion in the O+Ne+Mg core of 8 to 10 $M_\odot$ stars (section 2.4), i.e., electron captures. As the core mass ($M_C \simeq M_{He}$) increases due to triple shell-burnings, ρ_C becomes so high that the electron Fermi energy exceeds the threshold energies for electron captures on ^{24}Mg and ^{20}Ne in advance of the oxygen ignition (see Figure 1).

Effects of such electron captures on evolution of the degenerate core are summarized as follows (see Sugimoto and Nomoto 1980):
(a) Trigger of the collapse: Electron captures reduce the mean mole number of electrons per one gram, Y_e, and correspondingly the Chandrasekhar limit as $M_{Ch} \propto Y_e^2$. If M_{Ch} is reduced below the core mass, the core begins to collapse (Finzi and Wolf 1967).
(b) Entropy production: From the thermodynamic point of view, electron capture produces entropy by the distortion in the electron distribution function when the mean energy of the captured electrons is less than the electron chemical potential (Bisnovatyi-Kogan and Seidov 1970; Sugimoto 1970; Nakazawa et al. 1970). Also, the gamma-ray emission from the excited states of daughter nuclei produces entropy (Rudzski and Seidov 1974).

4.2 Collapse of O+Ne+Mg Core in 8 to 10 $M_\odot$

These two effects lead to a rise in temperature and therefore the

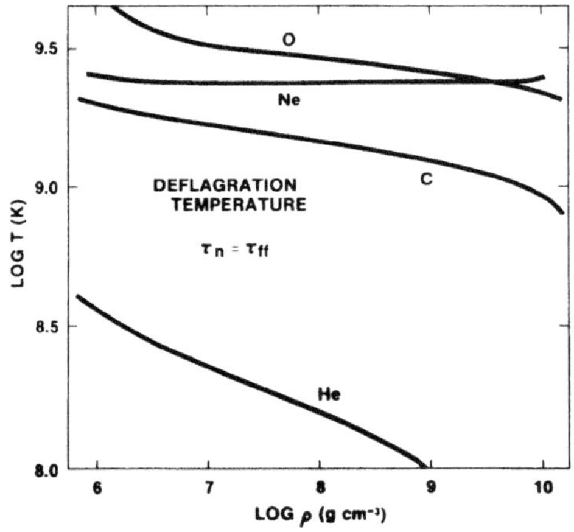

Figure 6. Deflagration temperature, T_{def}, defined at $\tau_n = \tau_{ff}$ for He, C, Ne, and O burnings.

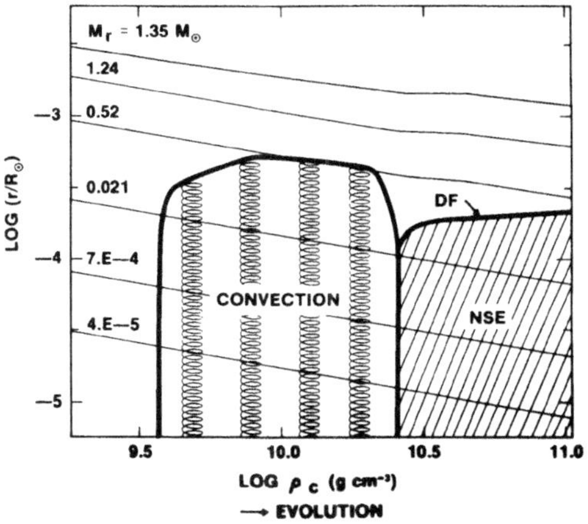

Figure 7. Quasi-dynamic contraction of O+Ne+Mg core due to electron captures on ^{20}Ne, ^{24}Mg, and NSE nuclei. Behind the oxygen deflagration front (DF), NSE region (shaded) is growing.

oxygen deflagration will be ignited. The question is which is the final fate of this core, collapse or explosion. In order to answer this question, we need to compute the transition phase from thermal through dynamical timescale because the electron capture is a rather slow process compared with the free-fall. Miyaji et al. (1980) was the first to make such a computation from the growing phase of the O+Ne+Mg core through its collapse.

The following sequence was found to take place.

(1) Formation of a convective core and quasi-dynamic collapse: As a result of electron captures on ^{24}Mg and ^{20}Ne, the central part of the core is heated and convective core appears as seen in Figure 7. Then Y_e decreases in the entire convective core, and M_{Ch} is reduced to become almost equal to the core mass. The contraction of the core is accelerated into a quasi-dynamic collapse.

(2) Competition between electron captures and oxygen deflagration: When ρ_c reaches 2.5×10^{10}g cm^{-3}, oxygen burning is ignited (Figure 1). It grows into a deflagration and the material is incinerated into NSE composition. However, it does not grow into a detonation because the nuclear energy release is only 0.04 u_0. The propagation of the deflagration front (DF in Figure 7) is too slow to halt the quasi-dynamic collapse of the core. Each contracting shell is deflagrated consecutively as it is compressed to $\rho = 2.5 \times 10^{10}$g cm^{-3}. Therefore, the DF is almost stationary in Eulerian coordinates (Figure 7).

(3) Collapse to form a neutron star: Behind the DF, the incinerated NSE region grows, from which energy is being lost by the processes of the electron captures and the photodisintegration of nuclei. Therefore, the total energy of the core is decreasing at an appreciable rate and the core continues to collapse.

4.3 Type II Supernovae Leaving Neutron Stars and Quiet Supernovae

Though Miyaji et al. (1980) stopped computing at the stage with $\rho_c = 1 \times 10^{11}$g cm^{-3}, the core will continue to collapse and make a bounce around the nuclear density. It is likely that a neutron star is left because of the following reasons: The mass interior to the hydrogen-burning shell, i.e., O+Ne+Mg core plus thin C+O and He layers, is smaller than the limiting mass of a neutron star. The extended hydrogen-rich envelope is so weakly bounded that it would be easily blown off by the shock wave which is formed at the bounce and propagates outward to the envelope (Van Riper 1979).

The electron capture supernovae, then, would be observed as SN II leaving neutron stars. Since the death rate of stars in this mass range is appreciable, the progenitors of the observed pulsar would mostly be these stars. On the other hand, they would contribute little to nucleosynthesis, because C+O zone contains only 10^{-3} - $10^{-2}M_\odot$.

When the 8-10 $M_\odot$ star is a primary star in a close binary system, it becomes a O+Ne+Mg white dwarf by losing its hydrogen-rich envelope and then also helium envelope at their expansions (Figure 3). When this white dwarf accretes matter, it becomes an electron capture supernova to leave a neutron star (Nomoto et al. 1979b). Since such a supernova ejects mass of only 10^{-1} - $10^{-2}M_\odot$, it is called as a quiet supernova which may be required to interpret some low mass X-ray binaries.

5. WHITE DWARF MODELS FOR TYPE I SUPERNOVAE

Although several mechanisms and progenitors for SN I have been suggested, there is little agreement on the correct one (see e.g. Wheeler 1980). White dwarf is one of the candidates for SN I (e.g. Whelan and Iben 1973). Recently, Arnett (1979) and Colgate et al. (1980) have shown that the light curves of SN I can be reproduced well by the radioactive decay model ($^{56}Ni \rightarrow {}^{56}Co \rightarrow {}^{56}Fe$). This has renewed interest in white dwarf models for SN I, because it can produce a large amount of ^{56}Ni by the process of nuclear detonation or deflagration.

Supernova explosion in a white dwarf is triggered by mass accretion from a companion star in a binary system. Although a hydrogen shell flash does not grow into a supernova explosion (Starrfield et al. 1975), accretion of helium as a result of such flashes is important in terms of supernovae. Such accumulation of helium takes place when the hydrogen shell-burning is stable and steady (Sienkiewicz 1980) or when the flash is so weak that the accreted matter does not expand appreciably

Table 1. Supernovae in Accreting White Dwarfs

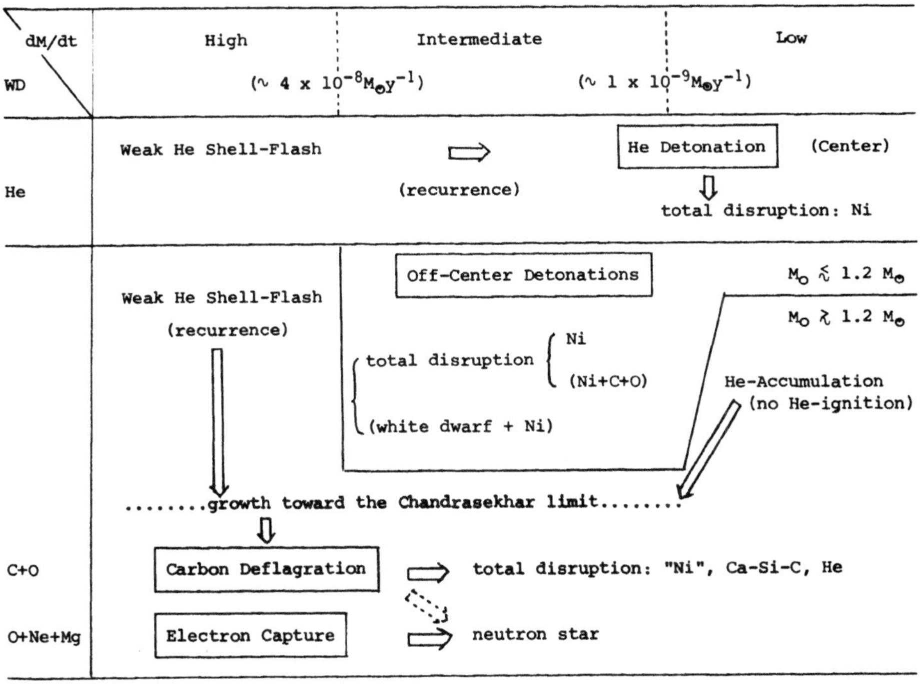

(Shara et al. 1978). These cases are realized when the accretion rate is as high as roughly $10^{-7}M_\odot y^{-1}$ but is lower than at most $6 \times 10^{-7}M_\odot y^{-1}$ (Paczyński and Żytkow 1978; Nomoto et al. 1979a). Another possible way to build up the helium zone would be with a slow accretion of $dM/dt < 10^{-10}M_\odot y^{-1}$; Starrfield et al. (1980) has suggested that diffusion of CNO nuclei out of the accreted matter could then occur and lead to the stable hydrogen burning by the p-p chain.

Recently, such an accretion of helium has been investigated extensively. Several types of supernova explosions which could occur depending on the accretion rate of helium dM/dt, and initial mass M_0 and composition of the white dwarf are summarized in Table 1. Comparison with the observational data of SN I will be made in section 5.5.

5.1 Helium Detonation in Helium White Dwarfs

The triggering mechanism of the supernova explosion depends on the thermal structure in the accreting white dwarf which is set by the accretion process of helium. Such a thermal history of helium accretion was first computed by Nomoto and Sugimoto (1977) for a helium white dwarf of initial mass 0.4 $M_\odot$. The results depend on dM/dt as follows.

When the accretion is as high as $dM/dt = 4 \times 10^{-8}M_\odot y^{-1}$, helium is ignited off-center because the compressional heating due to accretion is faster than the radiative cooling in the outer layer. The shell flash is too weak to induce any dynamical effect because of such a low ignition density as $\rho = 2.7 \times 10^5 g\ cm^{-3}$.

When the accretion is as slow as $dM/dt = (1 - 2) \times 10^{-8}M_\odot y^{-1}$, on the other hand, the white dwarf is well thermally relaxed due to heat diffusion. Ignition of helium occurs in the center and is delayed until the white dwarf mass grows to (0.8 - 1.0) $M_\odot$. The helium flash is so strong that it grows into detonation, and the white dwarf is disrupted completely; the entire mass of mostly ^{56}Ni is ejected (see also Mazurek 1973). The difference from the carbon deflagration is ascribed to the difference in the energetics. In the case of helium detonation, the ratio q/u_0 is as large as 10, while $q/u_0 = 0.2$ for the carbon deflagration. Resultant shock strength is as strong as $P_{sh}/P_0 \simeq 5$ which is large enough to overcome the spherical damping and to initiate a detonation.

5.2 Off-Center Dual Detonations in C+O White Dwarfs

When the accreting white dwarf is composed of carbon-oxygen, the helium zone grows on the C+O core. Fujimoto and Sugimoto (1979) and Taam (1980a, b) computed such an accretion. They found three types of evolution depending mainly on dM/dt as schematically summarized in Table 1: When the accretion is rapid, helium flash is weak and will recur many times. For the intermediate accretion rate, the helium flash is so strong that it grows into detonation. If the accretion is slow, the final fate depends on the initial mass M_{C+O} (M_0 in Table 1) as well as dM/dt, which will be discussed in sections 5.3 and 5.4.

Hydrodynamical stages for the cases of intermediate and low accretion rates were investigated by Nomoto (1980a, b; Cases A-D) and Woosley et al. (1980; Case W). In Table 2, summarized are the parameters chosen and the numerical results, i.e., dM/dt, the initial mass of the white dwarf M_{C+O}, the mass of the accreted helium ΔM_{He} at the ignition, the density of the ignited shell ρ_{He}, q/u_0, and the explosion energy E_{expl}. In Figure 8 of ρ-T plane, evolutionary paths in the bottom of the accreted helium zone and in the center are shown for Cases A-D; when the accretion is slower, the ignition density of helium ρ_{He} is higher.

First of all, we discuss the cases of intermediate accretion rate (Cases A, B, and W). The hydrodynamic behavior for Case A is as follows. (1) Formation of the dual detonation waves: In Figure 9, changes in the temperature and pressure profiles show the formation and development of the detonation waves. Stages #1 and #4 show the post-deflagration conditions, and #2 and #3 are the post-shock pre-deflagration stages which are denoted by subscript sh. Because of high ρ_{He}, the temperature in the burning shell rises above T_{def} (He), and the materials are incinerated into NSE matter. Since q/u_0 is as high as 17, the resultant overpressure is 9 times higher than the initial pressure P_0 (#1). A shock front is formed and propagates both outward and inward (#2 and #3). Both shock waves are so strong that they grow into the helium detonation wave (He-

Table 2. Physical Quantities at the Ignition and Explosion.

Case	dM/dt ($M_\odot y^{-1}$)	M_{C+O} ($M_\odot$)	ΔM_{He} ($M_\odot$)	ρ_{He} (g cm^{-3})	q/u_0	E_{expl} (erg)
A	3×10^{-8}	1.08	0.08	2.8×10^{6}	17	1.5×10^{51}
W	1×10^{-8}	0.50	0.62	1.4×10^{7}	--	2.2×10^{51}
B	3×10^{-9}	1.08	0.23	3.6×10^{7}	3	1.9×10^{51}
C	7×10^{-10}	1.28	0.12	2.4×10^{8}	1	---
D	4×10^{-10}	1.28	0.12	(carbon ignition)		

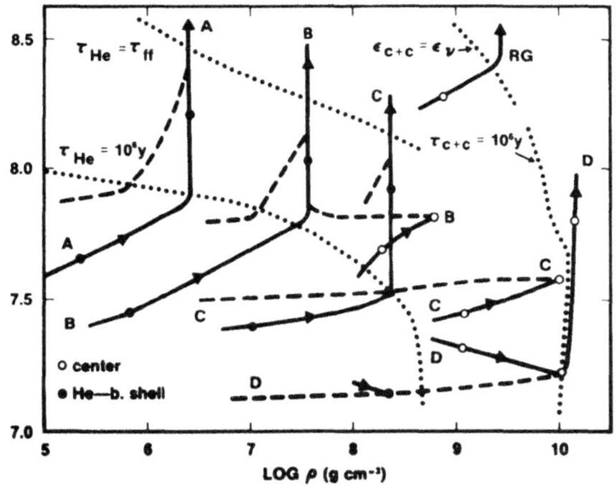

Figure 8. Evolutionary paths (solid) of the density and temperature in the bottom of the accreted helium zone (filled circle) and in the center of the white dwarfs (open circle) for Cases A-D and also for red giant stars (RG). Dashed lines are structure lines of the white dwarfs from the center toward the surface. Dotted lines are the ignition lines for He and C defined by $\tau_n = 10^6$ y and $\varepsilon_n = \varepsilon_\nu$. T_{def} (He) is also indicated by a dotted line.

DW) and carbon detonation wave (C-DW), respectively; i.e., T_{sh} is higher than T_{def} (He) and T_{def} (C), respectively.

(2) Propagation of the detonation waves: In the developing phase, both detonation waves (DWs) become stronger as seen from the increasing Psh in Figure 9. Afterwards, the dual DWs propagate self-consistently with the shock strength P_{sh}/P_0 close to that of the Chapman-Jouguet detonation; even the inward C-DW was found to be strong enough to overcome the pressure gradient. In Figure 10, their propagation and the accompanying changes in the radius of Lagrangian shells are shown. The outward He-DW gives the material kinetic energy which leads to the rapid expansion. On the other hand, the materials in the C+O layers are pushed inward by the C-DW but then begin to expand. Such a pushing effect is strengthened in the very central region because the C-DW converges due to the spherical geometry. However, it is not strong enough to induce the collapse by electron captures.

(3) Disruption of the white dwarf: Both DWs reach the surface and the center, respectively, and incinerate almost all materials into NSE elements except for $10^{-3} M_\odot$ near the surface. Then, the white dwarf is totally disrupted, i.e., the entire mass of 1.16 $M_\odot$ of mostly ^{56}Ni is ejected, because the released nuclear energy is much larger than the initial binding energy. This is a supernova with $E_{expl} = 1.5 \times 10^{51}$ ergs.

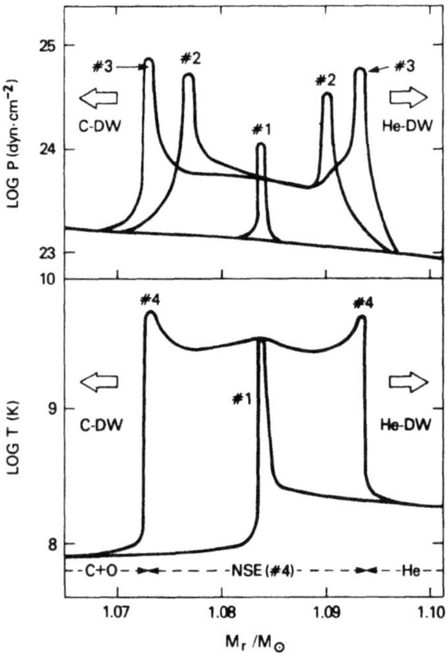

Figure 9. Formation and development of the dual detonation waves (He-DW and C-DW) for Case A.

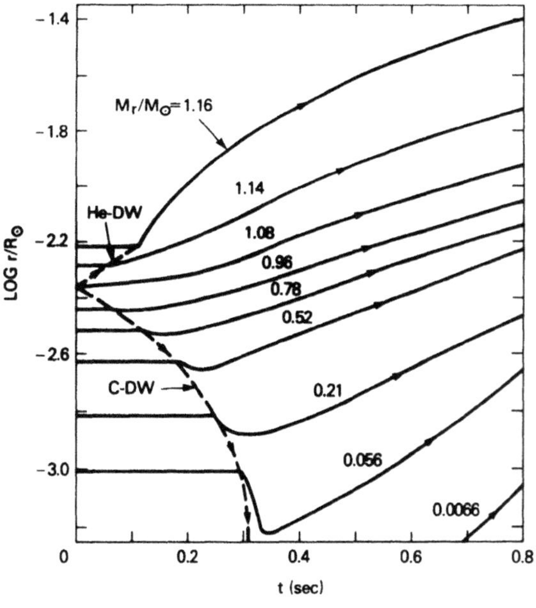

Figure 10. Propagation of the dual DWs (dashed) and the accompanying changes in the radial distance of the Lagrangian shells (solid) for Case A.

The white dwarfs in Cases B and W also become the same type of off-center detonation supernovae. Almost all the material is incinerated into ^{56}Ni by the dual DWs except for $10^{-3}M_\odot$ (Case W) and $10^{-4}M_\odot$ (Case B). The entire mass then ejected with E_{expl} as given in Table 2.

5.3 Helium Envelope-Detonation

The dual DWs does not always form. In Case C of slow accretion, ignition of helium is delayed to a stage with $\rho_{He} = 2.4 \times 10^8$g cm^{-3} (Figure 8). At this density, the shock strength generated by the incineration of helium is as low as $P_{sh}/P_0 = 1.8$ because of the low q/u_0 ($\simeq$1). The resultant T_{sh} is 5 x 10^9K which is higher than T_{def} (He) but lower than T_{def} (C) as seen in Figure 6. Accordingly, the He-DW forms and propagates toward the surface while incinerating helium into NSE products. On the other hand, the C-DW does not form; in order to initiate the Chapman-Jouguet C-DW at this density, energy release, q, is required to be as much as 3 times u_0 (Mazurek et al. 1977). During the propagation of the He-DW, the C+O core remains unburned (Nomoto 1980b).

Although the computation has not yet been completed for the stages after the arrival of the He-DW at the surface, the final fate of Case C is probably as follows. Since the white dwarf mass of 1.40 $M_\odot$ is close to M_{Ch}, the initial gravitational binding energy in the helium zone is as large as 3.8 x 10^{50}erg which is slightly larger than the nuclear energy release of 3.6 x 10^{50}erg. This implies that some incinerated material could remain bound on the C+O core, while a part of it could be ejected. During the subsequent long phase of helium accretion, ^{56}Ni will decay into ^{56}Co and ^{56}Fe. Therefore, this white dwarf would have a Fe/He envelope above the C+O core when a supernova explosion is triggered by the carbon ignition in the center.

If the initial mass of M_{C+O} is somewhat lower than in Case C with the same accretion rate, the total energy of the white dwarf will become positive as a result of the helium incineration. Then the unburned C+O will also be ejected following after ^{56}Ni by converting the internal energy into kinetic energy. In some cases, a white dwarf remnant could be left after a rather weak supernova explosion (Nomoto 1980a).

5.4 Carbon Deflagration

There exist two possible cases of accretion for which the C+O white dwarf grows toward the Chandrasekhar limit without suffering from the off-center detonations (Table 1).

One is the case of rapid accretion. Helium flash does not grow into the detonation, and the mass of the C+O white dwarf increases as a result of the many cycles of hydrogen and helium shell-flashes. Such a white dwarf is very similar to the C+O core in red giant (RG) stars evolving through the double shell-burnings. Therefore the white dwarf evolves along a similar path as a RG in the ρ_c-T_c plane (Figure 8). Then it will explode as a carbon deflagration supernova whose outcome is similar to the models discussed in section 3. The ejected materials would be the mixture of iron peak elements, Ca, S, Si, O, and C.

The other is the case in which the accretion rate of helium is lower than about $1 \times 10^{-9} M_\odot y^{-1}$, and the initial mass M_{C+O} is larger than a certain critical mass (roughly 1.1 - 1.4 $M_\odot$ depending on dM/dt). Carbon is ignited in the center due to pycnonuclear reaction before helium can ignite off-center. Case D in Figure 8 shows a development of such a carbon flash (see also Taam 1980b). For smaller M_{C+O}, ρ_{He} reaches the ignition line before ρ_c reaches the ignition density of carbon (Figure 8).

Although the development of the carbon flash under such a high ignition density as $\rho_{ign} \sim 10^{10} g\ cm^{-3}$ remains to be investigated, the carbon deflagration supernova would be a likely outcome because ρ_{ign} is lower than the critical density ($3 \times 10^{10} g\ cm^{-3}$) for the implosion by electron captures. In the outer layers of this type of carbon deflagration, at most 0.3 $M_\odot$ of unburned helium could be contained together with elements shown in Figure 5. In Case C, moreover, Fe could result as a product of helium envelope-detonation.

5.5 Observational Constraints on Models for SN I

Now, we will summarize the observational constraints on the theoretical models. They are the light curves and chemical compositions in the ejecta.
(1) In order to explain the peak luminosity of SN I by ^{56}Ni decay only, it is required that the amount of ^{56}Ni is as much as 1.0 - 1.4 $M_\odot$ (Arnett 1979; Chevalier 1980).
(2) Chemical compositions in the outer layers of SN I 1972e is being analyzed by Branch (1980) who compared the photoelectric scans of spectra during the thermal peak of the light curve with synthetic spectra. According to his preliminary results, the most plausible interpretation is that the outer layers of SN 1972e consisted of a mixture of Na, Ca, Si, Fe, and probably He. Abundances of such heavy elements seems to be higher than solar mass fraction. Also the upper limit to Co/Fe ratio was estimated to be about 0.1. This implies that Fe had been synthesized in the envelope before the explosion, which seems to be rather difficult to explain. (Branch (1980) noted that such a Co/Fe ratio is not yet a firm constraint in view of the present uncertainties.) Since the spectra of SN I are very homogeneous (Branch 1980), the compositions in the outer layers of most of SN I would be similar to SN 1972e.

From the theoretical point of view, there could be two types of supernova explosions in white dwarfs, i.e., detonation type (He and He/C) and deflagration type.

(1) Light curve: Both detonation and deflagration types produce large amounts of ^{56}Ni which can provide a sufficient amount of radioactive energy for the peak luminosity and exponential tail of SN I. Chevalier (1980) computed a theoretical light curve of an exploding white dwarf choosing a carbon deflagration model of $v_{def} \simeq 0.2\ v_s$ (Nomoto et al. 1976)

as a standard model, i.e., $M = 1.4\ M_\odot$ and $M_{Ni} = 1.0\ M_\odot$. He showed that the theoretical light curve fits well with the observations.

The detonation type models have larger explosion energies than the deflagration models though the ejected mass is smaller, i.e., expansion velocity is larger. Then their light curves would have steeper rise and decline, i.e., shorter width for the thermal peak if the same mean opacity is assumed (Chevalier 1980). According to Barbon et al. (1973), there may exist two subclasses of SN I, i.e., slow and fast; the slow SN I has wider width and longer decline time in the light curves than the fast SN I. Therefore the deflagration and detonation type might correspond to the slow and the fast SN I, respectively. However, the light curve shape depends also on opacity and density distribution in the ejected matter so that the above suggestion needs more investigation.

(2) Chemical Composition: The ejecta of the detonation type SN is composed almost exclusively of iron peak elements; the mass of the other elements is only $10^{-4} - 10^{-3} M_\odot$. This is in conflict with the existence of S, Si, Ca, etc. in the outer layers of SN 1972e (Branch 1980). On the other hand, appreciable amount of such elements exist in the ejecta of carbon deflagration supernova. Moreover, these elements are mixed together with Ni and unburned C+O (Figure 10). Therefore the abundance feature of SN 1972e can be explained qualitatively by the carbon deflagration model (Chevalier 1980).

If we take into account the probable existence of He and high Fe abundance with a low Co/Fe ratio, the progenitor of such a SN I would be a slowly accreting white dwarf. Such a white dwarf would explode as a carbon deflagration supernova with at most $0.3\ M_\odot$ helium in the outer layers. In some cases of accretion, He and Fe could exist as a result of helium envelope-detonation to produce ^{56}Ni which decays to ^{56}Fe during the subsequent accretion of helium.

In conclusion, the carbon deflagration would be more plausible model at least for SN 1972e and probably for most of SN I rather than the detonation type from the point of view of the chemical compositions in the ejecta (Chevalier 1980). Since the white dwarf mass could grow only when the hydrogen shell-burning is stable or a weak flash, supernovae could occur only for the case of rapid or very slow accretion. Such ranges of accretion rates might preclude the detonation type but correspond to the ranges for which the carbon deflagratin occurs (Table 1).

In the white dwarf models for SN I, about $1\ M_\odot$ iron peak elements are ejected into space by each SN I. One might worry about the overproduction of iron in the Galaxy. According to Tinsley (1980), however, the production of even $1.4\ M_\odot$ iron by each SN I is not ruled out in view of the uncertainties in the star formation rate and the SN I rate.

6. CONCLUDING REMARKS

New developments of the supernova models in degenerate stars have been made for electron capture supernovae in 8 to $10\ M_\odot$ stars and the white dwarf models for SN I. In order to advance these models, the following problems should be clarified.

Electron Capture Supernovae: Presupernova evolution should be fully investigated: The evolution of the growing O+Ne+Mg core starting from the phase of dredging up of helium zone up to the onset of electron captures remains to be computed. Moreover, it is important to confirm the anticipation that all stars in the mass range of 8 to 10 $M_\odot$ (± 1 $M_\odot$) evolve into the phase of commonly growing O+Ne+Mg core and also that several types of supernova explosion could occur for 10 to 12 $M_\odot$ stars depending sensitively on the stellar mass.

Hydrodynamical stages due to electron captures need to be studied by treating nuclear reactions and convective mixing in more detail. Problems in the core collapse and its bounce would be in common with the collapsing iron core in massive stars, although the case of a bare white dwarf might be somewhat different.

White dwarf models for SN I: Presupernova evolution, i.e., the link between the hydrogen shell-flashes and the accumulation of helium is poorly known. It depends on the several parameters in binary system (mainly on accretion rates). The relationship to the cataclysmic variables and the symbiotic stars is interesting (Nomoto 1980b).

Since the carbon deflagration supernovae seem to be the most plausible models for SN I, it is important to study the detailed nucleosynthesis by the partial burnings of Si, O, and C in the decaying deflagration wave for comparison of the chemical compositions with the observations. Appropriate treatment of the propagation of the deflagration wave due to convection is quite important. Despite the plausibility of such a deflagration model, whether the C+O white dwarf collapses or explodes (for the slow accretion in particular) is still debated; effects of convective URCA processes, critical ignition density for the implosion due to electron captures under the deflagration regime, effects of crystallization etc. are yet to be clarified.

Also other classes of models for SN I, i.e., $\sim$4 $M_\odot$ helium stars in binary systems (Arnett 1979) and $\sim$2 $M_\odot$ extended helium stars (Wheeler 1978a), need more investigation. The extended helium star should have a degenerate C+O core and explode as a carbon deflagration supernova, so that the discussions in sections 3 and 5.5 can be applied as well.

I would like to thank Prof. D. Sugmioto, and Drs. D. Branch, R.A. Chevalier, M.Y. Fujimoto, S. Miyaji, and J.C. Wheeler for stimulated discussions. Discussions with the participants in the 1980 Workshops on "Atomic Physics and Spectroscopy for Supernovae Spectra" (La Jolla), "Type I Supernova" (Austin), and "Origin and Distribution of the Elements" (Santa Cruz) were helpful to advance the present study. It is a pleasure to thank Dr. W.M. Sparks for the reading of the manuscript, useful comments, and encouragement during my stay in NASA/GSFC. This work has been supported in part by the Scientific Research Fund of the Ministry of Education, Science, and Culture of Japan (274062) and by NRC-NASA Research Associateships in 1979-1980.

REFERENCES

Alcock, C. and Paczyński, B.: 1978, Astrophys. J. 223, p.244.
Arnett, W.D.: 1969, Astrophys. Space Sci. 5, p.180.
Arnett, W.D.: 1979, Astrophys. J. Letters 230, p.L37.
Arnett, W.D.: 1980, Astrophys. J. 237, p.541.
Barbon, R., Ciatti, F., and Rosino, L.: 1973, Astron. Astrophys. 25, p.241.
Barkat, Z., Reiss, Y., and Rakavy, G.: 1974, Astrophys. J. Letters 193, p.L21.
Becker, S.A. and Iben, I. Jr.: 1979, Astrophys. J. 232, p.831.
Becker, S.A. and Iben, I. Jr.: 1980, Astrophys. J. 237, P.111.
Bisnovatyi-Kogan, G.S. and Seidov, A.F.: 1970, Astron. Zh. 47, p.139 (Soviet Astron. 14, p.113).
Boozer, A.H., Joss, P.C., and Salpeter, E.E.: 1973, Astrophys. J. 181, p.393.
Branch, D.,: 1980, in Proc. Austin Workshop on Type I Supernova, ed. J.C. Wheeler.
Bruenn, S.W.: 1971, Astrophys. J. 168, p.203.
Buchler, J.-R. and Mazurek, T.J.: 1975, Mem. Soc. Roy. Sci. Liege, 8, p.435.
Buchler, J.-R., Wheeler, J.C., and Barkat, Z.: 1971, Astrophys. J. 167, p.465.
Chandrasekhar, S.: 1939, An Introduction to the Study of Stellar Structure, (Chicago: Univ. of Chicago Press).
Chevalier, R.A.: 1980, submitted to Astrophys. J.
Colgate, S.A., Petschek, A.G., and Kriese, J.T.: 1980, Astrophys. J. Letters 237, p.L81.
Couch, R.G. and Arnett, W.D.: 1975, Astrophys. J. 196, p.791.
Cox, J.P. and Salpeter, E.E.: 1964, Astrophys. J. 140, p.485.
Ergma, E.V. and Vilhu, O.: 1978, Astron. Astrophys. 69, p.143.
Finzi, A. and Wolf, R.A.: 1967, Astrophys. J. 150, p.115.
Fujimoto, M.Y. and Sugimoto, D.: 1979, IAU Colloq. No.53, White Dwarfs and Variable Degenerate Stars, eds. H.M. Van Horn and V. Weidemann, p.285.
Hachisu, I. and Sugimoto, D.: 1978, Prog. Theor. Phys. 60, p.123.
Hoyle, F. and Fowler, W.A.: 1960, Astrophys. J. 132, p.565.
Iben, I., Jr.: 1978, Astrophys. J. 226, p.996.
Ikeuchi, S., Nakazawa, K., Murai, T., Hōshi, R., and Hayashi, C.: 1972, Prog. Theor. Phys. 48, p.1890.
Ivanova, L.N., Imshennik, V.S., and Chechetkin, V.M.: 1974, Astrophys. Space Sci, 31, p.497.
Kippenhahn, R.: 1970, Astron. Astrophys. 8, p.50.
Kumar, S.S.: 1963, Astrophys. J. 137, p.1121.
Lee, J.H.: 1972, Astronaut. Acta 17, p.455.
Mazurek, T.J.: 1973, Astrophys. Space Sci. 23, p.365.
Mazurek, T.J. and Wheeler, J.C.: 1980, Fund. of Cosmic Phys., 5, p.193.
Mazurek, T.J., Meier, D.L., and Wheeler, J.C.: 1977, Astrophys. J. 213, p.518.
Miyaji, S., Nomoto, K., Yokoi, K., and Sugimoto, D.: 1980, Publ. Astron. Soc. Japan 32, p.303.
Murai, T., Sugimoto, D., Hōshi, R., and Hayashi, C.: 1978, Prog. Theor. Phys. 39, p.619.

Nakazawa, K., Murai, T., Hōshi, R., and Hayashi, C.: 1970, Prog. Theor. Phys. 44, p.829.
Nomoto, K.: 1980a, in Proc. Austin Workshop on Type I Supernova, ed. J.C. Wheeler.
Nomoto, K.: 1980b, in Proc. IAU Colloq. No.58, Stellar Hydrodynamics.
Nomoto, K. and Sugimoto, D.: 1977, Publ. Astron. Soc. Japan 29, p.765.
Nomoto, K., Sugimoto, D., and Neo, S.: 1976, Astrophys. Space Sci, 39, p.L37.
Nomoto, K., Nariai, K., and Sugimoto, D.: 1979a, Publ. Astron. Soc. Japan 31, p.287.
Nomoto, K., Miyaji, S., Yokoi, K., and Sugimoto, D.: 1979b, IAU Colloq. No.53, White Dwarfs and Variable Degenerate Stars, eds. H.M. Van Horn and V. Weidemann, p.56.
Ōno, Y.: 1960, Prog. Theor. Phys. 24, p.825.
Paczyński, B.: 1970, Acta Astron. 20, p.47.
Paczyński, B.: 1971, Acta Astron. 21, p.271.
Paczyński, B.: 1972, Astrophys. Letters 11, p.53.
Paczyński, B. and Żytkow, A.N.: 1978, Astrophys. J. 222, p.604.
Rakavy, G., Shaviv, G., and Zinamon, Z.: 1967, Astrophys. J. 150, p.131.
Rose, W.K.: 1969, Astrophys. J. 155, p.491.
Rudzskii, M.A. and Seidov, Z.F.: 1974, Astron. Zh. 51, p.936. (Soviet Astron. 18, p.551).
Shara, M.M., Prialnik, D., and Shaviv, G.: 1978, Astron. Astrophys. 61, p.363.
Sienkiewicz, R.: 1980, Astron. Astrophys. 85, p.295.
Starrfield, S., Truran, J.W., and Sparks, W.M.: 1975, Astrophys. J. Letters 198, p.L113.
Starrfield, S., Truran, J.W., and Sparks, W.M.: 1980, submitted to Astrophys. J. Letters.
Sugimoto, D.: 1970, Astrophys. J. 161, p.1069.
Sugimoto, D. and Nomoto, K.: 1975, Publ. Astron. Soc. Japan 27, p.197.
Sugimoto, D. and Nomoto, K.: 1980, Space Sci. Rev. 25, p.155.
Taam, R.F.: 1980a, Astrophys. J. 237, p.142.
Taam, R.F.: 1980b, submitted to Astrophys. J.
Tinsley, B.M.: 1980, in Proc. Austin Workshop on Type I Supernova, ed. J.C. Wheeler.
Van Riper, K.A.: 1979, Astrophys. J. 232, p.558.
Weaver, T.A. and Woosley, S.E.: 1979, Bull. AAS 11, p.724.
Weaver, T.A. and Woosley, S.E.: 1980, in Proc. La Jolla Workshop on Atomic Physics and Spectroscopy for Supernovae Spectra, ed. R.E. Meyerott.
Wheeler, J.C.: 1978a, Astrophys. J. 225, p.212.
Wheeler, J.C.: 1978b, Mem. della Soc. Astron. Italiana 49, p.349.
Wheeler, J.C. (ed.): 1980, Proceeding of the Workshop on Type I Supernova, 17-19 March 1980, Univ. of Texas at Austin, in press.
Wheeler, J.C., Miller, G.E., and Scalo, J.M.: 1980, Astron. Astrophys. 82, p.152.
Whelan, J. and Iben, I., Jr.: 1973, Astrophys. J. 186, p.1007.
Woosely, S.E., Weaver, T.A., and Taam, R.F.: 1980 in Proc. Austin Workshop on Type I Supernova, ed. J.C. Wheeler.

DISCUSSION

Sugimoto: You discussed a model in which the helium and carbon detonation waves propagate both ways. As the carbon detonation wave propagates inwards, the pressure of the ambient matter, i.e., the average pressure, increases very greatly and the shock strength should decrease considerably. How far inwards does the front continue to be detonation wave?

Nomoto: The inward-moving detonation wave is weakened by the initial pressure gradient and strengthened by the spherical geometry in the central region. In our model, the nuclear energy release is large enough compared with the initial internal energy of the matter to overcome the damping effect of the pressure gradient. Although the shock strength decreases, it is still as large as about 4 when the detonation wave reaches the shell with a mass fraction of $M_r/M \simeq 0.2$; the detonation wave is therefore self-consistent there. After that, the inward detonation wave grows due to the focusing effect of the spherical geometry. The pressure in the centermost region increases by a factor of 300. So, one might expect that a collapse would be triggered by the inward-moving shock wave. In our model, however, the mass of the highly compressed region is only $3 \times 10^{-4}\ M_\odot$ which is too small to trigger the collapse.

Mazurek: One small comment regarding your point that the less tightly bound outer envelope of your electron capture supernova model may help to give an explosion. The present problem in getting a supernova from collapse in hydrodynamic calculations is getting the shock to propagate to the surface of the compact core that initially collapses. An inner region of the initial core having less than one solar mass collapses homologously and bounces. The shock that forms at its surface is not able to travel through the remaining $0.4\ M_\odot$ of collapsing initial core before it dies out. The problem for mass ejection therefore seems to be at the higher densities within the original degenerate core ($M \sim 1.4\ M_\odot$). The less tightly bound envelope exterior to the core therefore may be of little help.

Nomoto: The propagation of the reflected shock wave may depend on the equation of state at high densities.

Wheeler: Although I think your model for producing iron on top of a carbon core has some problems with insufficient helium at explosion, I would like to congratulate you on your ingenuity in suggesting a picture that is at all plausible. When we last talked, this was a point of great confusion. Concerning the models of Type I supernovae, there are still many problems. For instance, Chevalier's model, in which the whole light curve is produced by radioactivity, is simultaneously too dim at maximum and too bright in the exponential tail. This problem can not be solved by scaling the amount of ^{56}Ni. Some shock energy at maximum, and hence an extended envelope, may be necessary.

A NEW APPROACH TO THE ANALYTIC EVALUATION OF THERMONUCLEAR REACTION RATES

H.J. Haubold and R.W. John
Zentralinstitut für Astrophysik der
Akademie der Wissenschaften der DDR,
15 Potsdam, German Democratic Republic

In our paper (Haubold and John 1978) we succeeded in a closed-form evaluation of the reaction rate by means of a special function, known as Meijer's G-function

$$r_{ij} = (1-\frac{1}{2}\delta_{ij})N_iN_j(\frac{8}{\pi^2\mu})^{1/2}(kT)^{\mu-1/2}\frac{S^{(\mu)}(0)}{\mu!}G^{3,0}_{0,3}([\frac{x}{2}]^2|_{1+\mu,\frac{1}{2},0}),$$

$$x = 2\pi(\frac{\mu}{2kT})^{1/2}\frac{Z_iZ_je^2}{\hbar} \quad . \qquad (1)$$

This representation of the rate is appropriate to perform analytical operations (e.g. for the computation of energy generation in a fusion plasma).

Furthermore, from (1) easily follow approximate expressions for small and large values of the characteristic parameter Coulomb barrier energy divided by thermal energy, which appears in the argument of the G-function. For large values we get the asymptotic representation :

$$G^{3,0}_{0,3}([\frac{x}{2}]^2|_{1+\mu,\frac{1}{2},0}) \sim ([\frac{x}{2}]^2)^{\frac{2\mu+1}{6}}\exp\{-3[\frac{x}{2}]^{2/3}\}\times$$

$$\times(B_0(\mu)+B_1(\mu)[\frac{x}{2}]^{-2/3}+B_2(\mu)[\frac{x}{2}]^{-4/3}+\ldots)$$

The G-function in (1) admits a convergent series expansion, which may be used for the approximate computation of the rate in the case of small values of the parameter:

$$G^{3,0}_{0,3}([\frac{x}{2}]^2|_{1+\mu,\frac{1}{2},0}) \sim \sum_{n=0}^{\infty}(a_n(\mu)\,x^n\,\ln x + b_n(\mu)\,x^{n+g\mu}) \quad .$$

Haubold, H.J. and John, R.W.: 1978, "Astron. Nachr." 229, pp. 225-232; 1979, 300, pp. 173.

D. Sugimoto, D. Q. Lamb, and D. N. Schramm (eds.), Fundamental Problems in the Theory of Stellar Evolution, 317.

ENHANCEMENT OF THERMONUCLEAR REACTION RATE DUE TO STRONG SCREENING

N. Itoh*, H. Totsuji**, S. Ichimaru***, H.E. DeWitt****
* Department of Physics, Sophia University, Tokyo, Japan
** Department of Electronics, Okayama University, Japan
*** Department of Physics, University of Tokyo, Japan
**** Lawrence Livermore Laboratory, University of California, U.S.A.

The enhancement factor for the rate of thermonuclear reactions which involve two kinds of nuclei with charges Z_i and Z_j in the strong-screening regime is given for general cases of surrounding nuclear plasmas by the formula, $\exp[1.25\Gamma_{ij} - 0.095\tau_{ij}(3\Gamma_{ij}/\tau_{ij})^2]$. Here, $\Gamma_{ij} = 2Z_iZ_je^2/(a_i+a_j)T$; $a_i = [3Z_i/4\pi\Sigma_k Z_k n_k]^{1/3}$; $\tau_{ij} = [(27\pi^2/4)(2\mu_{ij}Z_i^2Z_j^2e^4/T\hbar^2)]^{1/3}$; μ_{ij} is the reduced mass for the two reacting nuclei Z_i and Z_j; and n_k is the number density of nuclei Z_k. The calculation is based on the recent results of Monte Carlo computations for binary ion mixtures, which have shown that the screening functions $h_{ij}(r)$ at intermediate distances $[0.5 \leq r/[(a_i+a_j)/2] \leq 1.6]$ can be expressed to a good degree of accuracy by

$$\frac{h_{ij}(r)}{T} = \Gamma_{ij}[1.25 - 0.39\frac{r}{(a_i+a_j)/2}].$$

Application to the calculation of carbon ignition in the carbon-oxygen core of a highly evolved star is discussed. The carbon ignition temperature is found to be single-valued as a function of the density in contrast to the work of Graboske.

DISCUSSION

Schatzman: The results of Jancovici and Alastney concerning the increased rate factor confirms, within perhaps a factor 3, your results. R. Mochkovitch has also reconsidered binary mixtures and finds that, for a constant electron density, the increased rate factor is independent of the chemical composition.

Itoh: These confirmations are very gratifying.

Sugimoto: As you said at the beginning of your talk, your results change the rate by an order of magnitude. This occurs because the quantities involved are in the exponent. By the same token, the uncertainty in your results could be large. How large is it?

D. Sugimoto, D. Q. Lamb, and D. N. Schramm (eds.), Fundamental Problems in the Theory of Stellar Evolution, 318–319.

Itoh: Let us define the dense plasma parameter $\Gamma \equiv E_{Coul}/kT$. The error bar of our result caused by the noise in the Monte Carlo data for the binary ion mixture is $\sim \exp(0.05\Gamma)$.

Salpeter: When $\Gamma \equiv E_{Coul}/kT$ is very large, the value of the screening factor may not be unique: The matter is "almost" solid, and the structure may depend on the previous history.

Itoh: We are concerned with the region in which $\Gamma \lesssim 160$. In this region the matter is in the liquid state.

Nomoto: When accretion onto a white dwarf is slow, helium ignites at a density as high as 3×10^8 g cm^{-3}. At this density the matter is a quantum liquid, i.e., the temperature is lower than the Debye temperature. Could you calculate the correct screening factor even for this region?

Itoh: Near the Debye temperature, the classical turning point radius for thermonuclear reactions is comparable to the mean distance between nuclei. Therefore it would seem to be rather difficult to extend the theory to this region, but we might nevertheless try to solve it.

Wheeler: There is a particular astrophysical problem that depends very much on the ignition in just the region where you have re-calculated it. That is the problem of accretion and carbon ignition in a cold white dwarf, and whether the white dwarf collapses before it ignites carbon or not. Drs. Schatzman, Mazurek, Tutukov, and I, at least, have worked on this problem and have all disagreed. Have you considered the application of your results to this problem?

Itoh: Your problem appears to be extremely interesting. I have heard from Dr. Nomoto that he is in the process of attacking this problem by applying our theory.

THE NEUTRONIZATION OF THE MATTER AND HYDRODYNAMIC INSTABILITY AT FINAL STAGES OF STELLAR EVOLUTION

V.M. Chechetkin
Institute of Applied Mathematics, Moscow, USSR

In this report I consider dynamic instability when the non-equilibrium neutronization of matter plays the main role. The model of the deflagration supernova has been developed during more than ten year in our group. In 1973 it was shown that thermal flash evolves in the deflagration regime. In our computations the deflagration regime is stipuleted by joint effect of compression of matter and weak shock waves, obtained at the pulsing regime of burning. Fig. 1 shows the graduated growth of entropy in mass zone through passage of weak shock waves.

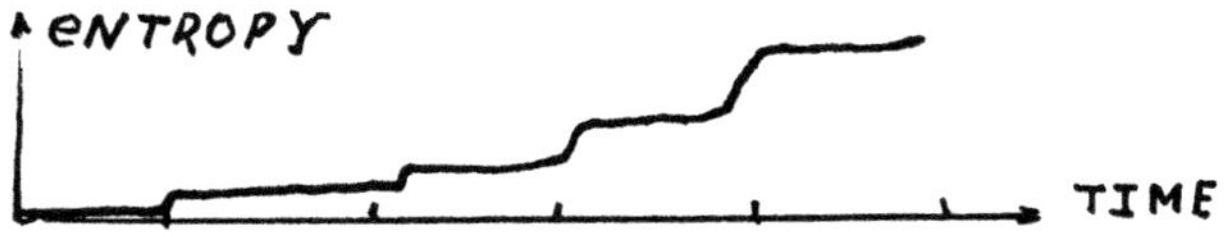

Figure 1. Entropy versus time for the mass zone.

The proof of the deflagration regime has the principal meaning for theory of supernova. The competitions between explosion and collapse in the flash of CO-core develops from start. Thermal explosion results in a hydrodynamic explosion. The neutronization of burning matter leads to the collapse of the core. The deflagration regime leads to the higher tendency towards collapse in the case $\rho_c < 3\times10^{10}$ g/cm^3. In 1975 the neutrino ignition of CO-cores was suggested. The neutrino ignition results in deflagration with subsonic propagation. After compression up to $\rho_c \gtrsim 3\times10^{10}$ g/cm^3 does it involve detonation with supersonic propagation. We obtained in calculations two versions of explosions: at $2\times10^9 < \rho_c < 9\times10^9$ g/cm^3 there is disruption of the star with 10^{50}-10^{51} erg kinetic energy and at $\rho_c > 9\times10^9$ g/cm^3 the stellar core collapses into a neutron star with outburst of the envelope with a kinetic energy of 10^{49}-10^{50} erg. The first version corresponds to SN II, and the second, supplemented by mechanism of slow energy release into the envelope, corresponds to SN I.

D. Sugimoto, D. Q. Lamb, and D. N. Schramm (eds.), Fundamental Problems in the Theory of Stellar Evolution, 320–321.

DISCUSSION

Sugimoto: As discussed in our paper of 1976, the fate of the deflagration wave depends sensitively on the speed with which the deflagration front propagates, and thus on the mode of heat transport. In the deflagration regime, the effect of heat transport by convection or by the Rayleigh-Tayler instability is more effective than the effect of the weak shock; this is even the definition of deflagration in contrast to detonation. If you take account of the effects of such heat transport, how do your results change?

Chechetkin: The propagation of the deflagration front by the weak shock wave is faster than the mechanism of convection. Therefore the first mechanism is more important.

Sugimoto: How long is the time scale involved in the propagation of the detonation front?

Chechetkin: The detonation front doesn't exist when the burning begins. If a detonation front arises, then the time scale of its propagation is less than 1 sec.

STELLAR COLLAPSE AND NASCENT NEUTRON STARS

T. J. Mazurek
Department of Physics
State University of New York
Stony Brook, N. Y. 11794, U.S.A.

The collapse of dense cores in massive stars proceeds as follows. Initially, leptons dominate the pressure because neutrinos become trapped at high densities. This results in the formation of a cool inner core that collapses homologously. At around nuclear densities the pressure from nucleons increases rapidly and halts the collapse, giving a core bounce. A shock forms at the surface of the inner core and propagates into the infalling envelope. These basic features have emerged in the hydrodynamic studies of collapse by various researchers. However, the question of final outcome of collapse is unresolved at present. The evolution of the core after the shock has propagated through the envelope has not been addressed in detail. This communication summarizes some current results of the author's ongoing study of stellar collapse.

The mass of the inner homologous core is found to vary with trapped lepton fraction. It typically is larger than the corresponding Chandrasekhar mass because of thermal and non-leptonic contributions to the pressure. Neutrino opacities determine the trapped lepton fraction. Calculations were performed with opacities that varied over an order of magnitude above their presently accepted values. The results were found to change only mildly. Masses of inner cores and trapped lepton fractions for different calculations all fell in the ranges: $.6 \lesssim M/M_\odot \lesssim .8$ and $.25 \lesssim X_\ell \lesssim .35$. The shock strength also varied with the mass of the inner core, being stronger for smaller core masses. Post-shock entropies ranged between 5k and 8k per baryon. In all cases, the shock died out before mass ejection could occur. However, this may be an artifact due to the finite zoning effects in the numerical study.

The collapsed core becomes essentially hydrostatic after one bounce. Its structure consists of a cool central region that collapsed homologously and an overlying shock-heated mantle of about half a solar mass. Neutrinos remain trapped on dynamic time scales for all densities above $\sim 10^{12}$g cm^{-3}. Below this density electron capture depletes the lepton fraction to values around 0.1. These results define a possible structure for neutron stars at birth. The evolution of this structure to the final state of cold neutron stars needs to be examined.

D. Sugimoto, D. Q. Lamb, and D. N. Schramm (eds.), Fundamental Problems in the Theory of Stellar Evolution, 322–325.

DISCUSSION

Schramm: If neutrinos oscillate (for example $\nu_e \leftrightarrow \nu_\tau$), how would that effect the X_e at the time of the bounce?

Mazurek: If neutrinos oscillate on time scales greater than those of dynamic collapse (~10 ms), it would have little effect on the trapped lepton fraction. If subsequent oscillations occur only between three types of neutrinos (e.g., ν_e, ν_μ, and ν_τ), then the effects on the equation of state would be moderate. The electron fraction X_e would decrease somewhat (from ~0.25 to perhaps ~0.20). However, I would guess that the effect on the pressure would be small, since this decrease in the electron fraction would not be sufficient to dissolve the heavy nuclei. On this basis, I do not think that neutrino oscillations would lead to new, dramatic effects. They could, however, aid the shock, in the sense that lower trapped lepton fractions give stronger shocks.

A further caveat is in order. Even if neutrino oscillations do occur in vacuum, I have been told by experts that such oscillations will be strongly supressed when the neutrinos find themselves in the presence of high density matter. The most pertinent point however, is that at present we do not know the properties of neutrino oscillations, if they occur. Such oscillations may have unforeseen properties that could have dramatic effects.

Sato: Previously, you pointed out the possibility that a supernova explosion could be produced when the degenerate neutrinos diffuse out of the core, because the energy of the degenerate neutrinos goes into thermal energy as they escape from the core. Now, you have carried out extensive numerical calculations of the collapse of the core. Have you been able to confirm the above possibility?

Mazurek: The mechanism you refer to can operate only on the long time-scale of neutrino diffusion. The situation is the following. The initial postshock core has a large fraction of its internal energy in the form of relativistically degenerate leptons. Being relativistic, the leptons exert only one-half of the pressure that a non-relativistic gas would exert if it has the same amount of energy. As the neutrinos leak out, reducing the lepton concentration, they take with them only a small fraction of their original energy (10 MeV out of ~70 MeV). Thus the energy in leptons will be transferred to nucleons which are non-relativistic. This will increase the pressure and will lead to a large, longterm expansion of the compact core. This longterm expansion may produce a piston-like shock that results in an explosion. However, to confirm this conjecture one needs to study the evolution of a core from its initial lepton-rich state through lepton depletion. To date, this is a fundamental problem that has not been addressed in adequate detail.

Tscharnuter: What are the reasons that the shock wave generally dies out? Is it, because some poorly understood physical processes are going on in the shock region or simply because the numerical grid is too coarse?

Mazurek: At the present time, it is not certain why the shock dies out.

There are reasons for believing this behaviour is physical. As I noted earlier, the core is essentially static after the "bounce". Its initial infall energy plus the work it does in a small expansion go into the shock. The shock loses energy to dissociation of nuclei and to neutrino losses. In the present numerical work, the energy that the core gives to the shock is roughly equal to the total nuclear energy of dissociation and of neutrino losses. However, the limits on precision imposed by the finite zoning are not good enough to firmly establish this. The discrepancy in energy between input and output is sufficiently large so that it could power a supernova. Thus models with much finer zoning than the calculations I described will have to be computed. Unfortunately, such models are extremely time-consuming in terms of the computing required. However, if semi-analytic solutions cannot be found, such calculations must be performed.

Wheeler: Your cold homologous core is below the expected maximum mass limit for a neutron star. Do you foresee a way to change the astrophysical conditions slightly and get an explosion with a black hole remnant, as opposed to a neutron star and an explosion?

Mazurek: At the present time we do not know how the shock ejects the matter, assuming that it indeed does so. However, you will recall that in the calculations I presented, the post-shock stationary core is composed of ~0.5 $M_{\odot}$ hot mantle in addition to the cold homologous core of ~0.7 $M_{\odot}$. In view of our lack of understanding of the shock ejection mechanism, one can envisage a situation where mass ejection occurs after the mantle is sufficiently large to result in a black hole on cooling. However, such an occurrence would demand addition of mass that is initially outside the ~1.5 $M_{\odot}$ degenerate core. In this case, the mechanism of explosion is likely to be somewhat more complicated than a "bounce" followed by shock ejection of the envelope as is usually discussed at present.

Tsuruta: Would you say that the expected core mass of ~0.6 - 0.8 $M_{\odot}$ is the final mass of the remnant neutron star, or that mass will keep falling onto the core? In the latter case, what is the expected final mass of the neutron star?

Mazurek: The final mass of neutron star will certainly be greater than the mass of the inner core that collapses homologously. The mass of this core falls in the range you quote. In the particular numerical example I presented, the inner core of ~0.7 $M_{\odot}$ had a hot a mantle (which would not be ejected even if the shock continued to propagate) of ~0.5 $M_{\odot}$. I believe that this particular case is representative of what can be expected in general when the core that initially collapses is an electron degenerate dwarf of around 1.5 $M_{\odot}$. On this basis, I would expect the initial mass of the neutron star to be greater than ~1.1 $M_{\odot}$.

Wheeler: Does the cold core you form have any affect on the subsequent rate of cooling of the neutron star, by setting up initial conditions which are different than assumed by Tsuruta, Lamb, or Sutherland?

Mazurek: No. The core configuration I have described will relax to the initial conditions assumed by Tsuruta, Lamb, and Sutherland on time scales of around a few seconds.

Cuyper: What will be the influence of rotation?

Mazurek: I don't know. Such effects are outside of the scope of the work I presented.

GENERAL RELATIVISTIC COLLAPSE OF AN AXIALLY SYMMETRIC STAR LEADING TO THE FORMATION OF NEUTRON STARS AND BLACK HOLES

Takashi Nakamura, Kei-ichi Maeda*, Shoken Miyama*, and Misao Sasaki*
Research Institute for Fundamental Physics, Kyoto University, Kyoto
*Department of Physics, Kyoto University, Kyoto

Using the [(2+1)+1]-dimensional representation of the Einstein equations, we have computed the general relativistic collapse of a rotating star. We adopt the cylindrical coordinate. The system is assumed to be axially and plane symmetric. The number of meshes is 28×28 in R and Z direction. The equation of state is $P=1/3\rho\varepsilon$ for $\rho<\rho^{*}\equiv 3\times 10^{14}$ g/cm^3 and $P=(\rho-\rho^{*})\varepsilon+1/3\rho^{*}\varepsilon$ for $\rho\geqq\rho^{*}$. We use the following initial conditions; $\rho\propto\exp(-(R^2+Z^2)/\lambda)$, $\Omega\propto\exp(-R^2/\lambda)$ where Ω and λ are angular velocity and a size parameter, respectively. We have calculated three models;

(1) Model 1 $M=10M_{\odot}$, $\rho_c=3\times10^{13}$g/cm^3, $\alpha=0.20$, $\beta=0.05$.
(2) Model 2 $M=10M_{\odot}$, $\rho_c=3\times10^{13}$g/cm^3, $\alpha=0.20$, $\beta=0.12$.
(3) Model 3 $M=10M_{\odot}$, $\rho_c=3\times10^{13}$g/cm^3, $\alpha=0.20$, $\beta=0.22$.

where $\alpha=E_{int}/|E_{grav}|$ and $\beta=E_{rot}/|E_{grav}|$. In all models, an apparent horizon was formed, that is, a black hole was formed. In Model 1, the final density distribution is oblate shape. In Model 2, there is a ringlike peak of the proper mass density distribution at the final stage. In Model 3, the determinant of the metric tensor goes to nearly zero at the ring in the equatorial plane, so that the proper mass density shows strong ringlike peak which is inside the apparent horizon. As the curvature invariant made from the Riemann tensor becomes very large at this ring, this may be a ring singularity of the space-time. These rotating black holes look like the Kerr black hole.

DISCUSSION

Gaffet: In the case of a black hole emerging from a pure gravitational wave, I would like to make the comment that, of course, the total mass has to be introduced as an initial condition, even though in this case there is no matter at all.

Nakamura: Even though there is no matter, the gravitational field does exist and it contains the energy. The total mass is determined by the asymptotic form of the metric tensor at large distance. So we don't have to introduce the total mass as an initial condition.

D. Sugimoto, D. Q. Lamb, and D. N. Schramm (eds.), Fundamental Problems in the Theory of Stellar Evolution, 326.

STABILITY AGAINST RADIAL PERTURBATIONS OF SLOWLY ROTATING NEUTRON STARS*

Kenzo ARAI and Keisuke KAMINISHI
Department of Physics, Kumamoto University

The dynamical equations governing pulsation in rotating neutron stars are derived in the framework of general relativity. Stellar models are constructed by using a realistic equation of state for cold neutron matter. Small radial displacement and slow rotation are treated as perturbations on spherically symmetric body. In these models the maximum masses are 1.761 $M_\odot$ at the central density 3.461×10^{15} g cm^{-3} for a sequence of nonrotating configurations and 2.165 $M_\odot$ for rotating models with the critical angular velocity $(GM/R^3)^{1/2}$.

We examine the stability of the three lowest quasi-radial modes in these models. The characteristic time scale of radial pulsation is of oder 1 ms for stable neutron stars. Effects of rotation do not stabilize the equilibrium configurations, but enhance the instability. This result presents a striking contrast to the case of the Tooper polytrope calculated by Hartle et al. (1975). The n = 3/2 polytropic equation has the value of the adiabatic index $\Gamma = 5/3$, so that the constructed model can sustain its stability upto higher central densities where the relativistic effects become significant. The centrifugal force may contribute to the restoring one on a radial perturbation, because there exist the stable regions in the middle and outer layers of their model. On the other hand, for the realistic equation of state used here, we have $\Gamma \cong 0.4$ at the neutron drip point. This small value of Γ makes the model unstable. In the models with higher central density the unstable outermost layer may be blown up by the centrifugal force and/or deformed significantly by the Coriolis force. The fundamental mode is broken away rapidly as the angular velocity increases.

* This paper was presented by K. Kaminishi.

D. Sugimoto, D. Q. Lamb, and D. N. Schramm (eds.), Fundamental Problems in the Theory of Stellar Evolution, 327.

CONCLUDING REMARKS

R. J. Tayler
Astronomy Centre
University of Sussex

In commencing this personal review of what we have heard in this Symposium, it is a great pleasure to congratulate Professor Hayashi on his 60th birthday today. For those participants like myself who were working on stellar evolution more than twenty years ago, the appearance of the famous review article by Hayashi, Hoshi and Sugimoto was a landmark in our subject. Together with Schwarzschild's book a few years earlier it marked the close of the second stage of the subject in which quite elaborate calculations had been made, but in which the contribution of the electronic computer was still small. The first stage in the subject perhaps ended with the publication of Chandrasekhar's book in 1939. What is obvious in this Symposium is that Professor Hayashi has created a very flourishing research school and that young Japanese astronomers are making many important contributions to the present third stage in the subject.

The principal emphasis in the subject has changed in the past twenty years. Then we were mainly concerned with the evolution of single stars and we used stellar systems, such as globular clusters, to give us information about single stars which we could not obtain if they were truly isolated. Now the emphasis has switched to the problem of galactic evolution; can we understand the variation of properties from place to place inside one galaxy and from one galaxy to another in terms of star formation, stellar evolution and nucleosynthesis and stellar mass loss, either quiet or catastrophic? The study of stellar birth and stellar death remains difficult and we continue to be grateful that the long main sequence lifetime enables us to discuss stellar evolution without understanding stellar birth.

There has been essentially nothing in this Symposium about main sequence and post main sequence evolution of single stars except a discussion of the influence of rotation and magnetic fields. This is not because there are no fundamental problems. The topic of mass loss was excluded from the Symposium at the request of the IAU Executive Committee, because it has recently been discussed at other IAU supported meetings. There are still some uncertainties in the opacity, nuclear

D. Sugimoto, D. Q. Lamb, and D. N. Schramm (eds.), Fundamental Problems in the Theory of Stellar Evolution, 329–334.

energy release and equation of state, although we may believe that these are relatively unimportant in early evolutionary phases. We cannot forget the solar neutrino problem even though we may hope that this will be solved by something outside stellar evolution such as the recently reported neutrino mixing. Above all there are serious problems involving stellar hydrodynamics. For example, UV observations are telling us that chromospheres, coronae and mass loss are much more common than was previously believed. We still lack a good theory of convection and of semi-convection. Mixing processes and mass loss which may seem relatively unimportant at the time that they occur may nevertheless have significant effects on later stages of evolution.

One fundamental problem that has not been raised in this Symposium, except perhaps in Mirzoyan's discussion of the formation of stars from very compact protostars, is whether the presently established laws of physics are adequate for the subject of stellar evolution. We have heard nothing of such things as variable G. Although we must be ready to accept changes in the laws of physics if they prove necessary, I am personally happy that the present laws provide all the necessary complications.

I now turn to points which have been specifically discussed in the Symposium. For single stars we have mainly been concerned with birth and death. The development of large computers has meant that it has been possible for much more ambitious calculations to be performed and for more physical processes to be included. However this has not always led to greater agreement amongst different workers either about the method of star formation or about the mechanism of supernova explosions. It is in fact quite easy to understand why both of these topics should present serious problems.

Consider first star formation in the solar neighbourhood. There remains between ten per cent and twenty per cent of matter in the form of interstellar gas. If we assume that stars are primarily formed as a result of compression in spiral shocks, it is clear that the fraction of gas used up in each passage through a shock is very small. If we had neither observations of dense clouds nor calculations of cloud collapse and fragmentation, we could consider two extreme possibilities. In the first we could assume that only a small amount of gas is compressed but that star formation is very efficient; in that case it might be easy to understand star formation once a dense cloud had formed. In the second case we could assume that all the gas is compressed but that star formation is very inefficient. In fact, both the observations of dense clouds and the calculations presented at the Symposium suggest that star formation is very inefficient. This means that it may be very difficult to obtain a definite result. We have seen that different workers are getting quite different answers, but what is impressive is that they are unanimous in saying that their results may not be correct. That in itself implies that the problem is difficult. There are obvious numerical problems involved with spatial resolution and with the existence of very different timescales. Because of these difficulties and of the variation of results from author to author, I hope that those working in the field will try to tell us what can be done and what cannot be done with the present, and indeed the next, generation of computers.

Consider next the final stages of stellar evolution. Despite many attempts to calculate the explosions of supernovae, there is no clear agreement about how a supernova is formed and what is its end product, although Wheeler and Nomoto gave us very impressive accounts of the highly detailed work which is at present being undertaken. What we particularly wish to know is what is the degree of mass loss in the explosion, what is its chemical composition and whether the end product is a neutron star, a black hole or nothing. Again it is obvious that the solution of the supernova problem must be difficult. If a stable neutron star is to be formed, the binding energy of such a star must be dissipated. However, the binding energy of a neutron star is several orders of magnitude greater than the energy obviously released in a supernova explosion. Most of the energy is probably lost in the form of neutrinos and gravitons and it is hardly surprising that it is difficult to determine whether the relatively small amount of energy required to remove the outside of the star is deposited in the correct place at the correct time.

A topic which was not discussed in the Symposium was the relation between the total numbers of neutron stars, visible supernovae and supernova remnants. The production rate of neutron stars appears to be significantly greater than that of supernovae and supernova remnants and we have heard that the explosions of supernovae may produce black holes or nothing. It is important to know whether a neutron star can ever be produced without either a visible supernova display or a subsequent optical or radio supernova remnant. Wheeler suggested that in one mass range neutron star production might not be accompanied by a visible supernova but that case would not make much difference to the total numbers. It is possible that, as has been suggested, many supernovae explode deep inside dense clouds and that neither they nor their subsequent remnants are observed and this possibility requires further investigation.

We heard from Lucy that there are at present difficulties in a numerical study of either of the more popular ideas concerning the formation of binary stars, fission and fragmentation. Numerical studies do indicate the formation of binary stars but they do not at present give the mass ratios most commonly observed. However we know that binary stars are common and that many of the most exciting objects in the Galaxy are included in binary systems. This is perhaps one of the two major developments in the theory of stellar evolution since the review by Hayashi, Hoshi and Sugimoto, the other being the recognition of the importance of mass loss. A further major development is in the attitude of astronomers. Although neutron stars had been discussed in 1934 and black holes in 1939, it was still a common belief amongst astronomers in the early 1960's that all stars must end their lives below the

Chandrasekhar mass as white dwarfs. At that time astronomers only believed what they saw and they did not see neutron stars and black holes; now they are much more likely to believe what they are told.

As we have heard, the number of stars which will become close binaries at some stage in their evolution is very large indeed. It is no longer sufficient to say that single spherical stars are the rule and close binaries the exception. Tutukov and van den Heuvel presented us with a variety of scenarios for the evolution of binary systems to form objects of all of the observed types and probably some others as well, but I think that they would agree that there are many difficult details to be discussed before the scenarios become real theories. We have heard about the problems of mass and angular momentum loss from the binary system and there may also be important effects due to departures from sphericity of the component stars. Once again we perhaps need a realistic assessment of what is the most that can possibly be included in numerical calculations in the foreseeable future and whether it is inevitable that progress must be made by ad hoc assumptions and comparison with observation.

When we consider planetary formation we have an uncertainty of a different order of magnitude. We know that there are very many binary stars, so that binary star formation must be an almost natural event. We have no idea how many planetary systems there are. We may have a post-Copernican prejudice that there is nothing very special about the Sun, but whether almost every G dwarf has a planetary system or whether less than one per cent have planets will make considerable difference to the difficulty of numerical work on the formation of planets. If we believe that the observed rotation speeds of main sequence stars tell us that late type stars have discs and planetary systems, that may be a clue; however the loss of angular momentum by stars with convective envelopes is probably very different from that by stars with radiative envelopes even in the absence of disc formation. The early stages in the formation of a planetary system were not discussed in this Symposium but Hayashi presented a very detailed theory of how the planets are formed once the protoplanetary disc is in existence.

We have heard two very interesting review talks about shell flashes in accreting degenerate stars. For someone like myself who almost runs away if the word computer is mentioned, it was very refreshing to hear Sugimoto's semi-analytical discussion of the occurrence of shell flashes. He did well to remind us that, if we do not have a physical understanding of what comes out of a computer, we shall have great difficulty in making progress. It is however equally true that, as Kippenhahn commented, it may be difficult to estimate the ultimate effect of an extremely large number of flashes, either by semi-analytical methods or by direct computation. The study of non-linear systems with two or more very disparate timescales is inevitably difficult. Having said that, it is very gratifying that the theories and observations of X-ray bursters appear to be in such good qualitative agreement.

Most stars rotate and contain magnetic fields. There are various effects related to them. The first is simply that of departures from spherical symmetry, if the rotation velocity or magnetic field strength is sufficiently great. However, even when departures from sphericity are small, we have problems related either to the lack of genuine equilibrium or to the occurrence of new instabilities. I do not think that either Kippenhahn or Mestel will complain for long if I say that we gave them an impossible task to discuss the effects of rotation and magnetic fields on stellar evolution and that they did not succeed. What they did do was very relevant to the subject of this Symposium. They demonstrated that there are very important fundamental problems related to rotation and magnetic fields which must be understood before they can be included with confidence in calculations of stellar evolution.

What for example is the natural rotational state of a star? Is it one of solid body rotation or a state nearer to constant angular momentum per unit mass? Given that many configurations are unstable, is the time for significant redistribution of angular momentum short compared to evolutionary timescales or is it long? Kippenhahn expressed a personal preference for a rather long time which would mean that in at least some cases what appears to be unstable differential rotation might survive. In the case of magnetic fields, we again have the prediction of many instabilities for magnetic fields of simple topology but as yet there are no good discussions of their non-linear development. At present I do not think that we have any convincing evidence in favour of "fossil" magnetic fields but we do have evidence that some fields must be produced by dynamos. Although both white dwarf and neutron star fields may be produced by approximate flux freezing in the collapse of a normal star, it is also possible that, as has been suggested in the case of pulsar formation, there might have been a dynamo process at the time of the collapse. In that case, even though the field is not at present being maintained by a dynamo, it will not be possible to relate it to the field at earlier stages of evolution. If magnetic fields do exist at all evolutionary phases, they couple different layers of a star in a manner which must strangely inhibit purely rotational effects.

In addition to the fluid dynamical effects related to rotation and magnetic fields, we cannot forget convection. Convection was not made a central theme of the present conference because it is only four years since an IAU colloquium was devoted specifically to this topic. This does not however mean that the problem has been solved. We continue to use the mixing length theory with its free parameter because we lack a better theory and there are particular uncertainties related to time dependent convection in variable stars. One role of convection and semi-convection is to mix different layers in a star. As several speakers have indicated there are less violent mixing processes caused by instabilities which may also have important effects.

To conclude, I select two fundamental problems which I believe to require particular attention before we can hope to be fully satisfied with our knowledge of stellar evolution. These are:

1) Numerical methods. We need to know what can be done and what cannot be done with the present and immediately foreseeable computers. There is no point in doing ever more elaborate calculations unless we can be certain that, given the physical input, the results are reliable;

2) Fluid dynamics. A major source of uncertainty at most stages of stellar evolution is mass loss and mixing. Can we hope to develop a true theory of mass loss and mixing or must we continue to have parametrised models which we try to fit to observation?

COMMENTS AND IMPRESSIONS

E.E. Salpeter
Physics and Astronomy Departments, Cornell University,
Ithaca, N.Y. U.S.A.

I will give no summaries or substantive remarks, partly because Roger Tayler has already done so and largely because I have been away from the field of stellar evolution for a long time. However I will give some personal impressions of a "prodigal son returned to the field" — or "how do things look 15 or 20 years after Professor Hayashi's pioneering work".

(A) PHYSICS INPUT AND MATHEMATICAL TECHNIQUES

(1) Rotation and Magnetic Fields: My first impressions are "A for Effort, but not necessarily for Achievement". Twenty years ago angular momentum and magnetic fields were simply ignored — not because they are unimportant but because they are difficult. They are still difficult today, but at least they are being tackled. I am particularly happy to see a "Two-pronged attack" — actual "honest and detailed" calculations on a few topics on the one hand, the establishment of semi-quantitative features for many topics on the other. Dissemination of expertise is still a bottleneck, but the Proceedings of this Symposium will help.

(2) Computer Codes for Evolutionary Calculations: One's first naive impression is that this area was much further advanced 15 years ago than today: Computational techniques were adequate then for one-demensional calculations (based on the unwarranted assumption of spherical symmetry), whereas current techniques are not fully adequate for more realistic calculations. Although a start has been made in this direction, a more systematic comparison of different techniques would be reassuring - to show that the computational schemes neither suppress real phenomena nor create phantom ones. There is an interesting parallel with the methodology of fundamental turbulence theory: Neither analytical nor numerical methods alone can solve a number of problems of principle (the determinacy of circulation modes, the non-Gaussian behavior of "intermittancy", etc.). The mixture, used there, of systematic numerical experiments with abstract theory may give some

D. Sugimoto, D. Q. Lamb, and D. N. Schramm (eds.), Fundamental Problems in the Theory of Stellar Evolution, 335–337.

useful hints by analogy.

(3) Physics of Dust Grains: I have a warning and/or question about dust grains in the formation period of the solar system. The warning relates to our ignorance of the dust grains in the interstellar medium from which the early solar system contracted. For instance, it is not clear whether collisions between dustgrains result in sticking or shattering or evaporation. Even for particles made of ordinary Materials there is surprisingly little experimental data. Furthermore, the solid-state structure of interstellar dust grains is likely to be quite unusual, because of radiation damage (U.V. and cosmic rays) and of chemical processing by radicals. My question to solar system theorists is: which of various uncertainties in the grain physics are really important. There are few groups working in experimental laboratory — astrophysics (astrochemistry), but some progress might be possible on the one or two key questions.

(4) The Solar Neutrino Problem: There is generally little interaction between work on solar physics and stellar evolution. In particular, I hope that stellar theorists will revisit the solar neutrino puzzle in the near future. It is likely that a second solar neutrino experiment will be carried out in a few years, employing a Gallium-detector. The two experiments are sensitive to different uncertainties and progress should be possible. However, it is important that different stellar theorists, with different views on the present solar neutrino puzzle, make very specific predictions for the Ga-experiment before it is carried out.

(B) ASTRONOMICAL TOPICS

(1) The Initial Mass Function: There has been real progress in at least a qualitative understanding of the intermediate stages of star formation, just before the Hayashi track. The even earlier stages seem to be elusive still and it is not clear what determines the distribution of masses to which individual stars settle down. In other words, a genuine a priori calculation of the Initial Mass Function (IMF) has not succeeded yet, although not for lack of trying. That's a pity, especially since the other two points I will make are closely connected to the IMF.

(2) The Extreme "Ends" of the Main Sequence: Main sequence stars of the lowest and highest masses are still somewhat shrouded in mystery, partly because observations are difficult and partly because the rate of massloss is appreciable but not well known. Theoretically it is not clear if we really understand the mechanism of massloss and observationally it is not clear if we know the IMF. These two uncertainties are connected, even if the luminosity function ψ is known observationally: If a star loses an appreciable fraction of its mass in a much shorter period than the main sequence lifetime, then the IMF derived by the standard methods from ψ is a gross underestimate.

I stress the uncertainty in the massloss rate, because of

recent indirect indications that it might be particularly large: (a) Infrared observations of dense molecular cloud complexes seem to require the frequent release of energy, packaged in relatively "small chunks". Supernovae come in "large chunks", but massive young stars are still likely to be involved. Although there is no direct evidence the following hypothesis is a possibility: The IMF may extend to larger masses than previously thought, but these stars lose mass rapidly and intermittently. Appreciable chunks of energy could be released during this evolution "down the main sequence" towards ordinary O-stars and these supermassive stars are difficult to find outside of molecular clouds because of their short lifetime. (b) X-ray observations from the Einstein satellite indicate a surprisingly large number of X-ray flares from cool M-dwarfs. Although there is no direct observational data, massloss is likely to accompany flares. The fact that cool M-dwarfs burn nuclear fuel in their deep interior and are convective from there all the way to the surface raises a theoretical challenge: Does convenction provide a close enough coupling between nuclear reactions and the surface, so that new modes of dynamical instability come into play?

(3) "Stellar Population III": In the stellar theory literature this phrase usually means stars in "Standard" regions of a galaxy, but formed early out of material with a very low abundance Z of heavier elements. I reserve the phrase "stellar population III" for an even more mysterious class of objects: We now have good evidence from rotation curves of spiral galaxies, that most galaxies have an "almost invisible" halo extending very far out. Even if massive neutrinos exist, they are not able to provide the mass for such a halo, but stars of very low mass and/or the neutronstar (or black hole) remnants of very massive stars have a sufficiently large mass-to-light ratio. One challenge to the theorist lies in the fact that the mass-density in these halos is very low but the total mass is very large. Another puzzle is that the gas in galaxy clusters, which was probably ejected from these galaxy haloes, has almost solar values for the abundance Z. This stellar population III, which contains more total mass than stellar populations I and II put together, presents a sufficiently large challenge that I hope Professor Hayashi will turn his brilliant talents into this direction!

GENERAL DISCUSSIONS

Hayashi: The summary by Professor Tayler is so complete and clear that there is almost nothing left for me to talk about. Perhaps I will say something about my impression of the Symposium. Study of stellar evolution began about thirty to forty years ago. In those days the problems were rather simple but nowadays they are so complex. However, computers have made great progress possible. But for them, the progress in the stellar evolution theory would have been very little. This is because nature is fond of complexities and diversities. With the aid of computers we have reached a very detailed understanding of the structure and evolution of the stars.

Although there are many differences and disputes in computational results, I am not worried about them because time will shortly solve them. In the IAU General Assembly 1961, Professor M. Schwarzschild gave an invited discourse. He divided stellar evolution into three phases: almost-hydrostatic, slowly contracting, and dynamical phases. He stressed the importance of the dynamical phases. Also in this Symposium, most of the fundamental problems other than elementary processes lie in the dynamical phase including magnetohydrodynamics, turbulence and convection. They have more degrees of freedom than those lying in stellar models. In the near future, I hope, much more progress will be made in hydrodynamic problems. In this Symposium it is the "problems" which are presented. When the problems are presented clearly, then the solution is not so far from being realized.

Nevertheless, I would offer a remark on computational work. Though sometimes the computer discovers new facts, the most important task is to construct a "theory" from the computational results. In the old days, I used computers somewhat by myself. These days I do not use them by myself, but rather I try to construct theory. I think it is a role to be played by an older generation because I started my career without computers. Before closing my talk on my impression of the Symposium, I would say once more that I, as a member of an older generation, like simplicity even though nature is fond of complexities.

Hayakawa (Chairman): Is there any comment, particularly from the younger generation who has grown up with computers?

Sugimoto: Of course both computational and theoretical works are equally important. In many cases simple interpolation formulae are usually confused with theories. The latter should explore the reasons why such numerical relations result. In developing theories, numerical results are very helpful. However, they are helpful only when important quantities are properly described in papers. In order to know what are the important quantities, a theory is required. Recently many numerical

D. Sugimoto, D. Q. Lamb, and D. N. Schramm (eds.), Fundamental Problems in the Theory of Stellar Evolution, 339–342.

results have been published, but very often, however, they can be used only for constructing interpolation formulae but not for constructing theory.

Tayler: When I was calculating stellar evolution on a desk calculator, I calculated the evolution of massive stars away from the main sequence. I considered the possibility of evolving low mass stars without convective cores but estimated that it would take six hours a day, five days a week for twenty months. I did not do it.

Miyaji: Perhaps the chairman nominated me because I have just finished my doctoral thesis and during the last two years I have been one of the top users of the computer at the Tokyo Astronomical Observatory. In my computations of electron-capture supernovae, convection due to entropy production by electron capture plays an essential role. Before my computation, such a possibility had not been pointed out, and of course I neglected it at first. Thus I had to spend another year in order to take such convection into account. From my limited experience above, I would like to agree with the comments from theorists of the older generation that the physics is sometimes recognized only after the computations.

Mouschovias: Concerning the effect of mass loss on stellar evolution, it may be important to remember that there are a large number of results available about our own solar wind, from which we should benefit. Skylab has shown that the solar wind (1) may originate mostly from coronal holes (which have open magnetic field lines); (2) may not exist as a quiescent wind; and (3) coronal transient events seem to contribute significantly to the mass in the solar wind. It is hoped that the Solar Maximum Mission, now in progress, will give definitive answers to these issues. In any case, in introducing mass loss into stellar evolution calculations, it is important to draw from the knowledge of our own sun.

Van den Heuvel: An important question, that is especially relevant to the X-ray binaries and bursters, is whether neutron star magnetic fields decay or not. Could anyone tell what the present status of the thinking about this problem is?

Lamb: Observations of pulsars suggest that they turn off after about 10^7 years (Manchester and Taylor 1977). This has frequently been taken as evidence that neutron star magnetic fields decay on such a time scale. Indeed, Flowers and Ruderman (1976) have shown that any magnetic field arising from currents in the crust may decay on such a time scale due to ohmic dissipation. However, Baym, Pethick, and Pines (1969) have shown that any magnetic field arising from currents in the core of the neutron star will decay on a time scale that is longer than the present age of the universe because the conductivity there is so high. Flowers and Ruderman therefore proposed that an instability may lead to fluid motions which cancel out the magnetic field, much as one can do so by flipping over one of two bar magnets. However, this model is a very simple one, and it is not known whether it has any relevance to the real physical situation in the cores of neutron stars. In fact, there are many other alternative ways in which pulsars may turn off. For example, the magnetic and rotation axes may align due to the radiation torque. Finally, recent observations of X-ray (Oda et al. 1980) and gamma-ray (Mazets et al. 1980) bursts indicate the presence of strong magnetic fields. In particular, the gamma-ray burst spectra appear to show cyclotron absorption or emission features. If the burst sources do, in fact, involve very old neutron stars, these observations may indicate that neutron star magnetic fields do not decay.

Van den Heuvel: There have recently been rumours that the neutrino might have a rest mass. Is there anyone who could comment on the effects which this might have for the solar neutrino problem?

Sato: I am not a specialist in the solar neutrino problem. On the basis of the measurement of neutrino flux, it is said that the electron neutrino may be oscillating among other types of neutrinos, for example the tau neutrino. The solar neutrino flux would then be decreased to half of that previously estimated. If neutrinos oscillate among three states of neutrinos, it will be decreased by a factor of three. This, I think, is the main effect of massive neutrinos on the solar neutrino problem. I think its more important effects in astrophysics are expected in the problems of "missing mass" in galactic halos, binary galaxies, rich cluster of galaxies and so on. They might be explained in terms of the rest mass of neutrinos.

Salpeter: I want to emphasize the point that has just been made that, even if there are massive neutrinos and neutrino oscillations, the flux of Reines' experiment is at most decreased by a factor of three. So it is not at all clear that even if there are massive neutrinos the astrophysicists can go home. There might still be a solar neutrino problem.

Salpeter: What would be the role of neutrino oscillations in stellar collapse?

Mazurek: If the oscillations occur on a time scale that is long compared to that of stellar collapse, they probably would not affect things too drastically. However, if oscillations occur on a much shorter time scale, the efficiency of neutrino trapping could be decreased appreciably. It is true that the Rines' results indicate a much shorter time scale for the oscillations. In this sense, we may have to worry about their effects. However, in dense matter the amplitude of such oscillations will be strongly suppressed. Thus it is not clear at present what the effects on stellar collapse will be. At present the most pertinent point, however, is that we do not know the detailed properties of neutrino oscillations, if they occur. Until we do, their effects on stellar collapse can only be conjectured.

Massevitch: Concerning the structure of the sun, another problem facing modern theory is the recently discovered solar oscillations with a period $2^h\ 40^{min}$. This result, which was obtained several years ago at the Crimean Observatory (USSR) by Prof. Severny and his collaborators, contradicts the generally adopted model of the sun (and its oscillation modes). Several attempts have been made to change the solar model in a way that would make such a large period possible. There have also been many doubts expressed about the reliability of the observational deta.

Recent observations carried out in the USA and UK have confirmed the period originally obtained and it is now up to theorists to provide its interpretation.

Osaki: I am sorry that I cannot say anything important concerning the $2^h\ 40^{min}$ solar oscillation. However, it is very difficult to explain because the fundamental period of the sun is about an hour. In order to explain a period of this length, we must consider a high order gravity mode. Even if this were the case, another difficult question arises, namely, how and why is such a high order gravity mode excited and why is a single mode selected from the dense spectrum of high order g-modes.

AUTHOR INDEX

SUBJECT INDEX

GPSR Compliance

The European Union's (EU) General Product Safety Regulation (GPSR) is a set of rules that requires consumer products to be safe and our obligations to ensure this.

If you have any concerns about our products, you can contact us on ProductSafety@springernature.com

In case Publisher is established outside the EU, the EU authorized representative is:

Springer Nature Customer Service Center GmbH
Europaplatz 3
69115 Heidelberg, Germany

Batch number: 10824476

Printed by Printforce, the Netherlands